Physics of Heavy Fermions

Heavy Fermions and Strongly Correlated Electrons Systems

Physics of Heavy Fermions

Heavy Fermions and Strongly Correlated Electrons Systems

Yoshichika Ōnuki
University of the Ryukyus, Japan

World Scientific

NEW JERSEY • LONDON • SINGAPORE • BEIJING • SHANGHAI • HONG KONG • TAIPEI • CHENNAI • TOKYO

Published by

World Scientific Publishing Co. Pte. Ltd.
5 Toh Tuck Link, Singapore 596224
USA office: 27 Warren Street, Suite 401-402, Hackensack, NJ 07601
UK office: 57 Shelton Street, Covent Garden, London WC2H 9HE

Library of Congress Cataloging-in-Publication Data
Names: Ōnuki, Yoshichika, 1947– author.
Title: Physics of heavy fermions : heavy fermions and strongly correlated electrons systems / Yoshichika Ōnuki (University of the Ryukyus, Japan).
Description: Singapore ; Hackensack, NJ : World Scientific, [2018] | Includes bibliographical references and index.
Identifiers: LCCN 2017041020| ISBN 9789813232198 (hardcover ; alk. paper) | ISBN 9813232196 (hardcover ; alk. paper)
Subjects: LCSH: Fermions. | Electron configuration. | Superconductivity. | Condensed matter.
Classification: LCC QC793.5.F42 O58 2018 | DDC 539.7/21--dc23
LC record available at https://lccn.loc.gov/2017041020

British Library Cataloguing-in-Publication Data
A catalogue record for this book is available from the British Library.

For any available supplementary material, please visit
http://www.worldscientific.com/worldscibooks/10.1142/10769#t=suppl

Guest editor Prof. L C Gupta

Typeset by Stallion Press
Email: enquiries@stallionpress.com

Printed in Singapore

Contents

Preface

A large variety of materials proves to be fascinating in solid state and condensed matter physics. New materials create new physics. Among them, the f electrons of rare earth and actinide compounds typically exhibit a variety of characteristic properties, including spin and charge orderings, spin and valence fluctuations, heavy fermions, and anisotropic superconductivity. These are mainly due to competitive phenomena between the RKKY interaction and the Kondo effect. The present text is written for the reader to understand these phenomena. For example, superconductivity was once regarded as one of the more well-understood many-body problems. However, superconductivity is still an exciting phenomenon in new materials. Magnetism complements and supports superconductivity in heavy fermion superconductors. The understanding of anisotropic superconductivity and magnetism is a challenging problem in solid state and condensed matter physics.

Furthermore, new measuring systems and/or new techniques correspond to new materials. The electronic states of the f electron systems, together with d electron systems containing the transition metals, can be changed by temperature, magnetic field, and pressure. Magnetic materials are changed to new superconductors by applying pressure. Readers are encouraged to participate in single crystal growth and various measurements described in this text. My belief on solid state and condensed matter physics is as follows:

> Solid state and condensed matter physics is a science with universal truth.
> To clarify it and make progress,
> it matters neither whether you are a man or a woman,
> nor which race and nation you belong to.
> An inexhaustible spring of solid state and condensed matter physics.
> New materials create new physics.

I would like to thank K. Miyake, K. Ueda, and H. Shiba for helpful discussions, and T. Komatsubara, A. Hasegawa, H. Sato, Y. Kitaoka, H. Harima, H. Yamagami, T. Sakakibara, K. Sato, A. Sumiyama, R. Settai, K. Sugiyama, Y. Inada, T. Takeuchi, D. Aoki, F. Honda, Y. Haga, E. Yamamoto, T. D. Matsuda, N. Tateiwa, M. Nakashima, A. Nakamura, A. Teruya, S. Fujimori, M. Hedo, and T. Nakama for their contribution to many fruitful collaborations and discussions. I am greatly indebted to T. Nakama for his help in typing the manuscript, together with additional helpful comments, and to M. Kakihana and M. Hedo for providing figures and tables. Finally, I would like to express sincere thanks to Prof. L. C. Gupta, who kindly invited me to write this monograph.

Okinawa, April 2017
Yoshichika Ōnuki

Chapter 1

Conduction Electrons and Fermi Surfaces

Conduction electrons are described by the one-electron band theory in which the electron-electron interactions (correlations) are simply neglected at the start. The Schrödinger wave function is solved for an electron moving in a potential which varies periodically with distance. The corresponding Bloch state with a band width of a few eV develops in the whole crystal. The conduction electrons are thus well described in momentum (k) space. The conduction electrons have energies up to a Fermi energy ε_{F}, which is simply expressed by $\varepsilon_{\mathrm{F}} = \frac{1}{2}mv_{\mathrm{F}}^2 = (\hbar k_{\mathrm{F}})^2/2m$. The Fermi surface is thus defined as a constant energy surface in k space of which the energy is equal to ε_{F}. Even though the electron-electron interactions may change the topology of the Fermi surface, the volume of the Fermi surface is kept invariant. It is the boundary of the electronic momentum distribution in the ground state, dividing k space sharply into two regions which electrons occupy or do not occupy. Note that the mass of the correlated conduction electron is changed from the bare mass m or the band mass m_{b} to an effective mass m^*.

1.1 Non-interacting conduction electrons -Fermi gas-

Strong magnetism described in the present text is present in the compounds containing the first-line series of transition metals (Ti, V, Cr,

Mn, Fe, Co, Ni, Cu), lanthanoids (lanthanides) (Ce, Pr, Nd, Sm, Eu, Gd, Tb, Dy, Ho, Er, Tm, Yb), and actinoids (actinides) (U, Np, Pu, Am, Cm). Their electron configurations are as follows

transition metals [Ar core] $3d^n\mathbf{4s^2}$ $(n = 0 - 10)$

lanthanides [Kr core] $4f^n5s^25p^6\mathbf{5d^16s^2}$ $(n = 0 - 14)$

actinides [Xe core] $5f^n6s^26p^6\mathbf{6d^17s^2}$ $(n = 0 - 14)$.

Three series of elements play a fundamental role in magnetism because $3d$, $4f$, and $5f$ shells can remain unfilled, leading to magnetism in the crystal. Note that $4d$, $5d$, and $6d$ electrons are rather delocalized, participating in the conduction band, and their contribution to magnetism is generally very weak. The $4s^2$ electrons in transition metals, $5d^16s^2$ electrons in lanthanides, and $6d^17s^2$ electrons in actinides contribute to the energy band, and partially become the conduction electrons.

First, we consider the conduction electrons based on a free electron model; that is, we consider N number of conduction electrons in a cubic crystal of length L, as shown in Fig. 1.1(a). The Schrödinger wave equation is

$$\left[-\frac{\hbar^2}{2m}\nabla^2 + V(\boldsymbol{r})\right]\Psi(\boldsymbol{r}) = \varepsilon\Psi(\boldsymbol{r}), \tag{1.1}$$

where the conduction electrons move freely in the crystal, as in an ideal gas. The one-particle wave function Ψ is expressed as Eq. (1.1), where the position $\boldsymbol{r} = (x, y, z)$, the wave vector $\boldsymbol{k} = (k_x, k_y, k_z)$, the potential energy $V(\boldsymbol{r})$, and

$$\nabla^2 = \frac{\partial^2}{\partial x^2} + \frac{\partial^2}{\partial y^2} + \frac{\partial^2}{\partial z^2}.$$

If we assume $V(\boldsymbol{r}) = 0$,

$$\Psi(\boldsymbol{r}) = \frac{1}{L^{3/2}}e^{i\boldsymbol{k}\cdot\boldsymbol{r}} \tag{1.2}$$

$$\varepsilon = \frac{\hbar^2\boldsymbol{k}^2}{2m} = \frac{\hbar^2}{2m}\left(k_x^2 + k_y^2 + k_z^2\right) \tag{1.3}$$

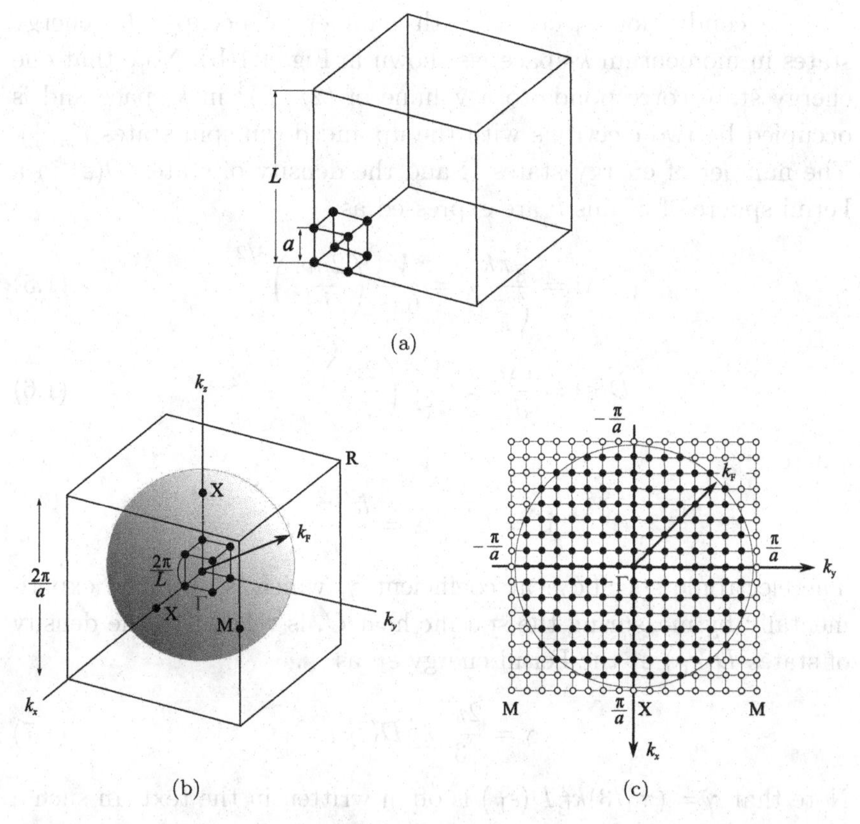

Figure 1.1: (a) Cubic crystal with a length L and the lattice constant a, (b) the corresponding spherical Fermi surface, and (c) the cross-section of the Fermi surface.

The wave function is conveniently required to be periodic in x, y, z with the length L, so that

$$\Psi(x,y,z+L)=\Psi(x,y,z),\quad \Psi(x,y+L,z)=\Psi(x,y,z),$$
$$\Psi(x+L,y,z)=\Psi(x,y,z), \tag{1.4}$$

which is called a periodic boundary condition. The wave vectors in x, y, z become

$$k_x=\frac{2\pi}{L}n_x,\quad k_y=\frac{2\pi}{L}n_y,\quad k_z=\frac{2\pi}{L}n_z$$
$$n_x,\ n_y,\ n_z=0,\ \pm1,\ \pm2,\ \cdots .$$

The conduction electrons with number N occupy the energy states in momentum k-space, as shown in Fig. 1.1(b). Note that one energy state corresponds to a volume of $(2\pi/L)^3$ in k-space and is occupied by two electrons with the up and down spin states ($\uparrow$, $\downarrow$). The number of energy states Ω and the density of states $D(\varepsilon)$ in a Fermi sphere of radius k are expressed as

$$\Omega = \frac{\frac{4}{3}\pi k^3}{\left(\frac{2\pi}{L}\right)^3} = \frac{V}{6\pi^2}\left(\frac{2m\varepsilon}{\hbar^2}\right)^{3/2} \tag{1.5}$$

$$D(\varepsilon) = \frac{d\Omega}{d\varepsilon} = \frac{V}{4\pi^2}\left(\frac{2m}{\hbar^2}\right)^{3/2}\varepsilon^{1/2} \tag{1.6}$$

where

$$V = L^3, \quad \varepsilon = \frac{\hbar^2 k^2}{2m}.$$

The electronic specific heat coefficient γ, which is obtained experimentally by measuring the specific heat C, is related to the density of states $D(\varepsilon_{\mathrm{F}})$ at the Fermi energy ε_{F} as

$$\gamma = \frac{2\pi^2}{3}k_{\mathrm{B}}^2 D(\varepsilon_{\mathrm{F}}). \tag{1.7}$$

Note that $\gamma = (\pi^2/3)k_{\mathrm{B}}^2 D(\varepsilon_{\mathrm{F}})$ is often written in the text. In such a case, the density of states $D(\varepsilon_{\mathrm{F}})$ contains a factor of 2 based on the up and down spin states. The present density of states in Eq. (1.6) does not contain the factor of 2.

We now consider N conduction electrons in the cube $V(= L^3)$. The number of conduction electrons per unit volume, n, is given as

$$\begin{aligned} n = \frac{N}{V} &= \frac{1}{V}\int_0^{\varepsilon_{\mathrm{F}}} 2D(\varepsilon)d\varepsilon \\ &= \int_0^{\varepsilon_{\mathrm{F}}} \frac{1}{2\pi^2}\left(\frac{2m}{\hbar^2}\right)^{3/2}\varepsilon^{1/2}d\varepsilon \\ &= \frac{1}{3\pi^2}\left(\frac{2m}{\hbar^2}\right)^{3/2}\varepsilon_{\mathrm{F}}^{3/2} \end{aligned} \tag{1.8a}$$

or

$$\varepsilon_{\mathrm{F}} = \frac{\hbar^2}{2m}(3\pi^2 n)^{2/3}. \tag{1.8b}$$

Here, we consider the simple cubic structure with a lattice constant $a(=4\,\text{Å})$, and assume one atom possesses one conduction electron, which is confined in the crystal of the cube of length $L(=1\,\text{cm})$. The carrier number $n(=1/a^3)$ is $1.6 \times 10^{22}\ \text{cm}^{-3}$, which leads to $\varepsilon_\text{F} = 2.3\,\text{eV}$ from Eq. (1.8b), or the Fermi temperature $T_\text{F}(=\varepsilon_\text{F}/k_\text{B}) = 27000\,\text{K}$, and the Fermi velocity $v_\text{F}(=\hbar k_\text{F}/m = \sqrt{2\varepsilon_\text{F}/m}) = 9.0 \times 10^7$ cm/s, where m is assumed to be m_0 (the electron rest mass).

Next, we consider the Schrödinger wave equation in Eq. (1.1) when the potential $V(\boldsymbol{r})$ exists but is weak. Both the wave function Ψ and potential V are periodic over a and therefore are expressed by Fourier expansions as

$$\Psi(\boldsymbol{r}) = e^{i\boldsymbol{k}\cdot\boldsymbol{r}}u(\boldsymbol{r}) = e^{i\boldsymbol{k}\cdot\boldsymbol{r}}\sum_{n'} u_{n'}e^{-i\boldsymbol{G}_{n'}\cdot\boldsymbol{r}} \tag{1.9}$$

$$V(\boldsymbol{r}) = \sum_{m} V_m e^{-i\boldsymbol{G}_m\cdot\boldsymbol{r}} \tag{1.10}$$

where $\boldsymbol{G}_m$ is the reciprocal lattice vector. From Eqs. (1.1), (1.9) and (1.10), we can obtain the wave function Ψ and the eigenvalue ε as

$$\Psi(\boldsymbol{r}) = u_0 e^{i\boldsymbol{k}\cdot\boldsymbol{r}} + \sum_{n\neq 0}\frac{V_n}{\frac{\hbar^2\boldsymbol{k}^2}{2m} - \frac{\hbar^2}{2m}(\boldsymbol{k}-\boldsymbol{G}_n)^2}u_0 e^{i(\boldsymbol{k}-\boldsymbol{G}_n)\cdot\boldsymbol{r}} \tag{1.11}$$

$$\varepsilon = \frac{\hbar^2\boldsymbol{k}^2}{2m} + V_0 + \sum_{n\neq 0}\frac{|V_n|^2}{\frac{\hbar^2\boldsymbol{k}^2}{2m} - \frac{\hbar^2}{2m}(\boldsymbol{k}-\boldsymbol{G}_n)^2} \tag{1.12}$$

where $V_{-n}V_n = |V_n|^2$.

When $\hbar^2\boldsymbol{k}^2/2m = (\hbar^2/2m)(\boldsymbol{k}-\boldsymbol{G}_n)^2$ in Eq. (1.12),

$$\varepsilon = V_0 + \frac{1}{2}\left\{\frac{\hbar^2\boldsymbol{k}^2}{2m} + \frac{\hbar^2}{2m}(\boldsymbol{k}-\boldsymbol{G}_n)^2\right\}$$

$$\pm\sqrt{\left\{\frac{1}{2}\left[\frac{\hbar^2\boldsymbol{k}^2}{2m^*} + \frac{\hbar^2}{2m^*}(\boldsymbol{k}-\boldsymbol{G}_n)^2\right]\right\}^2 + |V_n|^2}. \tag{1.13a}$$

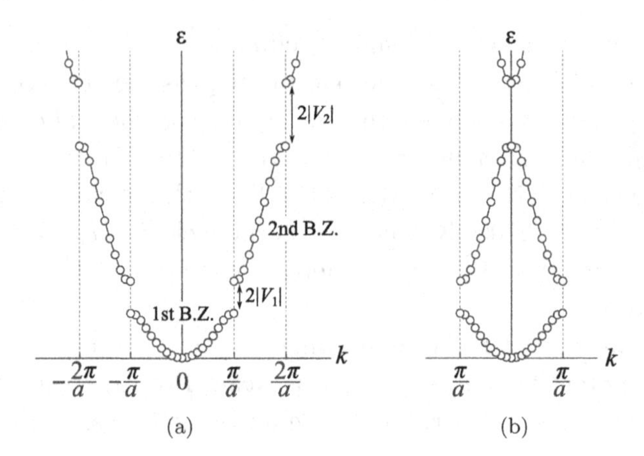

Figure 1.2: (a) Simplified energy band structure and (b) the corresponding one in the reduced zone scheme.

If we simplify the expansion by setting $G_n = (2\pi/a)n$ or $k = (\pi/a)n$,

$$\varepsilon = V_0 + \frac{\hbar^2}{2m}\left(\frac{\pi}{a}n\right)^2 \pm |V_n|. \tag{1.13b}$$

Figure 1.2(a) shows the simple energy band structure, where the energy band in the region of $k = -\pi/a$ to π/a is called the first Brillouin zone, while that in the region of $k = -2\pi/a$ to $-\pi/a$ and π/a to $2\pi/a$ is the second Brillouin zone. The energy gap at $k = (\pi/a)n$ is $2|V_n|$, which is expressed by the Fourier expansion coefficient V_n of the potential $V(\boldsymbol{r})$. The usual band structure is expressed in the reduced zone scheme, as shown in Fig. 1.2(b).

Note that the energy gap or the band gap is produced by the periodicity of the potential, which is called the Bragg reflection. The condition of the Bragg reflection, $\hbar^2\boldsymbol{k}^2/2m = (\hbar^2/2m)(\boldsymbol{k}-\boldsymbol{G}_n)^2$, is expressed as

$$\frac{|\boldsymbol{G}_n|}{2} = \boldsymbol{k}\cdot\frac{\boldsymbol{G}_n}{|\boldsymbol{G}_n|}. \tag{1.14}$$

This means that the zone boundary is normal to $\boldsymbol{G}_n$ at its midpoint. For example, the primitive lattice vector in the simple cubic crystal, as shown in Fig. 1.1(a), is

$$\boldsymbol{a}_1 = a\boldsymbol{i}, \quad \boldsymbol{a}_2 = a\boldsymbol{j}, \quad \boldsymbol{a}_3 = a\boldsymbol{k}$$

where $\boldsymbol{i}$, $\boldsymbol{j}$, and $\boldsymbol{k}$ are orthogonal unit vectors. The reciprocal lattice vector $\boldsymbol{b}_i$ is defined as $\boldsymbol{a}_i \cdot \boldsymbol{b}_j = 2\pi\delta_{ij}$, which leads to

$$\boldsymbol{b}_1 = \frac{2\pi}{a}\boldsymbol{i}, \quad \boldsymbol{b}_2 = \frac{2\pi}{a}\boldsymbol{j}, \quad \boldsymbol{b}_3 = \frac{2\pi}{a}\boldsymbol{k}.$$

The reciprocal lattice vector becomes $\boldsymbol{G} = n_1\boldsymbol{b}_1 + n_2\boldsymbol{b}_2 + n_3\boldsymbol{b}_3$ (where n_1, n_2, n_3 are integers). The zone boundaries of the first Brillouin zone in Fig. 1.1(b) are thus planes normal to $\pm(\pi/a)\boldsymbol{i}$, $\pm(\pi/a)\boldsymbol{j}$, $\pm(\pi/a)\boldsymbol{k}$, as shown in Figs. 1.1(b) and (c). The volume of the primitive reciprocal space is $\boldsymbol{b}_1 \cdot \boldsymbol{b}_2 \times \boldsymbol{b}_3 = (2\pi/a)^3$. The volume of the Fermi surface is $(2\pi/a)^3/2$ if one atom possesses one conduction electron, as in alkali metals of Li, Na, K, Rb, Cs and Fr.

Figure 1.3 shows the Fermi surfaces, based on the Harrison's free electron model, for the monovalent metal (Cu), divalent metal (Ca), trivalent metal (Al) and tetravalent metal (Pb) with the face-centered cubic structure [1]. The exact Fermi surfaces are slightly different from the Harrison's free electron model. For example, the neck Fermi surface exists along the $\langle 111 \rangle$ direction in Cu. The electron pocket Fermi surfaces in the fourth band of Pb do not exist. The first band of Pb is fully occupied by two of the electrons, and a closed Fermi surface in the second band is illustrated by the unoccupied (hole) Fermi surface, where another of its electrons occupies the second band. The last electron occupies the third band, revealing a "jungle gym" Fermi surface. In this case, the volume of the hole Fermi surface V_h in the second band is equal to the volume of the electron Fermi surface V_e in the third band. Pb is thus a compensated metal ($V_h = V_e$). On the other hand, Al is an uncompensated metal ($V_h \neq V_e$).

1.2 Crystal structure, crystal, and Fermi surface

For real metallic compounds, we will describe the framework of band theory. The usual band theory is based on the density functional formalism originating from the work of Hohenberg, Kohn and Sham [2, 3]. The wave function Ψ_i and the eigenvalue ε_i of an electron

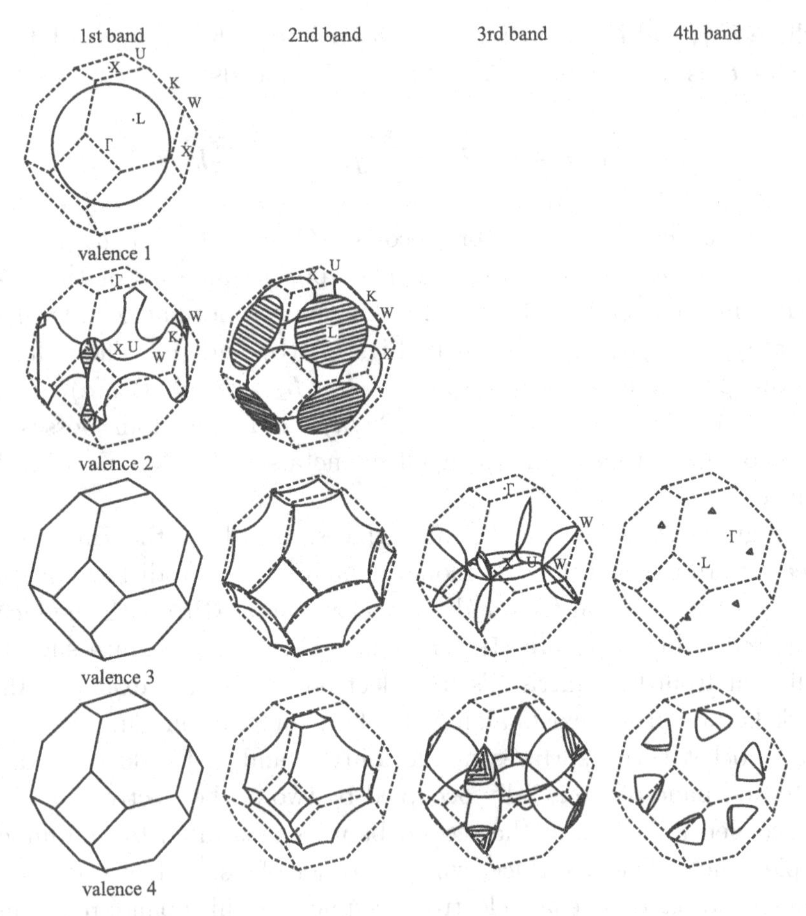

Figure 1.3: Fermi surfaces of Cu(valence 1), Ca(valence 2), Al(valence 3), and Pb(valence 4) with the face-centered cubic structure on the basis of the Harrison's free electron model, cited from Ref. [1].

in state i is given as a solution of the equation

$$\left\{-\frac{\hbar^2}{2m}\nabla^2 + U(\boldsymbol{r}) + e^2\int \frac{n(\boldsymbol{r}')}{|\boldsymbol{r}-\boldsymbol{r}'|}d\boldsymbol{r}' + v_{\rm xc}(\boldsymbol{r})\right\}\Psi_i(\boldsymbol{r}) = \varepsilon_i\Psi_i(\boldsymbol{r}) \tag{1.15}$$

where $n(\boldsymbol{r})$ is the electron density, given by

$$n(\boldsymbol{r}) = \sum_{i=1}^{N} |\Psi_i(\boldsymbol{r})|^2 \,. \tag{1.16}$$

In the left-hand side of Eq. (1.15), the first term is the kinetic energy, the second term the nuclear potential, the third term a direct Coulomb potential, and the fourth term $v_{\rm xc}(\boldsymbol{r})$, the exchange-correlation potential defined by

$$v_{\rm xc}(\boldsymbol{r}) = \frac{\delta E_{\rm xc}[n]}{\delta n(\boldsymbol{r})} \tag{1.17}$$

where $E_{\rm xc}$ is the exchange-correlation energy, which should be a complicated functional of the electron density n. Its exact form is unknown. The most essential point is how a reasonably good and simple form of the functional for the exchange-correlation potential can be found. The local density functional or the local density approximation (LDA) proposed by Gunnarsson and Lundqvist [4] is a drastic approximation. This is based on the assumption that the electron density is a slowly varying function of space. The exchange-correlation potential is thus expressed as

$$v_{\rm xc}(\boldsymbol{r}) = \alpha e^2 \left[\frac{3}{\pi} n(\boldsymbol{r})\right]^{1/3} \tag{1.18}$$

where α is constant, for example $\alpha \simeq 0.7$.

In order to calculate the energy band structure of rare earth and uranium compounds, relativistic effects should be taken into account because the extremely strong nuclear potential extends into the core region of the atom. Note that lanthanides and actinides have high atomic numbers. In this case, the spin-orbit interaction is taken into account self-consistently for all the valence electrons as in a second variational procedure. Alternatively, we can use the Dirac one-electron wave function instead of Eq. (1.15). Some standard techniques to calculate the band structure self-consistently within a required accuracy are the Green's function or Korringa-Kohn-Rostoker (KKR) method, the linearized muffin-tin orbital (LMTO) method, and the augmented plane wave (APW) method or linearized APW (LAPW) method, etc [5].

Here, we show the exact Fermi surfaces. The topology of the Fermi surface changes as a function of the number of valence electrons. This is well known as the Harrison's free electron model mentioned above. A compensated metal ($V_e = V_h$) is extremely different from an uncompensated metal ($V_e \neq V_h$) in topology. Here, V_e refers to the volume of an electron Fermi surface and V_h is the volume of a hole Fermi surface. For example, Fermi surface properties of rare-earth and actinide compounds with the $AuCu_3$-type cubic and $HoCoGa_5$-type tetragonal crystal structures have been shown to be functions of the number of valence electrons. See ref. [6] for details.

The dimensionality of the electronic state is another factor that changes the topology of the Fermi surface. If conduction electrons can move freely in real space, the topology of the Fermi surface is spherical, and this can be described as $\varepsilon_F = (\hbar^2/2m^*)(k_x^2 + k_y^2 + k_z^2)$, as shown in Fig. 1.4(a). If conduction electrons can move only in the x-y plane and not along the z-axis, the topology of the Fermi surface changes from a sphere to a cylinder: $\varepsilon_F = (\hbar^2/2m^*)(k_x^2+k_y^2)$, as shown in Fig. 1.4(b), revealing a two-dimensional electronic state. High-T_c cuprates are typical examples. If instead, the conduction electrons can move only along the z-axis [$\varepsilon_F = (\hbar^2/2m^*)k_z^2$] the topology of the Fermi surface can be changed into two plates, as shown in Fig. 1.4(c). This is one-dimensional in the electronic state. In this case, the well-known Peierls instability is realized, and a one-dimensional conductor becomes an insulator. Organic conductors are typical examples.

When the electronic state at $\boldsymbol{k}$ is transferred by a propagation vector $\boldsymbol{q}$, the electronic or Fermi surface instability occurs at $\varepsilon_{\boldsymbol{k}} = \varepsilon_{\boldsymbol{k}+\boldsymbol{q}}$. Here, $\varepsilon_{\boldsymbol{k}}$ and $\varepsilon_{\boldsymbol{k}+\boldsymbol{q}}$ are the electronic energies at $\boldsymbol{k}$ and $\boldsymbol{k}+\boldsymbol{q}$, respectively. An overlapping region is, however, only one point for the spherical Fermi surface, as shown in Fig. 1.4(d), and one line along the k_z-axis for the cylindrical Fermi surface. In these cases, the Fermi surface instability is not realized. The overlapping of the Fermi surface for the propagation vector $\boldsymbol{q}$ is called "nesting". The nesting of the Fermi surface occurs for a characteristic two-dimensional compound, 1T-TaS_2, for example. Figure 1.4(e) shows the characteristic cylindrical Fermi surface in 1T-TaS_2 [7, 8]. The nesting is realized in a wide region of the Fermi surface. The overlapping region of the Fermi surface disappears in the two-(or three-)dimensional case of the Fermi surface instability, which is called "charge density wave" (CDW) instability. Correspondingly, the lattice is distorted with a

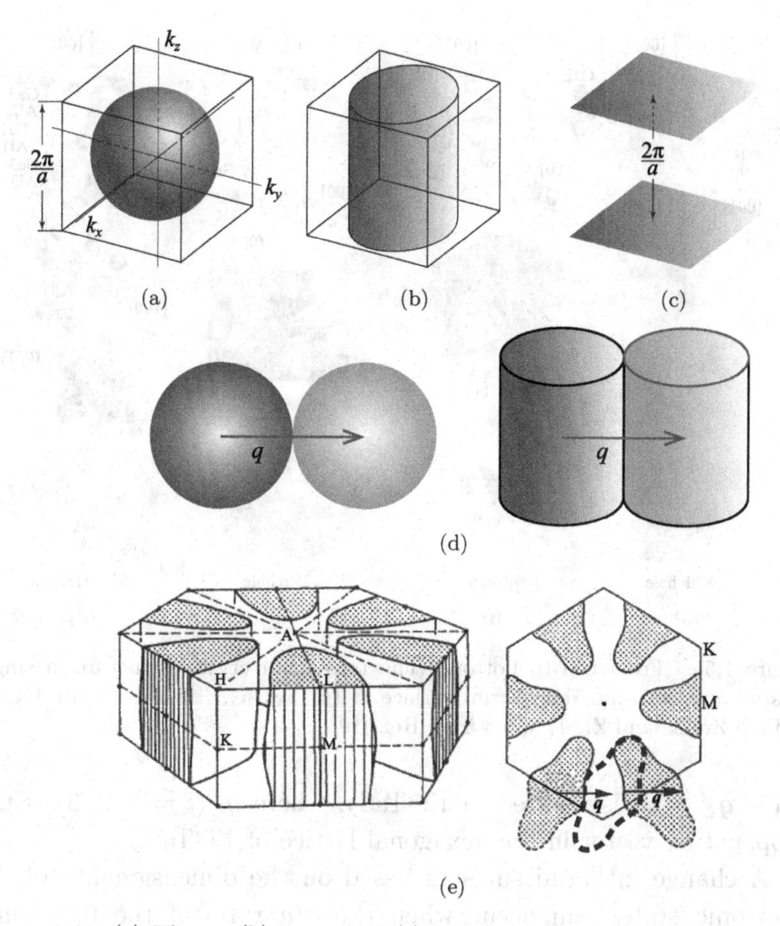

Figure 1.4: (a) Three-, (b) two-, and (c) one-dimensional Fermi surfaces, (d) nesting of the spherical and cylindrical Fermi surfaces, and (e) nesting of the Fermi surface in 1T-TaS$_2$, cited from Refs. [7, 8].

wavelength $2\pi/q$. A metallic state of 1T-TaS$_2$ at high temperatures with a carrier number of 10^{22} cm^{-3} is changed into an insulator at low temperatures with a carrier number of 10^{18} cm^{-3}(carriers of impurities), demonstrating an incommensurate CDW at 350 K with a partial disappearance of the Fermi surface, and a commensurate CDW at 180 K with a complete disappearance of the Fermi surface [9]. In this case, the term "commensurate" applies to the case where there is an integer number ratio of the reciprocal lattice vector $\boldsymbol{a}^*$ to the propagation vector $\boldsymbol{q}$, such as $3\boldsymbol{q} = \boldsymbol{a}^*$. A relation of

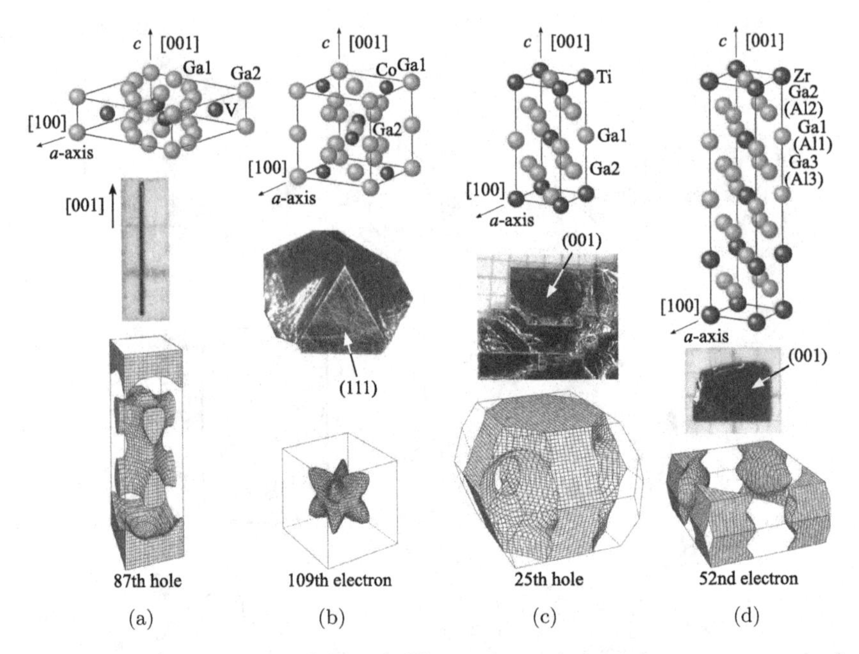

Figure 1.5: (From top to bottom) The tetragonal crystal structure, a single crystal, and corresponding Fermi surface in (a) V_2Ga_5, (b) $CoGa_3$, (c) $TiGa_3$, and (d) $ZrGa_3$ and $ZrAl_3$, cited from Ref. [10].

$3\boldsymbol{q}_1 - \boldsymbol{q}_2 = \boldsymbol{a}^*$ is realized in 1T-TaS_2, where $\boldsymbol{q}_i$ (i = 1, 2, 3) is the propagation vector in the hexagonal lattice of 1T-TaS_2.

A change in Fermi surfaces based on the dimensionality of the electronic states can occur when the c/a ratio of the tetragonal structure is continuously changed. Figure 1.5 shows several different tetragonal structures in T-Ga (T: transition metal) binary compounds [10]. The following is a main point in this study: the smaller the tetragonal c-value, the more enhanced the one-dimensionality of the electronic state; the larger the c-value, the more enhanced the two-dimensionality. We expect that nearly one-dimensional characteristic Fermi surfaces can be realized in V_2Ga_5 (space group No. 127, $a = 8.936$ Å, $c = 2.683$ Å, formula units per cell $Z = 2$), and two-dimensional Fermi surfaces in $ZrGa_3$ (No. 139, $a = 3.965$ Å, $c = 17.461$ Å, $Z = 4$) and $ZrAl_3$. Fermi surfaces in $CoGa_3$ (No. 136, $a = 6.230$ Å, $c = 6.431$ Å, $Z = 4$) are three-dimensional because the c-value is almost the same as the a-value. In contrast to these compounds, $TiGa_3$ (No. 139, $a = 3.789$ Å, $c = 8.734$ Å, $Z = 2$) crystallizes in a typical tetragonal structure as in $ThCr_2Si_2$.

Single crystals of these compounds have been grown by the Ga (Al) self-flux method, and de Haas-van Alphen (dHvA) experiments carried out, together with the full-potential linearized augmented plane wave (FLAPW) energy band calculations.

The characteristic shapes of single crystals in these compounds, as shown in Fig. 1.5, are interesting. The characteristic features of crystal structures and their corresponding single crystals are summarized below:

1) In V_2Ga_5, the crystal structure is flat along the tetragonal [001] direction (c-axis), and the corresponding single crystal is of needle shape along the [001] direction.
2) $CoGa_3$ is tetragonal in the crystal structure, but is approximately cubic, as shown in Fig. 1.5(b). Correspondingly, $CoGa_3$ is of pyramidal shape, which is characteristic of the crystals with the fcc structure, and also with the diamond-type structure. The flat plane of the pyramid corresponds to the (111) plane.
3) The shape of a single crystal in $TiGa_3$ is typically tetragonal. The flat plane of a rectangular single crystal corresponds to the tetragonal (001) plane (c-plane).
4) In $ZrGa_3$ and $ZrAl_3$, the crystal structure is elongated along the tetragonal [001] direction, and the corresponding single crystal is of thin-plate shape, of which the flat plane corresponds to the tetragonal (001) plane.

Reflecting the crystal structures or the corresponding Brillouin zones, the Fermi surfaces have characteristic properties which have been clarified through dHvA experiments and energy band calculations. They are summarized as follows:

1) A nearly one-dimensional plate-like Fermi surface is obtained in the band calculation, as shown in Fig. 1.5(a). The plate is, however, wavy in shape, and the electronic state is not

Table 1.1: Electronic specific heat coefficients γ in V_2Ga_5, $CoGa_3$, $TiGa_3$, $ZrGa_3$, and $ZrAl_3$ and the corresponding theoretical values γ_b, cited from Ref. [10].

	γ mJ/(K^2·mol)	γ_b mJ/(K^2·mol)	γ/γ_b
V_2Ga_5	16.7	10.41	1.60
$CoGa_3$	2.5	2.32	1.08
$TiGa_3$	3.6	3.57	1.01
$ZrGa_3$	2.5	2.41	1.04
$ZrAl_3$	2.7	2.50	1.08

one-dimensional but nearly three-dimensional. Therefore, Peierls instability is not realized in this compound.

2) The Fermi surfaces in $CoGa_3$ are very similar to those of Ni_3Ga with the $AuCu_3$-type cubic structure. The band 109th electron Fermi surface is typical between the two compounds, as shown in Fig. 1.5(b).
3) The Fermi surfaces in $TiGa_3$ are also very similar to those of YCu_2Si_2 with the $ThCr_2Si_2$-type tetragonal structure. The band 25th hole Fermi surface is typical between the two compounds, as shown in Fig. 1.5(c).
4) The flat Brillouin zone often produces the cylindrical Fermi surface, which is realized in $ZrGa_3$ and $ZrAl_3$. The band 52nd electron Fermi surface is typical, possessing concave and convex shapes in the cylinder.

Finally, we summarize in Table 1.1 the γ values of these compounds and the corresponding theoretical γ_b values. The γ value of 16.7 mJ/(K^2·mol) of V_2Ga_5 is slightly larger than γ_b of 10.41 mJ/(K^2·mol), which is characteristic in the $3d$-electron system. The experimental and theoretical values of the others compounds are approximately the same under consideration of a small mass enhancement based on the electron–phonon interaction.

1.3 Interacting conduction electrons -Fermi liquid-

The concept of the Fermi surface has already been introduced earlier in the one-electron band theory of metals in which the electron-electron interactions (correlations) are treated to some extent. It is important to check experimentally the magnitude of the electron-electron interaction in metals. This was obtained by measuring the electrical resistivity below 1.5 K, for example, for Al [11]. The resistivity ρ is expressed as

$$\rho = \rho_0 + AT^2.$$

The term $AT^2 (= \rho_{e\text{-}e})$ is due to the electron-electron scattering. The corresponding two electrons are occupied within a small energy region of $k_\mathrm{B}T$ from the Fermi energy ε_F. This is because the initial states of two scattering electrons are occupied in a Fermi surface while the final states after the electron-electron scattering correspond to two unoccupied states by the Pauli principle. The AT^2 term is simply obtained from a relation of $\rho_{e\text{-}e} \sim \tau^{-1} \sim (k_\mathrm{B}T/\varepsilon_\mathrm{F})^2 \sim T^2$, where $\varepsilon_\mathrm{F} \sim {m^*}^{-1}$ from Eq. (1.8b) and τ is the scattering lifetime. Note that the mass of Eq. (1.8b) is m. This is changed to the effective mass m^*, considering the mass enhancement of the electron-electron interaction. The low-temperature (below 1.5 K) electrical resistivity is of the form $\rho = \rho_0 + AT^2$ ($A = 3 \times 10^{-7}$ $\mu\Omega\cdot\mathrm{cm/K^2}$) in Al. This is very small in value. Experimentally, it is very difficult to detect the electron-electron interaction in the usual metal. In the heavy fermions of some rare earth and uranium compounds, the AT^2 term is extremely large and not negligible — for example, $A = 10^2$ $\mu\Omega\cdot\mathrm{cm/K^2}$ in $YbCo_2Zn_{20}$, which is shown later. Note that $\sqrt{A}$ is proportional to m^*, or $A \propto m^{*2}$.

Landau investigated the interacting fermion system of ^{3}He. The theoretical results are known as Landau's Fermi liquid theory, which are applied to interacting electrons in metals [12]. Landau arrived at a surprising conclusion. A system of interacting electrons can be attained when we start from a system of non-interacting electrons and slowly turn on the interaction. The ground state and low-lying excitations of Fermi liquid are therefore in one-to-one correspondence to those of the non-interacting Fermi gas. Note that the distribution function in the ground state for the interacting electrons $n(\boldsymbol{k}\sigma)$ is changed, shown schematically in Fig. 1.6(a), which is compared with the usual Fermi-Dirac distribution function of $n(\boldsymbol{k}) = 1/[(\varepsilon_{\boldsymbol{k}} - \mu)/k_\mathrm{B}T + 1]$ in Fig. 1.6(b). Here, μ is the chemical

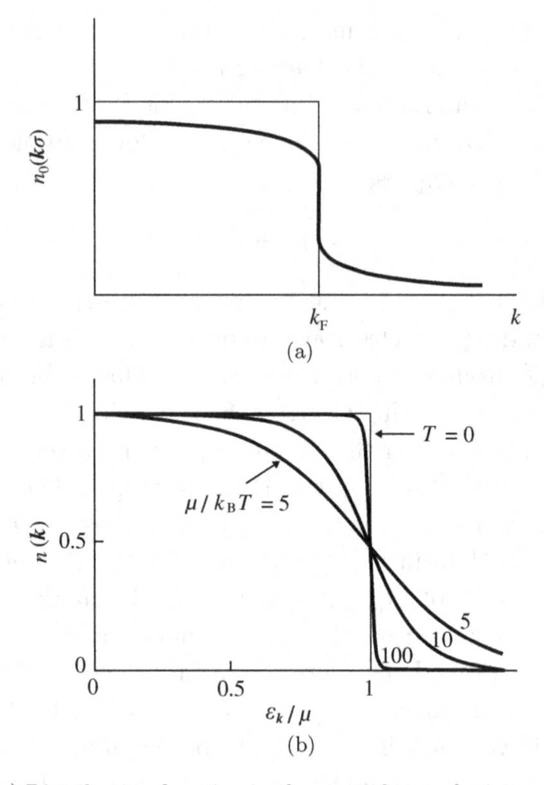

Figure 1.6: (a) Distribution function in the ground state for interacting electrons and (b) the Fermi-Dirac distribution function as a function of temperature.

potential. However, a sharp step at k_F remains from the one-to-one correspondence mentioned above. A change in the energy δE caused by a change in the distribution function of $n(\boldsymbol{k}\sigma) = \boldsymbol{n_0}(\boldsymbol{k}\sigma) + \delta n(\boldsymbol{k}\sigma)$ is given by

$$\delta E = \sum_{\boldsymbol{k}} \sum_{\sigma} \varepsilon_{\boldsymbol{k}} \delta n(\boldsymbol{k}\sigma) + \frac{1}{2} \sum_{\boldsymbol{k}} \sum_{\sigma} \sum_{\boldsymbol{k}'} \sum_{\sigma'} \times f(\boldsymbol{k}, \sigma, \boldsymbol{k}', \sigma') \delta n(\boldsymbol{k}, \sigma) \delta n(\boldsymbol{k}', \sigma') \tag{1.19}$$

where the function $f(\boldsymbol{k}, \sigma, \boldsymbol{k}', \sigma')$, which was introduced by Landau, is unknown, but characterizes the electron-electron interaction. The theoretical results obtained from Landau's Fermi-liquid theory are as follows. The effective mass m^* of the interacting electrons is enhanced as

$$m^* = m\left(1 + \frac{F_1^s}{3}\right). \tag{1.20}$$

The interacting electrons are called "quasiparticles" in Landau's Fermi-liquid theory, where names correspond to "dressed electrons". F_1^s is the interaction parameter derived from $f(\boldsymbol{k}, \sigma, \boldsymbol{k}', \sigma')$, together with F_0^a shown later. The electronic specific heat coefficient γ is given as

$$\gamma = \frac{2\pi^2 k_{\rm B}^2}{3} D^*(\varepsilon_{\rm F}). \tag{1.21}$$

The spin susceptibility χ_s is

$$\chi_s = \frac{2\mu_{\rm B}^2 D^*(\varepsilon_{\rm F})}{1 + F_0^a}. \tag{1.22}$$

The factor $1/(1 + F_0^a)$ corresponds to the Stoner enhancement factor shown in Chap. 4.

The effective mass m_c^* determined by the de Haas-van Alphen experiment is usually different from the band mass m_b, particularly in Ce, Yb, and U compounds. Here, the experimental γ and m_c^* values are usually larger than the theoretical γ_b and m_b values, which are obtained by the energy band calculations. The mass enhancement factor λ is defined as:

$$\frac{\gamma}{\gamma_b} = \frac{m_c^*}{m_b} = 1 + \lambda.$$

Origins for λ are ascribed to the many-body effects, which cannot be taken into account in the usual band theory. As for most probable origins, the electron-phonon interaction and the magnetic interactions are considered, and their contributions are denoted by λ_p and λ_m, respectively. Therefore, λ is expressed as a sum of two contributions

$$\lambda = \lambda_p + \lambda_m.$$

The electron-phonon term λ_p in normal metals such as Pb, including its temperature dependence, is well understood at present. Its magnitude is smaller that 1. If it were large, it might cause lattice instability. In contrast to this small value of λ_p, the magnetic contribution λ_m can take a large value in some Ce, Yb, and U compounds.

The magnetic contribution λ_m can be divided into two cases according to its origins. The first case occurs in many lanthanide compounds in which the $4f$ electrons are localized at lanthanide ions and their spin fluctuations enhance the effective mass of the conduction electrons via c–f interactions such as the Ruderman-Kittel-Kasuya-Yosida (RKKY) interaction, where c stands for conduction electrons and f for $4f$ electrons. A small mass enhancement of $\lambda_m = 1$–2 is observed in lanthanide compounds.

Another magnetic contribution to λ_m occurs when the $3d$ electrons in the ion-series transition metal compounds are itinerant and their spins are fluctuating. The magnitude of λ_m is, however, not extremely large; $\lambda_m = 1$–5 in the $3d$ electron system. In some Ce, Yb, and U compounds, λ_m is extremely large; $\lambda_m = 60$ in $CeRu_2Si_2$, for example, which is described later. The f electrons in these Ce, Yb, and U compounds are localized at temperatures higher than room temperature, but become itinerant at low temperatures, via the Kondo effect. The heavy f-electron system is a main theme in this text.

Chapter 2

Local Magnetic Moment

The f electrons in rare earth and actinide compounds are localized at high temperatures. The so-called RKKY interaction leads to long-range magnetic ordering of the f electrons at low temperatures, where the indirect f-f interaction is mediated by the spin polarization of the conduction electrons. The corresponding ordered moment is based on the crystalline electric field (CEF) scheme of the f electrons. In some CEF cases, the quadrupole moment becomes dominant instead of the magnetically ordered moment, which leads to quadrupole ordering.

2.1 Origin of magnetic moment

Strong magnetism is realized in the compounds containing the first-line series of transition metals, lanthanoids (lanthanides), and actinoids (actinides), as mentioned in Chap. 1. Their electron configurations are as follows

transition metals [Ar core] $\mathbf{3d^n}4s^2 (n = 0 - 10)$
lanthanides [Kr core] $\mathbf{4f^n}5s^2 5p^6 5d^1 6s^2 (n = 0 - 14)$
actinides [Xe core] $\mathbf{5f^n}6s^2 6p^6 6d^1 7s^2 (n = 0 - 14)$.

We will consider the Schrödinger wave function $\Psi(r, \theta, \phi)$ which describes the motion of an electron around the nucleus. It is given as

$$\Psi(r, \theta, \phi) = R(r)Y(\theta, \phi)$$

where $R(r)$ and $Y(\theta, \phi)$ are the radial and angular parts of $\Psi(r, \theta, \phi)$, respectively, in a spherical potential $V(r)$. The radial charge density $|rR(r)|^2$ can be calculated using the radial part of the electronic wave function. The radial part of the wave function is obtained by solving the following Schrödinger equation

$$-\frac{\hbar^2}{2m}\frac{1}{r^2}\frac{d}{dr}\left(r^2\frac{dR}{dr}\right) + V(r)R + \frac{l(l+1)\hbar^2}{2mr^2}R = \varepsilon R \tag{2.1}$$

where the contribution to $V(r)$ is mainly the Coulomb potential and the third term is the centrifugal potential. The azimuthal quantum number l is 2 for $3d$ and 3 for $4f$ and $5f$ electrons. Their radial charge densities in Ni, Ce, and U atoms are shown in Figs. 2.1(a), (b), and (c), respectively. The $4f$ electron in the Ce atom, as shown in Fig. 2.1(b), is pushed deep into the interior of the $5d$ and $6s$ electrons due to a large centrifugal potential, which is compared with the $3d$ electrons in Ni in Fig. 2.1(a) [13]. This means that the $4f$ electrons are more localized than the $3d$ electrons. Magnetism in the rare earth metallic compounds arises from the $4f$ electrons with a local moment. Therefore, a direct magnetic interaction does not occur between the $4f$ electrons because of a non-overlap of wavefunctions. A magnetic order does occur via an indirect interaction, which is mediated by the spins of conduction electrons. This indirect interaction is the RKKY interaction. The $5f$ electrons in the U-based metallic compounds have a character between that of the itinerant-like $3d$ electrons and the localized $4f$ electrons.

Furthermore, we consider the relativistic effect of Ce and U atoms with large atomic numbers Z. This is because the velocity of a $1s$ electron is close to the velocity of light. Here, we calculate velocity v of an electron, for example, in a stationary circulating orbit of radius r. The centrifugal force mv^2/r is provided by the Coulomb force Ze^2/r^2. That is

$$\frac{mv^2}{r} = \frac{Ze^2}{r^2}.$$

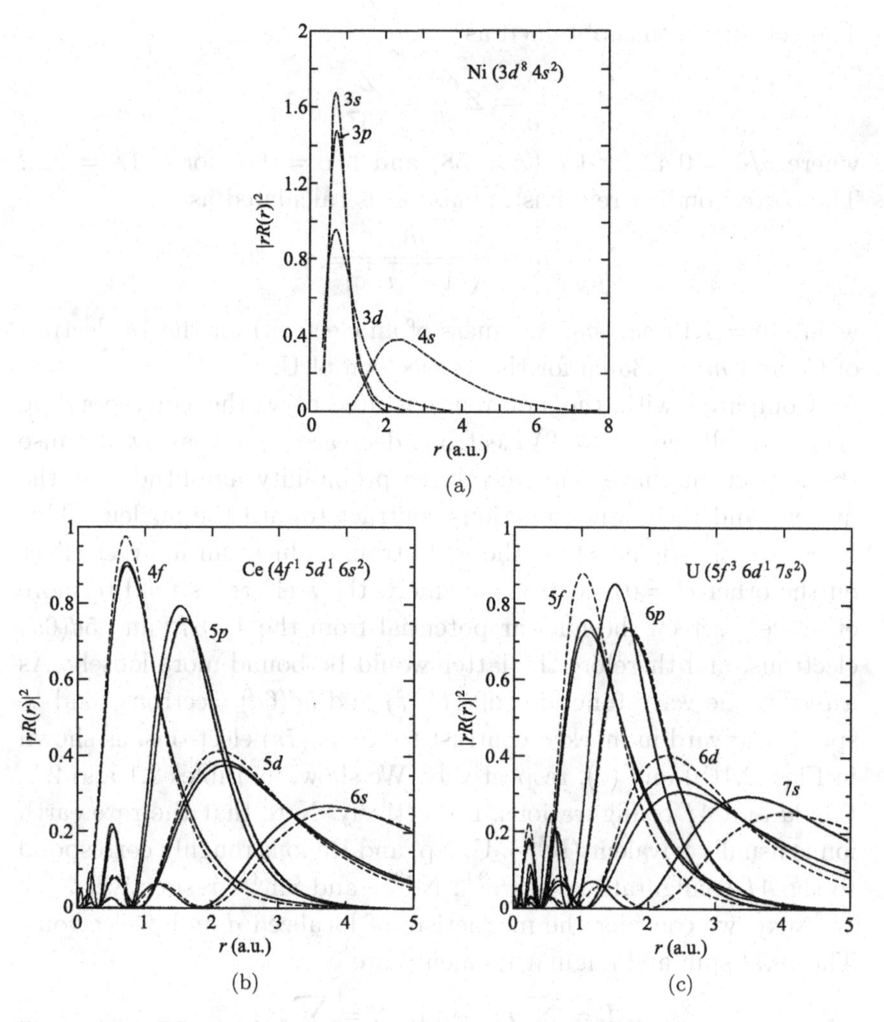

Figure 2.1: Effective radial charge densities of (a) Ni [13], (b) Ce, and (c) U, where dashed and solid lines are due to non-relativistic and relativistic calculations, respectively. Figures provided courtesy of Y. Higuchi and A. Hasegawa. The horizontal scale of a.u. refers to atomic unit.

The Bohr postulate for the $1s$ electron (the principal quantum number $n = 1$) is

$$mvr = \hbar.$$

The velocity is thus obtained as

$$\frac{v}{c} = Z\frac{e^2}{c\hbar} = \frac{Z}{137},$$

where $v/c = 0.42$ for Ce (Z = 58) and $v/c = 0.67$ for U (Z = 92). The corresponding relativistic mass m is calculated as

$$m = \frac{m_0}{\sqrt{1-(v/c)^2}}$$

where $m = 1.19m_0$ (m_0: rest mass of an electron) for the 1s electron of Ce and $m = 1.35m_0$ for the 1s electron of U.

Compared with the non-relativistic theory, the corresponding energy of all the s ($l = 0$) electrons decreases significantly, because the s electrons have relatively large probability amplitudes at the nucleus and their wave functions contract toward the nucleus. This direct relativistic effect on the s electrons induces an indirect effect on the other ($l \neq 0$) electrons. Namely, the s electrons tend to more effectively screen the nuclear potential from the $4f(5f)$ and $5d(6d)$ electrons, and therefore the latter would be bound more loosely. As a result, the wave functions of $4f(5f)$ and $5d(6d)$ electrons tend to spread outward in marked contrast to the $6s(7s)$ electrons, as shown in Figs. 2.1(b) and (c), respectively. We show, in Tables 2.1 and 2.2, the $3d$ and $4f$ configurations, respectively. Note that the rare earth ion is usually trivalent (R^{3+}). U, Np, and Pu ions roughly correspond to the $4f$ configurations of Pr^{3+}, Nd^{3+}, and Sm^{3+}, respectively.

Next, we consider the magnetism of localized d and f electrons. The total spin and angular momenta are

$$\boldsymbol{L} = \sum_i \boldsymbol{\ell}_i \quad \text{and} \quad \boldsymbol{S} = \sum_i \boldsymbol{s}_i.$$

The ground state determination is worked out with the help of Hund's rules:

(1) $\boldsymbol{S}$ has the maximum value consistent with the Pauli principle.
(2) $\boldsymbol{L}$ has the maximum value consistent with the Pauli principle and rule (1).

The multiplicity of each term is $(2L+1)(2S+1)$. Note that the maximum S is associated with a fully antisymmetric spatial wave

Table 2.1: 3*d* configuration.

Ion	Ground state	Electron number	m_l: 2	1	0	−1	−2	S	L	J	g_J	μ_{eff}: $g_J\sqrt{J(J+1)}$	$2\sqrt{S(S+1)}$
$Ti^{3+}V^{4+}$	$^2D_{3/2}$	1	↑					1/2	2	3/2	4/5	1.55	1.73
V^{3+}	$^3F_2$	2	↑	↑				1	3	2	2/3	1.63	2.83
$V^{2+}Cr^{3+}Mn^{4+}$	$^4F_{3/2}$	3	↑	↑	↑			3/2	3	3/2	2/5	0.77	3.87
$Cr^{2+}Mn^{3+}$	$^5D_0$	4	↑	↑	↑	↑		2	2	0	–	0	4.90
$Mn^{2+}Fe^{3+}$	$^6S_{5/2}$	5	↑	↑	↑	↑	↑	5/2	0	5/2	2	5.92	5.92
Fe^{2+}	$^5D_4$	6	↑↓	↑	↑	↑	↑	2	2	4	3/2	6.70	4.90
Co^{2+}	$^4F_{9/2}$	7	↑↓	↑↓	↑	↑	↑	3/2	3	9/2	4/3	6.54	3.87
Ni^{2+}	$^3F_4$	8	↑↓	↑↓	↑↓	↑	↑	1	3	4	5/4	5.59	2.83
Cu^{2+}	$^2D_{5/2}$	9	↑↓	↑↓	↑↓	↑↓	↑	1/2	2	5/2	6/5	3.55	1.73

Table 2.2: $4f$ configuration.

Ion	Ground state	Electron number	m_l 3	2	1	0	−1	−2	−3	S	L	J	g_J	$g_J J$	$\mu_{\mathrm{eff}} = g_J\sqrt{J(J+1)}$
Ce^{3+}	$^2F_{5/2}$	1	↑							1/2	3	5/2	6/7	2.14	2.54
Pr^{3+}	$^3H_4$	2	↑	↑						1	5	4	4/5	3.2	3.58
Nd^{3+}	$^4I_{9/2}$	3	↑	↑	↑					3/2	6	9/2	8/11	3.27	3.62
Pm^{3+}	$^5I_4$	4	↑	↑	↑	↑				2	6	4	3/5	2.4	2.68
Sm^{3+}	$^6H_{5/2}$	5	↑	↑	↑	↑	↑			5/2	5	5/2	2/7	0.71	0.85
Eu^{3+}	$^7F_0$	6	↑	↑	↑	↑	↑	↑		3	3	0	–	0	0
Gd^{3+}	$^8S_{7/2}$	7	↑	↑	↑	↑	↑	↑	↑	7/2	0	7/2	2	7	7.94
Tb^{3+}	$^7F_6$	8	↑↓	↑	↑	↑	↑	↑	↑	3	3	6	3/2	9	9.72
Dy^{3+}	$^6H_{15/2}$	9	↑↓	↑↓	↑	↑	↑	↑	↑	5/2	5	15/2	4/3	10	10.65
Ho^{3+}	$^5I_8$	10	↑↓	↑↓	↑↓	↑	↑	↑	↑	2	6	8	5/4	10	10.61
Er^{3+}	$^4I_{15/2}$	11	↑↓	↑↓	↑↓	↑↓	↑	↑	↑	3/2	6	15/2	6/5	9	9.58
Tm^{3+}	$^3H_6$	12	↑↓	↑↓	↑↓	↑↓	↑↓	↑	↑	1	5	6	7/6	7	7.56
Yb^{3+}	$^2F_{7/2}$	13	↑↓	↑↓	↑↓	↑↓	↑↓	↑↓	↑	1/2	3	7/2	8/7	4	4.54

function. Then, each electron (or hole if the shell is more than half filled) belongs to a different orbital, which lowers the repulsive electrostatic energy. Correspondingly, the correlations between electrons give rise to "ferromagnetism" inside the atom.

In the atom, there is a coupling between the spin and orbital magnetic moments. Each electron sees the nucleus moving and is therefore subject to a magnetic field acting on the spin of the atom. This perturbation, or the spin-orbit coupling, which is based on a relativistic interaction, can be written as

$$\mathcal{H}_{\rm so} = \lambda \boldsymbol{L} \cdot \boldsymbol{S}. \tag{2.2}$$

The L-S term is thus split into J-multiplets characterized by the new quantum number J ($\boldsymbol{J} = \boldsymbol{L} + \boldsymbol{S}$, where $\boldsymbol{J}$ is the total angular momentum). The ground state multiplet is such that $J = |L - S|$ if the shell is less than half filled and $J = L + S$ in the other case. The multiplicity of each multiplet is $2J + 1$.

The incomplete shells have orbital and spin angular momenta and thus lead to a magnetic moment $\boldsymbol{\mu}$ given by

$$\boldsymbol{\mu} = -\mu_{\rm B}(\boldsymbol{L} + 2\boldsymbol{S}) \tag{2.3a}$$

or

$$\boldsymbol{\mu}_J = -g_{\rm J}\mu_{\rm B}\boldsymbol{J} \tag{2.3b}$$

where $g_{\rm J}$ is the Landé g factor

$$g_{\rm J} = \frac{3}{2} + \frac{S(S+1) - L(L+1)}{2J(J+1)}. \tag{2.4}$$

The relation between $\boldsymbol{J}$ and $\boldsymbol{L}$+2$\boldsymbol{S}$ is schematically shown in Fig. 2.2. $\boldsymbol{S}$ and $\boldsymbol{L}$ rotate precessionally along $\boldsymbol{J}$, revealing that $\boldsymbol{J}$ and $g_J\boldsymbol{J}$ do not depend on time. In a magnetic field $\boldsymbol{H}$, $g_J\boldsymbol{J}$ rotates along the field direction.

When the magnetic field H is applied to the d or f electron systems, the so-called Zeeman energy is written as follows

$$\mathcal{H}_{\rm Z} = -\boldsymbol{\mu}_J \cdot \boldsymbol{H}$$

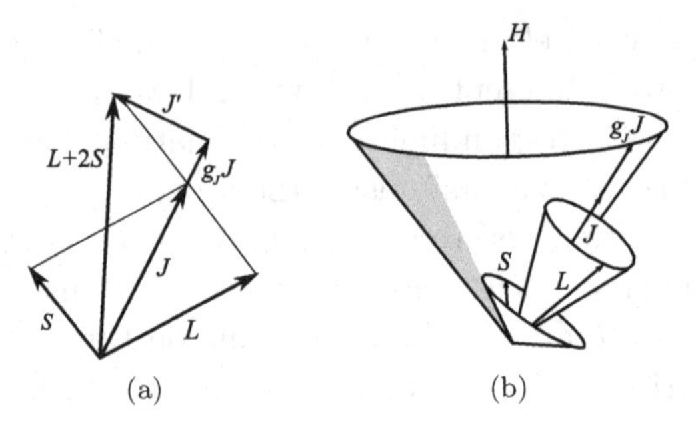

Figure 2.2: (a) Relation between $\boldsymbol{J}$ and $\boldsymbol{L}+2\boldsymbol{S}$ and (b) precessions of $\boldsymbol{S}$ and $\boldsymbol{L}$ along $\boldsymbol{J}$, together with $g_J\boldsymbol{J}$ in the magnetic field $\boldsymbol{H}$.

$$
\begin{aligned}
&= g_{\mathrm{J}}\mu_{\mathrm{B}}\boldsymbol{J}\cdot\boldsymbol{H} \\
&= g_{\mathrm{J}}\mu_{\mathrm{B}}J_zH \quad (J_z=-J,-J+1,\ldots,J-1,J).
\end{aligned} \tag{2.5}
$$

The magnetization M is given as

$$
\begin{aligned}
M = N\langle\mu_{J_z}\rangle &= N\frac{\displaystyle\sum_{J_z=-J}^{J} -g_{\mathrm{J}}\mu_{\mathrm{B}}J_z\, e^{-\frac{g_{\mathrm{J}}\mu_{\mathrm{B}}J_zH}{k_{\mathrm{B}}T}}}{\displaystyle\sum_{J_z=-J}^{J} e^{-\frac{g_{\mathrm{J}}\mu_{\mathrm{B}}J_zH}{k_{\mathrm{B}}T}}} \\
&= Ng_{\mathrm{J}}\mu_{\mathrm{B}}JB_J\left(\frac{g_{\mathrm{J}}\mu_{\mathrm{B}}JH}{k_{\mathrm{B}}T}\right)
\end{aligned} \tag{2.6}
$$

where N is the number of magnetic atoms (ions), and $B_J(\frac{g_{\mathrm{J}}\mu_{\mathrm{B}}JH}{k_{\mathrm{B}}T})$ is the Brillouin function. Under the condition of $(g_{\mathrm{J}}\mu_{\mathrm{B}}JH/k_{\mathrm{B}}T)\ll 1$, namely high temperatures and small magnetic fields, the magnetic susceptibility $\chi(=M/H)$ follows the Curie law of $\chi=C/T$, so that

$$
\begin{aligned}
&B_J\left(\frac{g_{\mathrm{J}}\mu_{\mathrm{B}}JH}{k_{\mathrm{B}}T}\right)\simeq\frac{J+1}{3J}\frac{g_{\mathrm{J}}\mu_{\mathrm{B}}JH}{k_{\mathrm{B}}T} \\
&M = N\frac{g_{\mathrm{J}}^2\mu_{\mathrm{B}}^2J(J+1)}{3k_{\mathrm{B}}T}H \\
&\chi=\frac{M}{H}=N\frac{g_{\mathrm{J}}^2\mu_{\mathrm{B}}^2J(J+1)}{3k_{\mathrm{B}}T}=N\frac{\mu_{\mathrm{eff}}^2\mu_{\mathrm{B}}^2}{3k_{\mathrm{B}}T}
\end{aligned} \tag{2.7}
$$

and

$$\mu_{\text{eff}} = g_{\text{J}}\sqrt{J(J+1)}. \tag{2.8}$$

Under the opposite condition $(g_{\text{J}}\mu_{\text{B}}JH/k_{\text{B}}T) \gg 1$, which corresponds to low temperatures and high magnetic fields,

$$B_J\left(\frac{g_{\text{J}}\mu_{\text{B}}JH}{k_{\text{B}}T}\right) = 1$$

and

$$M = N\mu_{\text{s}} = Ng_{\text{J}}\mu_{\text{B}}J. \tag{2.9}$$

The effective magnetic moment μ_{eff} in Eq. (2.8) is generally consistent with the experimental values of the rare earth compounds except Sm and Eu ions. On the other hand, it is close to $2\sqrt{S(S+1)}$ in the $3d$ compounds, which is only due to the contribution of the spin moment ($L = 0$ and $g_{\text{J}} = 2$). This is mainly due to the crystalline electric field (CEF), which results from the interaction between the non-spherical $3d$-orbital and the electrostatic fields of the environment. As shown in Fig. 2.3, the spin-orbit coupling is much larger in

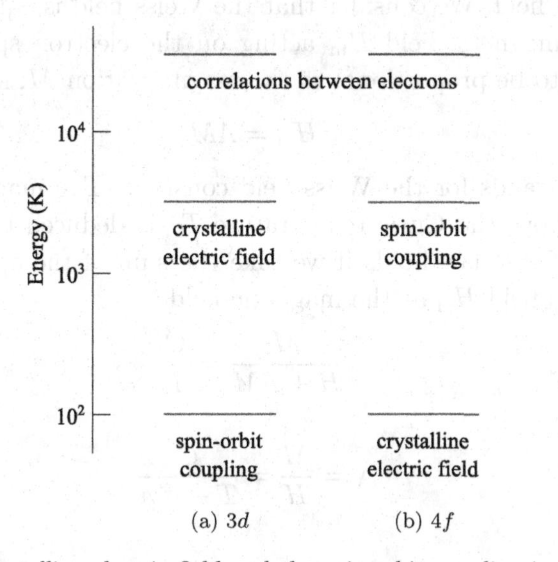

Figure 2.3: Crystalline electric field and the spin-orbit coupling in (a) $3d$ and (b) $4f$ electrons.

the $4f$ series than in the $3d$ ones since the rare earth elements are heavy. Conversely, the CEF is weaker in the $4f$ series, since the $4f$ shell is screened by the outer $5s^25p^6$ shell and $5d^16s^2$ electrons, as shown in Fig. 2.1. Note that $5f$ electrons are approximately similar to $4f$ electrons, but the relativistic effect of $5f$ electrons is much larger than that of $4f$ electrons, as shown in Figs. 2.1(b) and (c). This enhances the itinerant nature of $5f$ electrons in the crystal.

The spin S, angular L, and total J momenta, the Landé g_J factor, the effective magnetic moment $\mu_{\rm eff}$ and the saturated moment $g_J J$ for the $3d$ transition and $4f$ rare earth ions are summarized in Tables 2.1 and 2.2, respectively.

2.2 Weiss field and magnetic ordering

We call a compound ferromagnetic if it possesses a spontaneous (saturated or ordered) magnetic moment, that is, a magnetic moment even in absence of an applied magnetic field. A ferromagnetic compound corresponds to a paramagnetic compound with an added interaction that tends to make the ionic (atomic) magnetic moments line up the same way. Such an interaction of the molecular field is called the Weiss field. We consider that the Weiss field is equivalent to an effective magnetic field $H_{\rm m}$ acting on the electron spins, which is assumed to be proportional to the magnetization M, so that

$$H_{\rm m} = \Lambda M \tag{2.10}$$

where Λ stands for the Weiss field constant. The magnetic susceptibility above the Curie temperature $T_{\rm C}$ is deduced by postulating that the Curie law holds if we take the sum of the applied field H and Weiss field $H_{\rm m}$ as the magnetic field

$$\frac{M}{H + \Lambda M} = \frac{C}{T}$$

or

$$\chi = \frac{M}{H} = \frac{C}{T - C\Lambda}$$

or

$$\chi = \frac{C}{T - T_{\rm C}} \quad (T_{\rm C} = C\Lambda). \tag{2.11}$$

This equation, which is known as the Curie-Weiss law, describes the observed susceptibility in the paramagnetic region above the Curie temperature T_C quite well.

The origin of the Weiss field is due to the quantum-mechanical magnetic exchange interaction, as pointed out by Heisenberg [14]. The interaction between magnetic atoms i, j bearing spins $\boldsymbol{S}_i$, $\boldsymbol{S}_j$, known as the exchange energy, contains a term

$$\mathcal{H}_{\mathrm{ex}} = -2J_{\mathrm{ex}}\boldsymbol{S}_i \cdot \boldsymbol{S}_j \tag{2.12a}$$

where J_{ex} is the exchange integral and is related to the overlap of the charge distributions i, j. The exchange energy has no classical analogue, although it is of electrostatic origin. It expresses a difference in Coulomb interaction energies of the system when the spins are parallel or antiparallel. It is a consequence of the Pauli exclusion principle — that in quantum mechanics, one cannot usually change the relative direction of two spins without making changes in the spatial charge distribution in the overlap region. The resulting changes in the Coulomb energy of the system may conveniently be written in the form of Eq. (2.12a), so that it appears as if there were a direct coupling between the spins $\boldsymbol{S}_i$, $\boldsymbol{S}_j$.

We present here an approximate relation between the exchange integral J_{ex} and the Weiss field constant Λ. Supposing that the magnetic atom with the spin $\boldsymbol{S}_i$ has Z nearest neighbors with an average spin $\langle \boldsymbol{S} \rangle$ per atom, Eq. (2.12a) can be changed into

$$\begin{aligned} \mathcal{H}_{\mathrm{ex}} &= -2J_{\mathrm{ex}}\boldsymbol{S}_i \cdot Z\langle \boldsymbol{S} \rangle \\ &= -\boldsymbol{\mu}_i \cdot \boldsymbol{H}_{\mathrm{m}} \end{aligned} \tag{2.12b}$$

where $\boldsymbol{\mu}_i = -g_{\mathrm{J}}\mu_{\mathrm{B}}\boldsymbol{S}_i$, and $\boldsymbol{H}_{\mathrm{m}} = -2J_{\mathrm{ex}}Z\langle \boldsymbol{S} \rangle / g_{\mathrm{J}}\mu_{\mathrm{B}}$, considering that the exchange energy corresponds to the Zeeman energy. From a relation of $H_{\mathrm{m}} = \Lambda M$, we obtain the following relation between the Weiss field constant Λ and the magnetic exchange integral J_{ex}:

$$-\frac{1}{g_{\mathrm{J}}\mu_{\mathrm{B}}}2J_{\mathrm{ex}}Z\langle \boldsymbol{S} \rangle = \Lambda(-Ng_{\mathrm{J}}\mu_{\mathrm{B}}\langle \boldsymbol{S} \rangle)$$

$$\Lambda = \frac{2ZJ_{\mathrm{ex}}}{Ng_{\mathrm{J}}^2\mu_{\mathrm{B}}^2}. \tag{2.13}$$

At present, the magnetic exchange interaction, or the Heisenberg model is applied to the magnetic insulating $3d$-compounds. If the $3d$-compounds are metallic, observed magneton numbers are usually considerably smaller than the theoretical values calculated from the free ions and also are frequently non-integral, revealing itinerant magnetism (described in Chap. 4).

Note that the present theory is also applied to the antiferromagnetic state which is characterized by an ordered antiparallel arrangement of electron spins. The magnetic susceptibility is usually expressed as

$$\frac{1}{\chi} = \frac{T + \theta_{\rm p}}{C}. \tag{2.14}$$

The temperature dependence of the magnetic susceptibility is schematically shown in Fig. 2.4.

(1) If the paramagnetic Curie temperature $\theta_{\rm p}$ is positive, the magnetic state is antiferromagnetic. Note that the antiferromagnetic transition temperature is called the Néel temperature.
(2) If $\theta_{\rm p}$ is zero, it is paramagnetic.
(3) If $\theta_{\rm p}$ is negative, it is ferromagnetic.

Furthermore, note that the $4f$ electrons in rare earth compounds are principally localized in nature. The RKKY interaction plays a predominant role in magnetism of such compounds. Therefore, the mutual magnetic interaction between the $4f$ electrons occupying

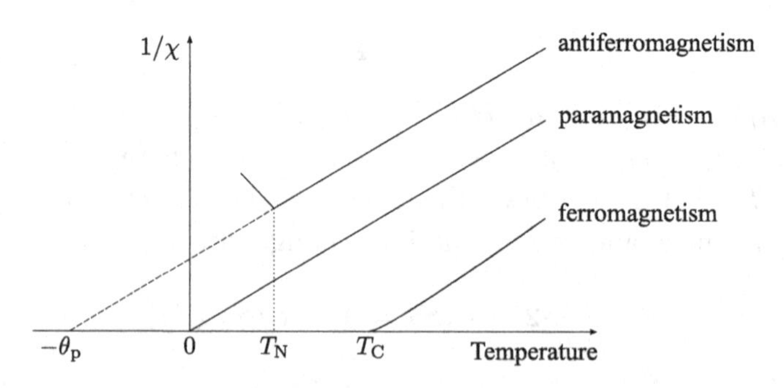

Figure 2.4: Temperature dependences of the inverse magnetic susceptibilities $1/\chi$ in antiferromagnetic, paramagnetic, and ferromagnetic compounds.

different atomic sites cannot be direct, such as in $3d$-magnetism, but should be indirect, which occurs via the conduction electrons [15–17].

In the RKKY interaction, a localized spin $\boldsymbol{S}_i$ of the $4f$ electrons (f) interacts with a conduction electron (c) that has a spin $\boldsymbol{s}$, which leads to a spin polarization of the conduction electron. The exchange interaction is expressed as

$$H_{\mathrm{cf}} = -J_{\mathrm{cf}}\boldsymbol{s} \cdot \boldsymbol{S}. \tag{2.15}$$

Note that the present exchange interaction is related not to $\boldsymbol{J}$, but to $\boldsymbol{S}$. This polarization interacts with another spin $\boldsymbol{S}_j$ localized on a magnetic ion (atom) j and therefore creates an indirect interaction between the spins $\boldsymbol{S}_i$ and $\boldsymbol{S}_j$, which is proportional to $-J(2k_{\mathrm{F}}R_{ij})$ $\boldsymbol{S}_i \cdot \boldsymbol{S}_j$. This indirect interaction extends to the far distance and damps with the Friedel oscillation, a sinusoidal $2k_{\mathrm{F}}R_{ij}$ oscillation, as shown in Fig. 2.5. Here, $J(2k_{\mathrm{F}}R_{ij}) = J(x)$ is expressed as follows:

$$J(x) \sim \frac{-x\cos x + \sin x}{x^4}. \tag{2.16}$$

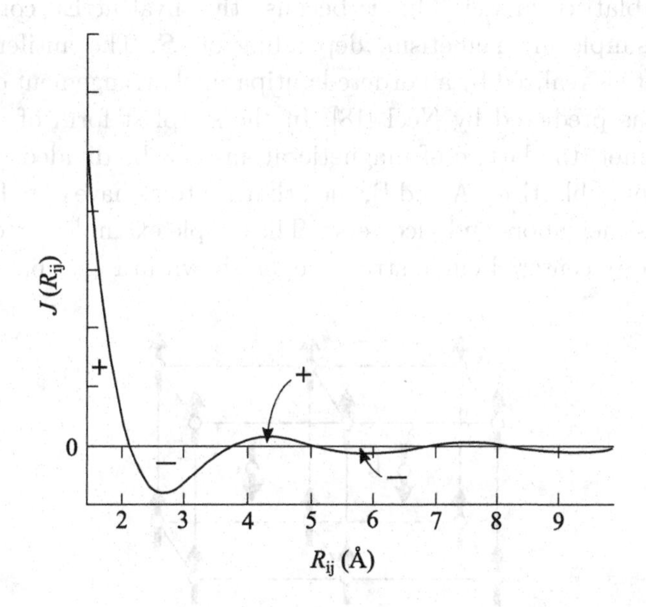

Figure 2.5: Friedel oscillation in the RKKY interaction, shown as an example.

When the number of $4f$ electrons increases in such a way that the lanthanide elements change from Ce to Gd or conversely from Yb to Gd in the compound, the magnetic moment becomes larger and the RKKY interaction stronger, leading to magnetic order with the ordering temperature roughly following the de Gennes relation $\boldsymbol{S}^2 = (g_\mathrm{J}-1)^2 J(J+1)$, where $\boldsymbol{S} = (g_\mathrm{J}-1)\boldsymbol{J}$.

2.3 Magnetism in Eu compounds

Most of the Eu compounds are in the Eu-divalent electronic state ($4f^7$ in Eu^{2+}: $S=7/2$, $L=0$, and $J=7/2$) and order magnetically. While generally rare earth compounds are trivalent, a Eu-divalent (Eu^{2+}) electronic state is more stable than a Eu-trivalent (Eu^{3+}) state. Large lattice parameters of the Eu compounds, compared with those of the other trivalent rare earth compounds, are a good indication of the divalence of Eu. Some Eu compounds such as $EuPd_3$ are, however, in the trivalent electronic state ($4f^6$ in Eu^{3+}: $S = 3$, $L = 3$, and $J = 0$).

First, we explain the Eu-divalent antiferromagnetism based on a two-sublattice model. This is because the divalent Eu compound is very simple in magnetism, depending on $\boldsymbol{S}$. The antiferromagnetic state is realized by an ordered antiparallel arrangement of spins, which was predicted by Néel [18]. In the simplest form of an antiferromagnet, the lattice of magnetic atoms can be divided into two equivalent sublattices, A and B, such that A atoms have only B atoms as nearest neighbors and vice versa. The simple example corresponds to the body-centered cubic structure, as shown in Fig. 2.6.

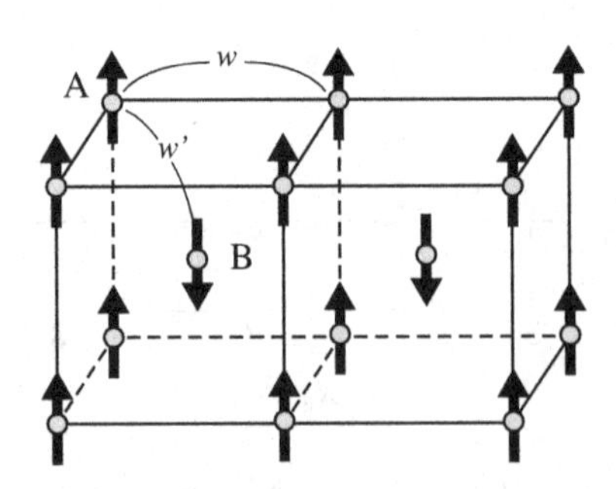

Figure 2.6: Arrangement of antiferromagnetic moments.

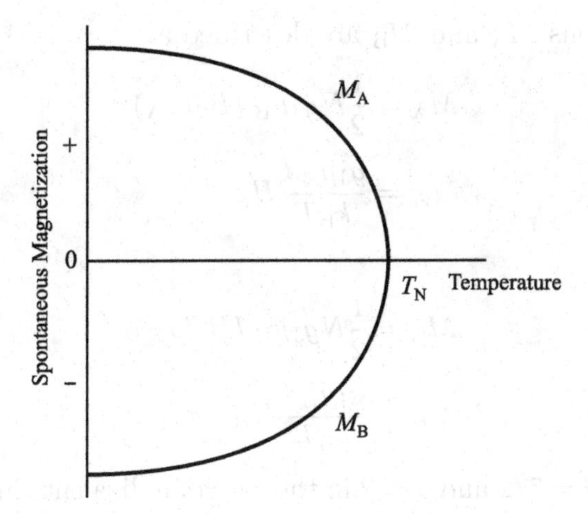

Figure 2.7: Spontaneous magnetizations M_A and M_B for an antiferromagnet.

At 0 K, each sublattice has maximum magnetization, and with increasing temperature, the thermal activation reduces the sublattice magnetizations in the same way as in ferromagnets. The net magnetization is, however, zero at all the temperatures because of the exact cancellation of the spontaneous magnetization of each sublattice, as shown in Fig. 2.7. The magnetic susceptibility is different because the magnetic field is applied parallel or perpendicular to the direction of the magnetic moments.

Let $w_{AA} = w_{BB} = w$ be the molecular field (Weiss field) coefficient inside each sublattice and $w_{AB} = w_{BA} = w'$ the molecular field coefficient between the two sublattices. The total fields acting on the A and B sublattices in an external magnetic field H are:

$$\begin{aligned} H_A &= H - w_{AA}M_A - w_{AB}M_B \\ H_B &= H - w_{BA}M_A - w_{BB}M_B \end{aligned} \tag{2.17}$$

or

$$\begin{aligned} H_A &= H - wM_A - w'M_B \\ H_B &= H - w'M_A - wM_B \end{aligned} \tag{2.18}$$

Note that w' is always positive ($w' > 0$) because of the antiferromagnetic interaction, but the sign of w becomes positive or negative depending on the compounds with various crystal structures.

Magnetizations $M_{\rm A}$ and $M_{\rm B}$ are described as

$$M_{\rm A} = \frac{1}{2} N g_{\rm J} \mu_{\rm B} J B_J(x_{\rm A})$$
$$x_{\rm A} = \frac{g_{\rm J}\mu_{\rm B} J}{k_{\rm B}T} H_{\rm A} \tag{2.19}$$

and

$$M_{\rm B} = \frac{1}{2} N g_{\rm J} \mu_{\rm B} J B_J(x_{\rm B})$$
$$x_{\rm B} = \frac{g_{\rm J}\mu_{\rm B} J}{k_{\rm B}T} H_{\rm B} \tag{2.20}$$

where $J = S = 7/2$ and $g = 2$ in the case of a divalent Eu compound.

When temperature is high, $x \ll 1$. Using $B_J(x) = \frac{J+1}{3J}x$, Eqs. (2.19) and (2.20) become

$$M_{\rm A} = \frac{(N/2)g_{\rm J}^2\mu_{\rm B}^2 J(J+1)}{3k_{\rm B}T} H_{\rm A} = \frac{N\mu_{\rm eff}^2}{6k_{\rm B}T} H_{\rm A} = \frac{C}{2T} H_{\rm A} \tag{2.21}$$

$$M_{\rm B} = \frac{(N/2)g_{\rm J}^2\mu_{\rm B}^2 J(J+1)}{3k_{\rm B}T} H_{\rm B} = \frac{N\mu_{\rm eff}^2}{6k_{\rm B}T} H_{\rm B} = \frac{C}{2T} H_{\rm B} \tag{2.22}$$

where $\mu_{\rm eff} = g_{\rm J}\mu_{\rm B}\sqrt{J(J+1)}$ and $C = N\mu_{\rm eff}^2/3k_{\rm B}$. The total magnetization M is thus given as

$$M = M_{\rm A} + M_{\rm B} = \frac{C}{2T}\{2H - (w + w')M\}. \tag{2.23}$$

The magnetic susceptibility becomes

$$\chi = \frac{M}{H} = \frac{C}{T + \frac{C}{2}(w + w')} = \frac{C}{T + \theta_{\rm p}} \tag{2.24}$$

and

$$\theta_{\rm p} = \frac{C}{2}(w + w'). \tag{2.25}$$

For $H = 0$, Eqs. (2.21) and (2.22) become

$$M_{\rm A} = \frac{C}{2T}(-wM_{\rm A} - w'M_{\rm B}) \tag{2.26}$$

$$M_{\rm B} = \frac{C}{2T}(-w'M_{\rm A} - wM_{\rm B}) \tag{2.27}$$

or

$$\left(1+\frac{Cw}{2T}\right)M_{\rm A}+\frac{Cw'}{2T}M_{\rm B}=0 \tag{2.28}$$

$$\frac{Cw'}{2T}M_{\rm A}+\left(1+\frac{Cw}{2T}\right)M_{\rm B}=0. \tag{2.29}$$

This set of equations has a non-trivial solution ($M_{\rm A}\neq 0$ and $M_{\rm B}\neq 0$) only below the ordering temperature, which is called the Néel temperature $T_{\rm N}$. $T_{\rm N}$ is such that

$$\begin{vmatrix} 1+\frac{Cw}{2T_{\rm N}} & \frac{Cw'}{2T_{\rm N}} \\ \frac{Cw'}{2T_{\rm N}} & 1+\frac{Cw}{2T_{\rm N}} \end{vmatrix}=0$$

or

$$T_{\rm N}=\frac{C}{2}(w'-w). \tag{2.30}$$

Next, we consider the magnetic susceptibility below the Néel temperature. When the magnetic field is applied parallel to the magnetic moments, namely along the antiferromagnetic easy-axis, the parallel susceptibility $\chi_{\parallel}$ decreases with decreasing temperature and vanishes at 0 K ($\chi_{\parallel}\to 0$ for $T\to 0$), as shown in Fig. 2.8(a). This is because the antiferromagnetic structure is stabilized with decreasing temperature, and there is no orientation of the magnetic moment with the field direction at 0 K.

On the other hand, the perpendicular susceptibility (antiferromagnetic hard-axis susceptibility) $\chi_{\perp}$ is unchanged below $T_{\rm N}$, as shown in Fig. 2.8(a). This is because the molecular field $w'M_{\rm B}$ due to the magnetic moment $M_{\rm B}$ is applied to the magnetic moment $M_{\rm A}$, as shown in Fig. 2.8(b), satisfying the relation

$$w'M_{\rm B}\sin\theta=\frac{H}{2}.$$

The magnetic moment along the field direction is

$$\begin{aligned} M &= M_{\rm A}\sin\theta+M_{\rm B}\sin\theta \quad (M_{\rm A}=M_{\rm B}) \\ &= 2M_{\rm B}\sin\theta \\ &= \frac{H}{w'} \end{aligned} \tag{2.31}$$

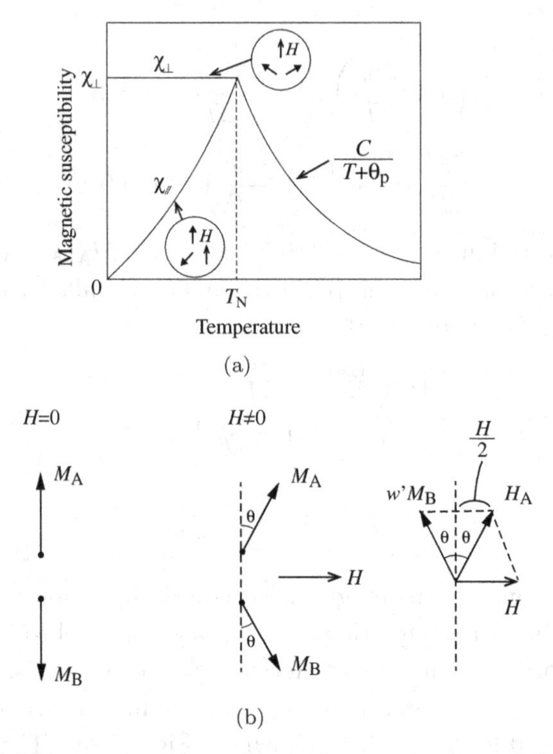

Figure 2.8: (a) Temperature dependence of magnetic susceptibility for an antiferromagnet. (b) Schematic views for the antiferromagnetic moments M_A and M_B for the applied magnetic field H along the hard-axis.

and $\chi_\perp = \frac{M}{H} = \frac{1}{w'}$, which is temperature-independent. The magnetic susceptibility in the antiferromagnetic state is also shown in Fig. 2.8(a).

When the magnetic moment saturates at H_c, via the canting process of magnetization, the following relation is obtained from Eq. (2.31):

$$H_c = w' M_{sat} = w' N g \mu_B J. \tag{2.32}$$

H_c can be described by θ_p in Eq. (2.25) and T_N in (2.30), so it follows that

$$H_c = N g \mu_B J \cdot \frac{1}{C}(T_N + \theta_p)$$

$$= \frac{3k_B}{g\mu_B(J+1)}(T_N + \theta_p)$$
$$= \frac{k_B}{3\mu_B}(T_N + \theta_p) \tag{2.33}$$

and H_c[kOe] = $4.96(T_N + \theta_p)$[K], where $J = S = 7/2$.

Here, we define again the above equation, following the usual definition. The magnetic susceptibility χ in $EuPb_3$ is obtained as $\frac{1}{\chi} = \frac{1}{C}(T + 57 \text{ K})$, for example. The θ_p value is 57 K, but is experimentally expressed as $\theta_p = -57$ K. This is because the antiferromagnet possesses a negative θ_p value, while positive θ_p corresponds to the case of a ferromagnet. In this particular case, Eq. (2.33) should be changed to

$$H_c = \frac{k_B}{3\mu_B}(T_N - \theta_p) \tag{2.34a}$$

or

$$H_c[\text{kOe}] = 4.96(T_N - \theta_p)[\text{K}]. \tag{2.34b}$$

The high-field magnetization was once obtained in $EuPb_3$ with $T_N = 20$ K, $\theta_p = -57$ K, and $H_c = 380$ kOe, as shown in Fig. 2.9 [19]. The relation indicates a very simple magnetization, which is realized in other Eu compounds such as $EuGa_4$, $EuBi_3$, and $EuCd_{11}$. It is characteristic that there is no anisotropy of the magnetic susceptibilities in the paramagnetic state even for these non-cubic Eu compounds. This is mainly due to the lack of any CEF effect in the divalent Eu state ($S = 7/2$ and $L = 0$).

Generally speaking, anisotropies of the magnetic susceptibility and magnetization in the f electron systems are based on the CEF effect and the magnetic exchange interaction. Usually, the CEF effect is the main contribution to the anisotropies of the magnetic susceptibility and magnetization. The simple canting magnetization for the magnetic field along the hard-axis, mentioned above, is realized under the following conditions: (1) no anisotropy of the magnetic susceptibilities in the paramagnetic state and (2) no successive magnetic transitions below the Néel temperature.

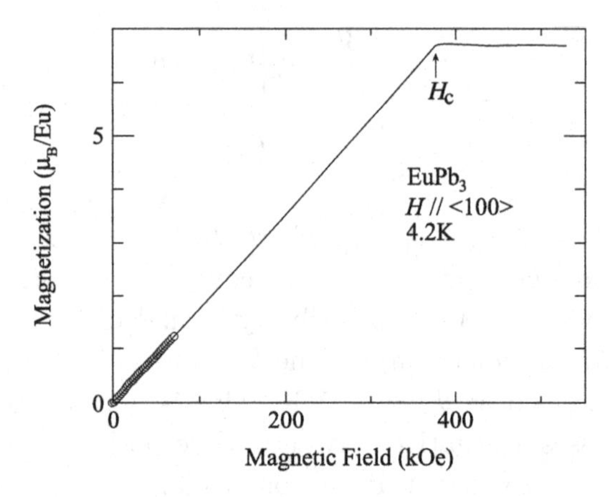

Figure 2.9: Canting magnetization of $EuPb_3$, cited from Ref. [19]. Data shown by circles and solid lines are from SQUID and pulse-field magnetization measurements, respectively.

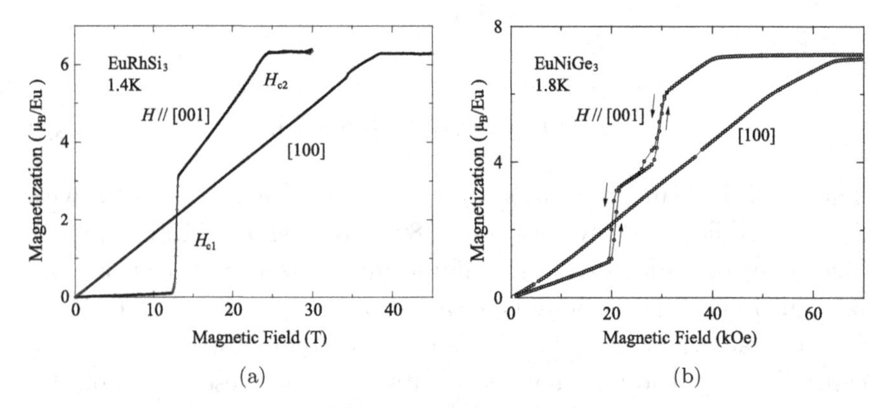

Figure 2.10: Magnetization curves in (a) $EuRhSi_3$ and (b) $EuNiGe_3$, cited from Refs. [20, 21].

In other Eu compounds with the divalent electronic state, characteristic magnetization curves are observed, for example, in $EuRhSi_3$ and $EuNiGe_3$ [20, 21], as shown in Fig. 2.10.

Next, we describe the Van Vleck paramagnetic susceptibility of the trivalent Eu ions [22]. The $L{-}S$ coupling (or spin-orbit coupling) scheme is expressed as $^{2S+1}L_J$, where L is $0(S)$, $1(P)$, $2(D)$, $3(F)$, ..., and J is $L+S$, $L+S-1$, $\cdots$ $|L-S|$. The trivalent Eu ions ($L = S = 3$) correspond to ^7F_J (J = 0, 1, 2, 3, 4, 5, and 6). The spin-orbit interaction $\lambda \boldsymbol{L} \cdot \boldsymbol{S}$ with the coupling constant λ forms the energy levels in the L–S coupling. The energy E_J and the energy difference $E_J - E_{J-1}$ are given as

$$E_J = \frac{\lambda}{2}\{J(J+1) - L(L+1) - S(S+1)\} \tag{2.35}$$

$$E_J - E_{J-1} = \lambda J. \tag{2.36}$$

When the separation of the multiplet components is sufficiently large compared to $k_B T$, only the lowest energy is taken into consideration for the magnetic susceptibility. This is applied to most of the rare earth compounds, except Sm^{2+} and Eu^{3+}. Here, we consider the case of Eu^{3+}. The paramagnetic susceptibility can be expressed using the Van Vleck theory [23] as

$$\chi = \frac{\sum_{J=0}^{6} \chi_J (2J+1) e^{-\frac{E_J}{k_B T}}}{\sum_{J=0}^{6} (2J+1) e^{-\frac{E_J}{k_B T}}} = N \frac{\sum_{J=0}^{6} \left[\frac{g^2 \mu_B^2 J(J+1)}{3 k_B T} + \alpha_J\right] (2J+1) e^{\frac{E_J}{k_B T}}}{\sum_{J=0}^{6} (2J+1) e^{-\frac{E_J}{k_B T}}} \tag{2.37}$$

$$\text{where } \alpha_J = \frac{\mu_B^2}{6(2J+1)} \left[\frac{F_{J+1}}{E_{J+1} - E_J} - \frac{F_J}{E_J - E_{J-1}}\right] \tag{2.38}$$

$$F_J = \frac{[(L+S+1)^2 - J^2][J^2 - (S-L)^2]}{J}. \tag{2.39}$$

The explicit paramagnetic susceptibility of Eu^{3+} ions can be derived in the forms

$$\chi_{\text{para}}(Eu^{3+}) = \frac{N \mu_B^2}{Z} \frac{A}{3\lambda} \tag{2.40}$$

$$Z = 1 + 3e^{-\frac{\lambda}{k_B T}} + 5e^{-\frac{3\lambda}{k_B T}} + 7e^{-\frac{6\lambda}{k_B T}} + 9e^{-\frac{10\lambda}{k_B T}} + 11e^{-\frac{15\lambda}{k_B T}} + 13e^{-\frac{21\lambda}{k_B T}} \tag{2.41}$$

$$A = 24 + \left(13.5\frac{\lambda}{k_B T} - 1.5\right) e^{-\frac{\lambda}{k_B T}} + \left(67.5\frac{\lambda}{k_B T} - 2.5\right) e^{-\frac{3\lambda}{k_B T}}$$

$$+\left(189\frac{\lambda}{k_{\mathrm{B}}T}-3.5\right)\mathrm{e}^{-6\frac{\lambda}{k_{\mathrm{B}}T}}+\left(405\frac{\lambda}{k_{\mathrm{B}}T}-4.5\right)\mathrm{e}^{-10\frac{\lambda}{k_{\mathrm{B}}T}}$$
$$+\left(742.5\frac{\lambda}{k_{\mathrm{B}}T}-5.5\right)\mathrm{e}^{-15\frac{\lambda}{k_{\mathrm{B}}T}}+\left(1228.5\frac{\lambda}{k_{\mathrm{B}}T}-6.5\right)\mathrm{e}^{-21\frac{\lambda}{k_{\mathrm{B}}T}}. \tag{2.42}$$

Since the formulae of Eqs. (2.40)–(2.42) include only one unknown parameter λ, the value of λ can be determined uniquely in contrast to the experimental result of the temperature dependence of the magnetic susceptibility. The magnetic susceptibility can be presented with the variation of one parameter λ. However, the exact energy levels of J multiplets are not described by one parameter λ, for example, in $EuPd_3$ with the $AuCu_3$-type cubic structure (No. 221) [24]. Figure 2.11 shows the temperature dependence of the magnetic susceptibility in $EuPd_3$ [25].

As mentioned above, most of the Eu compounds are different from the other rare earth compounds, including the cerium compounds in the trivalent electronic state as the Eu-divalent (Eu^{2+}) electronic state is more stable. The energy difference between Eu^{2+}

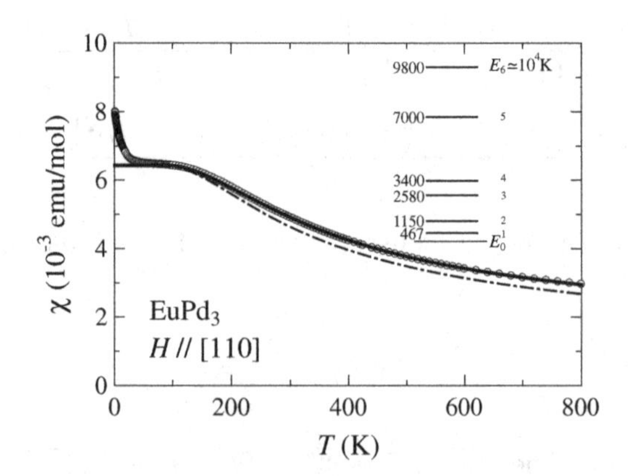

Figure 2.11: Comparison between the observed and calculated $\chi(T)$ for $H \parallel \langle 110\rangle$ of $EuPd_3$, cited from Ref. [25]. The observed data are plotted as open circles. The dot-dashed line is based on Eq. (2.37), with $\lambda = 467$ K. The solid line is the best fit curve calculated using $E_0 = 0$, $E_1 = 467$ K, $E_2 = 1150$ K, $E_3 = 2500$ K, and $E_4 = 3400$ K.

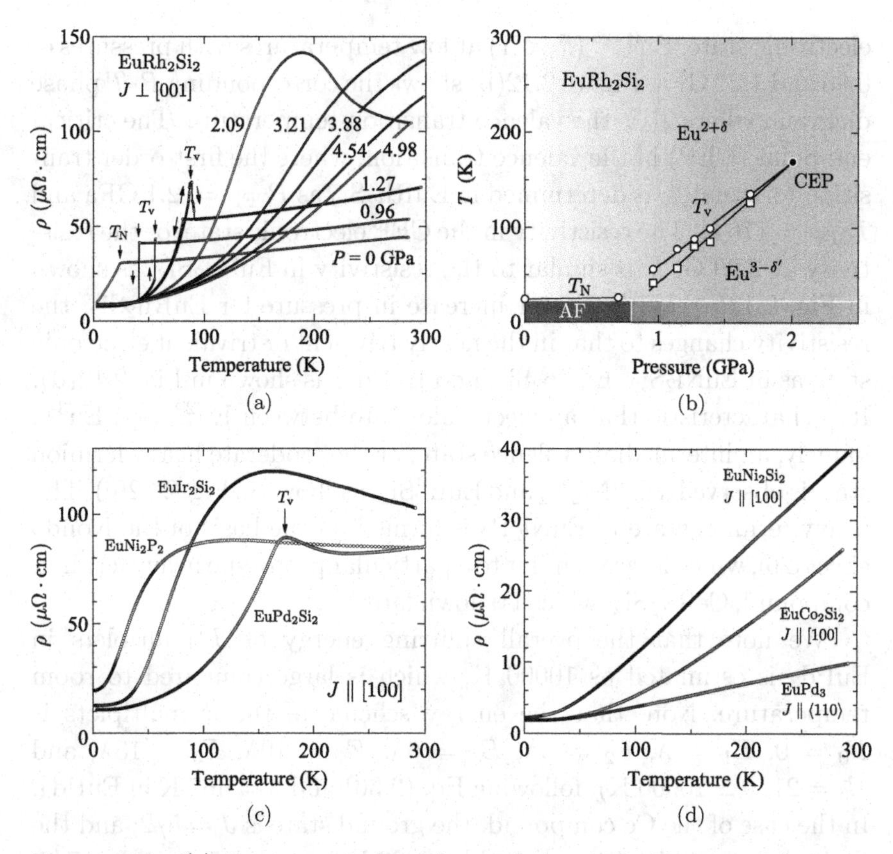

Figure 2.12: (a) Temperature dependence of electrical resistivity under pressure in an antiferromagnet of $EuRh_2Si_2$, indicating the first-order valence transition, (b) *P*-*T* phase diagram in $EuRh_2Si_2$, (c) temperature dependence of electrical resistivity in a temperature-induced valence transition compound of $EuPd_2Si_2$ and moderate heavy fermion compounds of $EuNi_2P_2$ and $EuIr_2Si_2$, and (d) nearly trivalent or trivalent compounds of $EuNi_2Si_2$, $EuCo_2Si_2$, and $EuPd_3$.

and Eu^{3+} states is, however, not extensively large. Therefore, the Eu^{2+} electronic state can be changed through application of high pressures.

Figure 2.12(a) shows the temperature dependences of electrical resistivity under pressure in $EuRh_2Si_2$ with the $ThCr_2Si_2$-type tetragonal structure. Valence instability occurs from a nearly divalent electronic state, $Eu^{2+\delta}$ ($\delta < 1$) at high temperatures to a nearly trivalent electronic state, $Eu^{3-\delta'}$ ($\delta' < 1$) at low temperatures with pressures of 0.96 and 1.27 GPa. Figure 2.12(b) shows the corresponding P–T phase diagram, where T_V is the valence transition temperature. The critical end point (CEP) in the valence transition, where the first-order transition terminates, is determined in $EuRh_2Si_2$ as $P_{CEP} = 2.1$ GPa and $T_{CEP} = 176$ K. The resistivity in the CEP electronic state, or the resistivity at 2.09 GPa, is similar to the resistivity in $EuPd_2Si_2$, as shown in Fig. 2.12(c). With further increase in pressure for $EuRh_2Si_2$, the resistivity changes to that in the nearly trivalent or trivalent electronic state as in $EuNi_2Si_2$, $EuCo_2Si_2$, and $EuPd_3$, as shown in Fig. 2.12(d). It is characteristic that an electronic state between Eu^{2+} and Eu^{3+}, namely, an intermediate valence state, or the moderate heavy fermion state is observed in $EuNi_2P_2$ and $EuIr_2Si_2$, as shown in Fig. 2.12(c). The heavy fermion state in $EuNi_2P_2$ is formed on the basis of the Kondo effect [26], which is very similar to a particular prototype heavy fermion compound, $CeRu_2Si_2$, which is shown later.

We note that the overall splitting energy of J multiplets in $EuPd_3$ is estimated as 10000 K, which is large compared to room temperature. Note that the energy scheme of the J multiplets is $E_0 = 0$, $E_1 = \lambda$, $E_2 = 3\lambda$, $E_3 = 6\lambda$, $E_4 = 10\lambda$, $E_5 = 15\lambda$, and $E_6 = 21\lambda$ ($\simeq 10000$ K), following Eq. (2.36) and $\lambda = 467$ K in $EuPd_3$. In the case of the Ce compound, the ground state is $J = 5/2$, and the excited state, which is separated by 3000 K from $J = 5/2$, is $J = 7/2$, where J consists of $J = |3 + 1/2| = 7/2$ and $J = |3 - 1/2| = 5/2$ for $S = 1/2$ and $L = 3$ in the spin-orbit coupling. We do not consider the $J = 7/2$ state in the usual magnetic properties below room temperature. Eu compounds with the trivalent state are a rare case, with a small value of $E_1 = 467$ K.

Sm compounds are another rare case. The ground state of $J = 5/2$ ($S = 5/2$ and $L = 5$) in the trivalent Sm compound is the same as that of the Ce compound. For example, the excited state of $J = 7/2$ is not largely separated from $J = 5/2$ by $\Delta E = 1300$ K in $SmCu_2Si_2$. The Van Vleck susceptibility between $J = 5/2$ and $J = 7/2$ is therefore included in the magnetic susceptibility, together with the Curie-Weiss magnetic susceptibility due to $J = 5/2$. Namely, the magnetic susceptibility in the trivalent Sm compound can be described as

$$\chi = \frac{\mu_{\rm eff}^2}{3k_{\rm B}(T+\theta_{\rm p})} + \frac{20\mu_{\rm B}^2}{7\Delta E}$$

where $\mu_{\rm eff}$(=0.85 $\mu_{\rm B}$/Sm for Sm^{3+}) is the effective moment. Note that most Sm compounds are trivalent but some are in the intermediate valence state between Sm^{3+} and Sm^{2+} ($4f^6$: $S = L = 3$, $J = 0$).

2.4 Crystalline electric field (CEF)

The $4f$ electrons in the Ce atom are pushed deeply into the interior of the closed $5s$ and $5p$ shells because of the strong centrifugal potential $l(l+1)/r^2$, where $l = 3$ holds for the f electrons. This gives rise to the atomic-like character of the $4f$ electrons in the crystal [5]. On the other hand, the tail of their wave function spreads to the outside of the closed $5s$ and $5p$ shells, which is highly influenced by the potential energy, the relativistic effect, and the distance between the Ce atoms. This results in the hybridization of the $4f$ electrons with the conduction electrons, causing various phenomena such as magnetic ordering, quadrupole ordering, valence fluctuations, Kondo lattice, heavy fermions, Kondo insulators and unconventional superconductivity [27].

The Coulomb repulsive force of the $4f$ electron at one atomic site, U, is so strong, e.g. $U \simeq 5$ eV in the Ce compounds (see Fig. 2.13), that occupancy of the same site by two $4f$ electrons is usually prohibited. In the Ce compounds, the tail of the $4f$ partial density of states extends to the Fermi level even at room temperature. This means that the $4f$ level approaches the Fermi level in energy and the $4f$ electron hybridizes strongly with the conduction electrons. This $4f$-hybridization coupling constant is denoted by V_{cf}. When U is strong and V_{cf} is ignored, the freedom of the charge in the $4f$ electron is suppressed, while the freedom of the spin is retained, representing the $4f$-localized state. Naturally, the degree of localization depends on the level of the $4f$ electron, E_f, where larger E_f accelerates the localization.

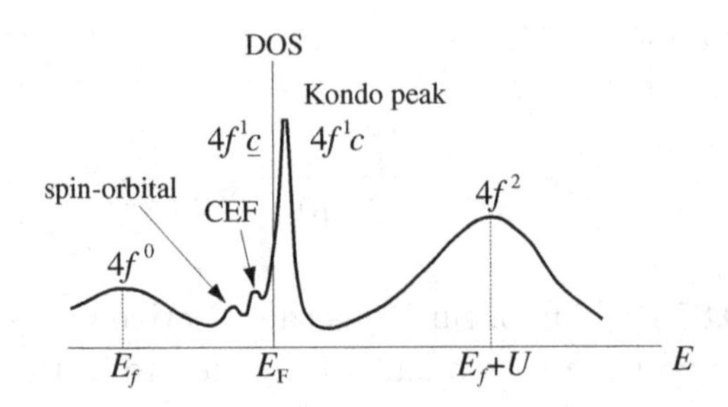

Figure 2.13: Density of states (DOS) of the $4f$ electron in the Ce compound (Ce^{3+}), cited from Ref. [27].

In the localized $4f$-electronic scheme, the $4f$ ground multiplets, which obey the Hund rule in the LS-multiplets, split into the J-multiplets — for example, $J = 7/2$ and $J = 5/2$ in Ce^{3+} — by the spin-orbit coupling. Moreover, the $J = 5/2$-multiplet splits into three doublets ($2J + 1 = 6$ for $J = 5/2$) based on the CEF as shown in Figs. 2.14 (a) for Ce^{3+} in the non-cubic crystal. It is noted that the $J = 7/2$ multiplet becomes the ground state in the spin-orbit coupling of Yb^{3+}, and eight fold $4f$-levels ($2J + 1 = 8$ for $J = 7/2$) split into four doublets in the non-cubic CEF scheme, as shown in Fig. 2.14(b). Note that the magnitude of spin-orbit coupling is 3000 K in Ce^{3+} and 15000 K in Yb^{3+} and the CEF $4f$-level splitting is in the range of 300 K. The CEF effect is explained below.

The $4f$ electronic state strongly depends on the electric field of the surrounding negative ions, the CEF effect. The electrostatic potential $\phi(\boldsymbol{r})$ can be expressed as [28, 29]

$$\phi(\boldsymbol{r}) = \sum_i \frac{q_i}{|\boldsymbol{r} - \boldsymbol{R}_i|} \tag{2.43}$$

where $\boldsymbol{r}$ is the positional vector of the $4f$ electron in Ce^{3+}, q_i is the charge of the six-coordinated negative ion and $\boldsymbol{R}_i$ is the positional vector of the corresponding ion.

For example, the negative ion with charge q is located at $(a, 0, 0), (-a, 0, 0), (0, a, 0), (0, -a, 0), (0, 0, a)$ and $(0, 0, -a)$, as shown in Fig. 2.15. We express Eq. (2.43) through Taylor expansion, and

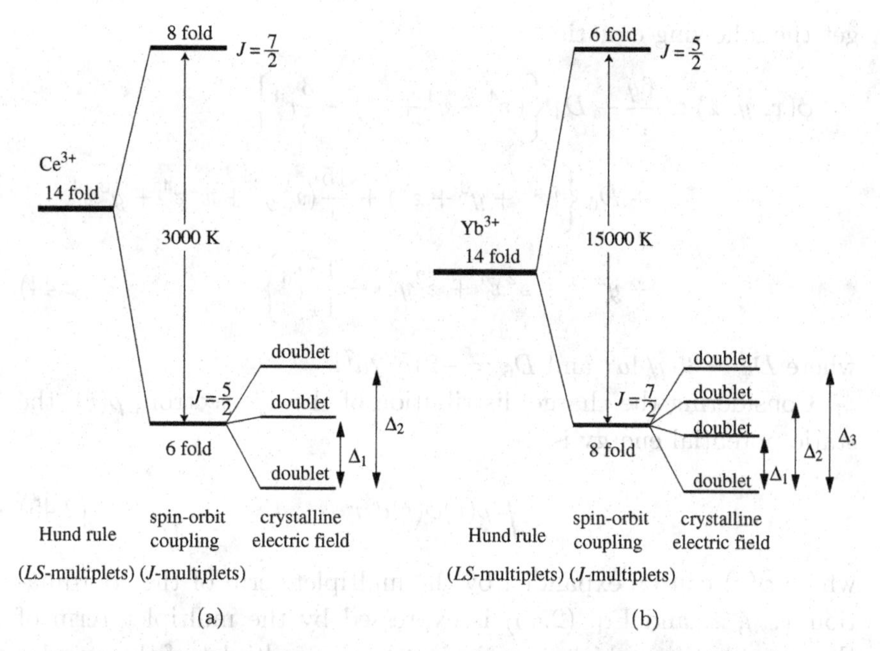

Figure 2.14: Level scheme of the $4f$ electrons for (a) Ce^{3+} and (b) Yb^{3+} in the non-cubic crystal.

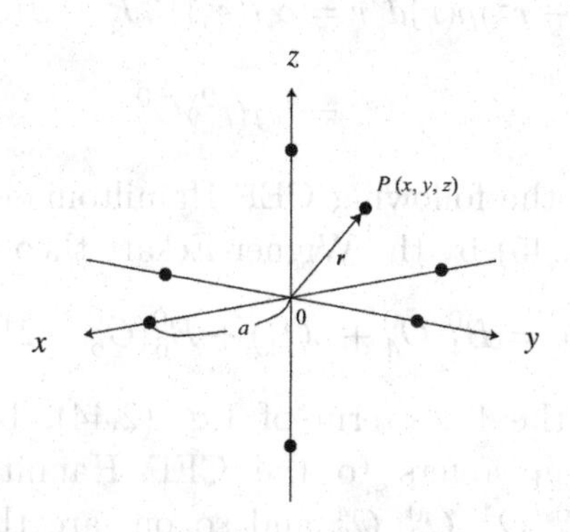

Figure 2.15: Six-coordinated negative ions and the $4f$ electron at the point P in the CEF scheme.

get the following equation

$$\phi(x, y, z) \simeq \frac{6q}{a} + D_4\left\{(x^4 + y^4 + z^4) - \frac{3}{5}r^4\right\}$$
$$+ D_6\left\{(x^6 + y^6 + z^6) + \frac{15}{4}(x^2y^4 + x^2z^4 + y^2x^4\right.$$
$$\left. + y^2z^4 + z^2x^4 + z^2y^4) - \frac{15}{14}r^6\right\} \quad (2.44)$$

where $D_4 = 35q/4a^5$ and $D_6 = -21q/2a^7$.

Considering the charge distribution of the $4f$ electron, $\rho(r)$, the static potential energy is

$$\int \rho(r)\phi(r)d^3r \quad (2.45)$$

where $\phi(r)$ can be expanded by the multiplet term of the coordination x, y, z, and Eq. (2.45) is expressed by the multiplet term of the coordination, which is equivalent to the multiplet of the angular momentum operator based on the Wigner-Eckart theorem in quantum mechanics. For example,

$$\int (3z^2 - r^2)\rho(r)d^3r = \alpha_J\langle r^2\rangle\{3J_z^2 - J(J+1)\}$$
$$= \alpha_J\langle r^2\rangle O_2^0. \quad (2.46)$$

We can represent the following CEF Hamiltonian corresponding to Eqs. (2.44) and (2.45) by the Wigner-Eckart theorem:

$$\mathcal{H}_{\mathrm{CEF}} = B_4^0(O_4^0 + 5O_4^4) + B_6^0(O_6^0 - 21O_6^4). \quad (2.47)$$

Here, we ignore the first term of Eq. (2.44), because it has no coordination. $\mathcal{H}_{\mathrm{CEF}}$ refers to the CEF Hamiltonian, while the operators O_n^m: O_4^0, O_4^4, O_6^0, O_6^4 and so on, are the Stevens operators. These operators are expressed in the matrix representation by Hutchings [28, 29].

Next, we consider the case where Ce^{3+} is sited in the cubic crystalline electric field: $L = 3$, $S = 1/2$, $J = 5/2$ and $M = \frac{5}{2}, \frac{3}{2}, \frac{1}{2}, -\frac{1}{2}, -\frac{3}{2}, -\frac{5}{2}$. Therefore, the multiplet with the $J = 5/2$ case (six fold degenerate of $2J + 1 = 6$) splits by the CEF effect. For $J = 5/2$, $O_6^0 = O_6^4 = 0$, and O_4^0 and O_4^4 can be expressed as

$$O_4^0 = 35J_z^4 - 30J(J+1)J_z^2 + 25J_z^2 - 6J(J+1) + 3J^2(J+1)^2 \tag{2.48}$$

$$O_4^4 = \frac{1}{2}(J_+^4 + J_-^4) \tag{2.49}$$

where $J_\pm = J_x \pm iJ_y$. The operator O_n^m can be expressed by the (6×6)-matrix. Therefore, the CEF Hamiltonian of the cubic Ce^{3+} is expressed as follows

$$\mathcal{H}_{\mathrm{CEF}} = \begin{array}{c} \\ \langle\frac{5}{2}| \\ \langle\frac{3}{2}| \\ \langle\frac{1}{2}| \\ \langle-\frac{1}{2}| \\ \langle-\frac{3}{2}| \\ \langle-\frac{5}{2}| \end{array} \begin{array}{c} \begin{array}{cccccc} |\frac{5}{2}\rangle & |\frac{3}{2}\rangle & |\frac{1}{2}\rangle & |-\frac{1}{2}\rangle & |-\frac{3}{2}\rangle & |-\frac{5}{2}\rangle \end{array} \\ \begin{pmatrix} 60B_4^0 & 0 & 0 & 0 & 60\sqrt{5}B_4^0 & 0 \\ 0 & -180B_4^0 & 0 & 0 & 0 & 60\sqrt{5}B_4^0 \\ 0 & 0 & 120B_4^0 & 0 & 0 & 0 \\ 0 & 0 & 0 & 120B_4^0 & 0 & 0 \\ 60\sqrt{5}B_4^0 & 0 & 0 & 0 & -180B_4^0 & 0 \\ 0 & 60\sqrt{5}B_4^0 & 0 & 0 & 0 & 60B_4^0 \end{pmatrix} \end{array}. \tag{2.50}$$

Next, we calculate the $4f$ level state $|i\rangle$ and its energy E_i as

$$\mathcal{H}_{\mathrm{CEF}}|i\rangle = E_i|i\rangle. \tag{2.51}$$

The following wave functions and energies are obtained

$$\left.\begin{aligned} |\Gamma_7^\alpha\rangle &= \tfrac{1}{\sqrt{6}}\left|\tfrac{5}{2}\right\rangle - \sqrt{\tfrac{5}{6}}\left|-\tfrac{3}{2}\right\rangle \\ |\Gamma_7^\beta\rangle &= \tfrac{1}{\sqrt{6}}\left|-\tfrac{5}{2}\right\rangle - \sqrt{\tfrac{5}{6}}\left|\tfrac{3}{2}\right\rangle \end{aligned}\right\} E_{\Gamma_7} = -240B_4^0 \tag{2.52}$$

$$\left.\begin{aligned} |\Gamma_8^\nu\rangle &= \sqrt{\tfrac{5}{6}}\left|\tfrac{5}{2}\right\rangle + \tfrac{1}{\sqrt{6}}\left|-\tfrac{3}{2}\right\rangle \\ |\Gamma_8^\kappa\rangle &= \sqrt{\tfrac{5}{6}}\left|-\tfrac{5}{2}\right\rangle + \tfrac{1}{\sqrt{6}}\left|\tfrac{3}{2}\right\rangle \\ |\Gamma_8^\lambda\rangle &= \left|\tfrac{1}{2}\right\rangle \\ |\Gamma_8^\mu\rangle &= \left|-\tfrac{1}{2}\right\rangle \end{aligned}\right\} E_{\Gamma_8} = 120B_4^0. \tag{2.53}$$

The energy state $-240B_4^0$ is Γ_7 and the energy state $120B_4^0$ is Γ_8. We show in Fig. 2.16 the charge distribution of Γ_7 and Γ_8 states [27]. The quartet Γ_8 wave function expands along the x, y, z directions, while the doublet Γ_7 expands along the $\langle 111\rangle$ direction. If negative

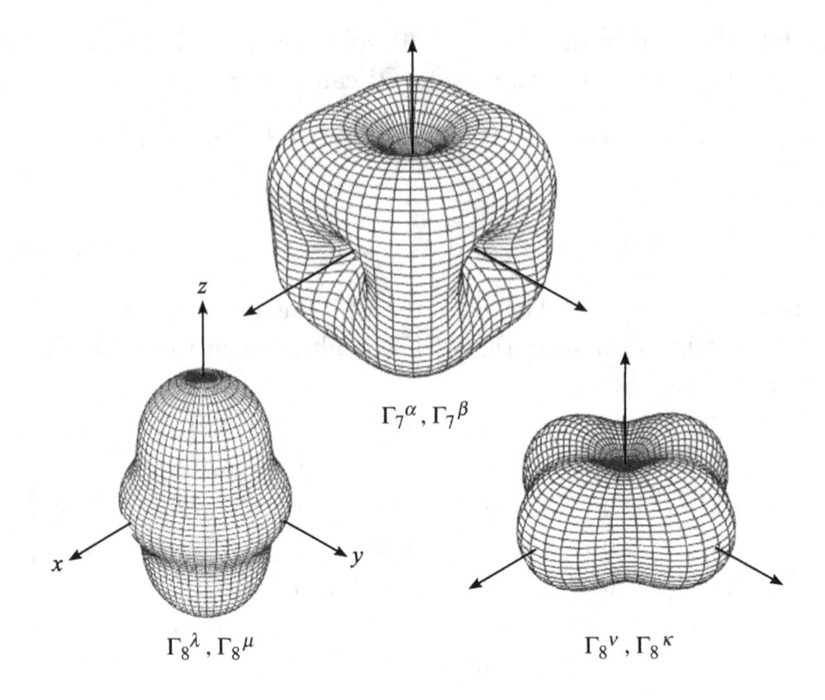

Figure 2.16: Charge distribution of Γ_7 and Γ_8 states, cited from Ref. [27].

ions exist along the principal axes, the Γ_7 ground state is preferable compared to the Γ_8 ground state, so that the latter becomes an excited state.

In general, the CEF Hamiltonian can be expressed as

$$\mathcal{H}_{\mathrm{CEF}} = \sum_{n,m} B_n^m O_n^m. \tag{2.54}$$

If the number of the $4f$ electrons is odd, J has a half-integer value for Ce^{3+}, Nd^{3+}, Sm^{3+}, Dy^{3+}, Er^{3+} and Yb^{3+} and the corresponding $4f$ energy levels always possess doublets. These doublets are called the Kramers doublets. Kramers degeneracy is based on time reversal symmetry. The magnetic properties of the rare earth compounds are very different, depending on whether the number of the $4f$ electrons is odd or even. When a magnetic field is applied to the system, all the degenerate $4f$ states, including the Kramers doublets, split into singlets.

We can obtain the magnetic moment of the $4f$ electrons by measuring the magnetic susceptibility or magnetization under magnetic field H, considering the Zeeman energy term, as follows

$$\mathcal{H} = \mathcal{H}_{\text{CEF}} - g_J \mu_B H J_z \qquad (\boldsymbol{H}//z). \tag{2.55}$$

Here, $|i\rangle$ is the wave function of the $4f$ energy level i, E_i the eigenvalue, and μ_i the magnetic moment of the energy level. The energy level is mixed with the other energy levels under the influence of a magnetic field. We represent the wave function and eigenvalue of this energy state as $|\tilde{i}\rangle$ and $E_i(H)$, respectively. Namely, we calculate the energy state under magnetic field, $|\tilde{i}\rangle$ and $E_i(H)$, by diagonalizing the matrix of the Hamiltonian in Eq. (2.50). Then, we calculate the magnetization and the magnetic susceptibility by $|\tilde{i}\rangle$ and $E_i(H)$. Here, the Helmholtz free energy F can be expressed in terms of the partition function Z as

$$F = -k_B T \ln Z \tag{2.56}$$

$$\text{where } Z = \sum_i e^{-\frac{E_i(H)}{k_B T}}. \tag{2.57}$$

The magnetization M is expressed as the differential of F by magnetic field $\frac{\partial F}{\partial H}$ as

$$\begin{aligned} M &= -\frac{\partial F}{\partial H} \\ &= \frac{\sum_i \mu_{z_i} e^{-E_i(H)/k_B T}}{\sum_i e^{-E_i(H)/k_B T}} \\ &\equiv \langle \mu_{z_i} \rangle \end{aligned} \tag{2.58}$$

where μ_{z_i} is the magnetic moment of the state $|\tilde{i}\rangle$

$$\begin{aligned} \mu_{z_i} &= -\frac{\partial E_i(H)}{\partial H} \\ &= g_J \mu_B \langle \tilde{i} | J_z | \tilde{i} \rangle. \end{aligned} \tag{2.59}$$

M corresponds to the average $\langle \mu_{z_i} \rangle$ of the magnetic moment μ_{z_i}.

The magnetic susceptibility χ is the field differential of magnetization $\partial M/\partial H(H \to 0)$

$$\chi = \frac{1}{k_{\mathrm{B}}T}\left(\left\langle\left(\frac{\partial E_i(H)}{\partial H}\right)^2\right\rangle - \left\langle\frac{\partial E_i(H)}{\partial H}\right\rangle^2\right) - \left\langle\frac{\partial^2 E_i(H)}{\partial H^2}\right\rangle. \tag{2.60}$$

In the calculation of χ, we can treat the Zeeman energy $-g_{\mathrm{J}}\mu_{\mathrm{B}}HJ_z$ as the perturbation. The energy $E_i(H)$ can be expressed by the second perturbation

$$E_i(H) = E_i - g_{\mathrm{J}}\mu_{\mathrm{B}}H\langle i|J_z|i\rangle - (g_{\mathrm{J}}\mu_{\mathrm{B}})^2 H^2 \sum_{j(\neq i)} \frac{|\langle j|J_z|i\rangle|^2}{E_j - E_i}. \tag{2.61}$$

By using Eq. (2.61), the magnetic susceptibility in Eq. (2.60) is obtained as

$$\chi = \frac{(g_{\mathrm{J}}\mu_{\mathrm{B}})^2 \sum_i e^{-E_i/k_{\mathrm{B}}T}\left(|\langle i|J_z|i\rangle|^2 + 2k_{\mathrm{B}}T\sum_{j(\neq i)}\frac{|\langle j|J_z|i\rangle|^2}{E_j - E_i}\right)}{k_{\mathrm{B}}T\sum_i e^{-E_i/k_{\mathrm{B}}T}}. \tag{2.62a}$$

Equation (2.62a) is the general expression of the magnetic susceptibility under consideration of CEF, but another expression is often used:

$$\chi = \frac{(g_{\mathrm{J}}\mu_{\mathrm{B}})^2}{\sum_i e^{-E_i/k_{\mathrm{B}}T}}\left(\frac{\sum_i |\langle i|J_z|i\rangle|^2 e^{-E_i/k_{\mathrm{B}}T}}{k_{\mathrm{B}}T} + \sum_i\sum_{j(\neq i)} |\langle j|J_z|i\rangle|^2 \frac{e^{-E_i/k_{\mathrm{B}}T} - e^{-E_j/k_{\mathrm{B}}T}}{E_j - E_i}\right). \tag{2.62b}$$

The first term is the Curie term which can be determined from the diagonal terms of the matrix J_z, and the second term is related to the non-diagonal terms. Namely, it is the Van-Vleck term, which is related to the transition between the $4f$-CEF states. It is known from Eqs. (2.62a and b) that the magnetic susceptibility can be determined from the $4f$-CEF states without magnetic field.

Next, we calculate J_z for the cubic Ce^{3+}. The J_z matrix can be expressed as

$$J_z = \begin{array}{c} \\ \langle\Gamma_7^\alpha| \\ \langle\Gamma_7^\beta| \\ \langle\Gamma_8^\nu| \\ \langle\Gamma_8^\kappa| \\ \langle\Gamma_8^\lambda| \\ \langle\Gamma_8^\mu| \end{array} \begin{array}{c} \begin{array}{cccccc} |\Gamma_7^\alpha\rangle & |\Gamma_7^\beta\rangle & |\Gamma_8^\nu\rangle & |\Gamma_8^\kappa\rangle & |\Gamma_8^\lambda\rangle & |\Gamma_8^\mu\rangle \end{array} \\ \begin{pmatrix} -\frac{5}{6} & 0 & \frac{2\sqrt{5}}{3} & 0 & 0 & 0 \\ 0 & \frac{5}{6} & 0 & -\frac{2\sqrt{5}}{3} & 0 & 0 \\ \frac{2\sqrt{5}}{3} & 0 & \frac{11}{6} & 0 & 0 & 0 \\ 0 & -\frac{2\sqrt{5}}{3} & 0 & -\frac{11}{6} & 0 & 0 \\ 0 & 0 & 0 & 0 & \frac{1}{2} & 0 \\ 0 & 0 & 0 & 0 & 0 & -\frac{1}{2} \end{pmatrix} \end{array}. \tag{2.63}$$

We obtain the magnetic moment to be $-5/7\,\mu_B$ for $|\Gamma_7^\alpha\rangle$ and $+5/7\,\mu_B$ for $|\Gamma_7^\beta\rangle$ from $g_J = 6/7$. The summation over the two degenerate states of the Γ_7 state is zero. The magnetic moments for $|\Gamma_8^\nu\rangle$, $|\Gamma_8^\kappa\rangle$, $|\Gamma_8^\lambda\rangle$ and $|\Gamma_8^\mu\rangle$ are $11/7\,\mu_B$, $-11/7\,\mu_B$, $3/7\,\mu_B$ and $-3/7\,\mu_B$, respectively. Equation (2.62b) can be expressed as follows (recall Γ_7 is the ground state, and Γ_8 is the excited state $E_{\Gamma_8} - E_{\Gamma_7} = \Delta$)

$$\chi_z = \frac{(g_J\mu_B)^2}{1 + 2e^{-\Delta/k_BT}} \left\{ \frac{\frac{25}{36} + \frac{65}{18}e^{-\Delta/k_BT}}{k_BT} + \frac{40\left(1 - e^{-\Delta/k_BT}\right)}{9\Delta} \right\}. \tag{2.64}$$

Figures 2.17(a) and (b) show the temperature dependence of the inverse magnetic susceptibility and magnetization, respectively, on the basis of Eqs. (2.58), (2.62) and (2.63), for three cases: no CEF, Γ_7 ground state, and Γ_8 ground state with the splitting energy $\Delta = 200\,\mathrm{K}$ between Γ_7 and Γ_8. If there is no CEF, $\Delta \to 0$ and $\chi_z = \frac{35}{4}(g_J\mu_B)^2/3k_BT$ [27]. This is equivalent to the case of $k_BT \gg \Delta$ and to the Curie law which ignores CEF. When Γ_7 is the ground state, the magnetization approaches the magnetic moment of 0.7–0.8 μ_B. Lastly, when Γ_8 is the ground state, the magnetization

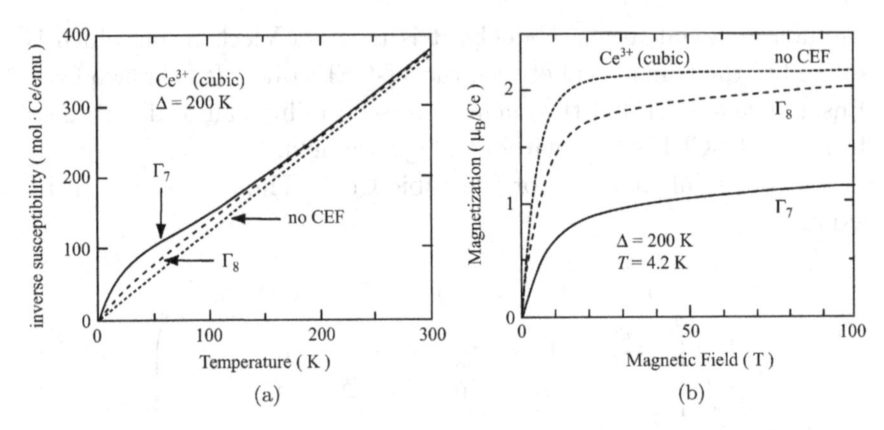

Figure 2.17: (a) Inverse magnetic susceptibility and (b) magnetization for $\Delta = 200\,\text{K}$ in cubic Ce^{3+}, cited from Ref. [27].

becomes $1.7\text{–}1.8\,\mu_\text{B}$. If the Zeeman energy is larger than the CEF splitting energy, the magnetization becomes the saturated magnetic moment $g_\text{J}J = 2.3\ \mu_\text{B}/\text{Ce}$.

2.5 Quadrupole moment

The quadrupole interaction is of basic importance in f-electron magnetism as well as in magnetic interactions such as the RKKY interaction and the many-body Kondo effect. The quadrupole moment is obtained from the electrostatic energy:

$$\mathcal{H} = \int \rho(r)\phi(r)d^3r$$
$$= Ze\phi(0) + \sum_j P_j \left(\frac{\partial \phi}{\partial x_j}\right)_0 + \frac{1}{2}\sum_{j,k} Q_{jk}\left(\frac{\partial^2 \phi}{\partial x_j \partial x_k}\right)_0 + \cdots \tag{2.65}$$

$$Ze = \int \rho(r)d^3r \tag{2.66}$$

$$P_j = \int \rho(r)x_j d^3r \tag{2.67}$$

$$Q_{jk} = \int \rho(r)x_j x_k d^3r \tag{2.68}$$

where $\rho(r)$ is the charge distribution of the f electrons in the rare earth and uranium compounds, $\phi(r)$ is the electrostatic potential, Ze is the charge of the f electrons, $P_{\rm j}$ is a dipole term and Q_{jk} is a quadrupole term. The independent terms in $x_{\rm j}x_{\rm k}$ of Eq. (2.68) are the following five: $2z^2 - x^2 - y^2$, $x^2 - y^2$, xy, yz and zx. These are converted into the quadrupole moments $(2J_{\rm z}^2 - J_{\rm x}^2 - J_{\rm y}^2)/\sqrt{3} = O_2^0$, $J_{\rm x}^2 - J_{\rm y}^2 = O_2^2$, $J_{\rm x}J_{\rm y} + J_{\rm y}J_{\rm x} = O_{\rm xy}$, $J_{\rm y}J_{\rm z} + J_{\rm z}J_{\rm y} = O_{\rm yz}$ and $J_{\rm z}J_{\rm x} + J_{\rm x}J_{\rm z} = O_{\rm zx}$, respectively, as shown in Fig. 2.18.

The quadrupole moment $O_{\Gamma\gamma}$ couples to the strain $\varepsilon_{\Gamma\gamma}$, as shown in Table 2.3, and distorts the crystal. In Fig. 2.19, we show the strain of the crystal in the simple case of the cubic structure. For example, the cubic structure in CeAg with the CEF-Γ_8 ground state is changed into a tetragonal structure when O_2^0 becomes the order parameter below 16 K [30]. This is because O_2^0 couples to the strain ε_u, which brings about expansion along the z-axis and shrinkage along the x- and y-axes. The Γ_8-quartet state is changed into two Kramers doublets and the ferromagnetic ordering occurs below $T_{\rm C} = 5.0$ K.

In the following, we discuss the metamagnetic transition based on the quadrupole interaction $\langle O_2^2 \rangle$ in $PrCu_2$ [31–33]. The crystal structure of $PrCu_2$ is orthorhombic (No. 74, $a = 4.410$ Å, $b = 7.050$ Å, $c = 7.454$ Å). This can be viewed as a distorted hexagonal AlB_2-type structure, as shown in Fig. 2.20(a). $PrCu_2$ is a Van Vleck paramagnetic compound in which the quadrupole interaction plays an important role in the magnetic behavior and the ferroquadrupole ordering is realized below $T_{\rm Q} = 7.6$ K, as shown in Fig. 2.20(d). A change of the strain is characteristic in the quadrupole ordering. The transverse mode C_{66} indicates a softening effect with decreasing temperature that decreases steeply below $T_{\rm Q} = 7.6$ K, revealing a ferroquadrupole ordering.

Figure 2.21 shows the magnetization curves at 15 K for three principal axes. The c-axis magnetization $M_{\rm c}$ is the hard-axis magnetization, while the a-axis magnetization $M_{\rm a}$ corresponds to the easy-axis magnetization. Both magnetizations should show almost the same curves if $PrCu_2$ has a hexagonal structure, but they indicate highly different behavior. This is due to the CEF effect caused by the orthorhombic symmetry.

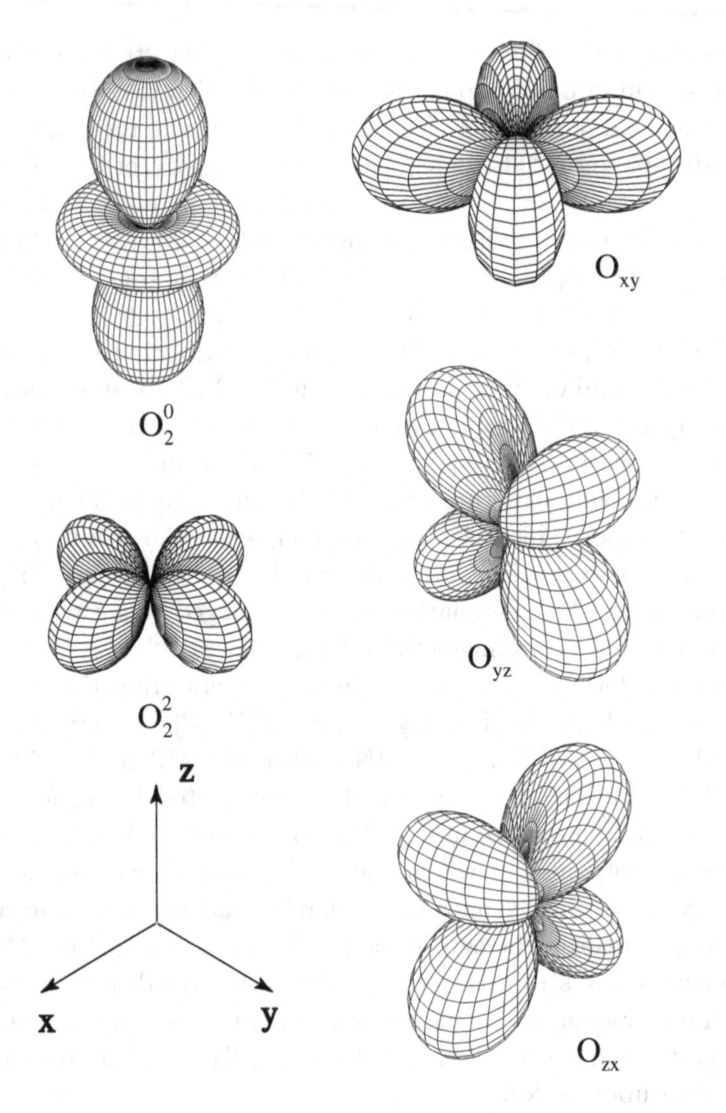

Figure 2.18: Schematic view of the quadrupole moments, cited from Ref. [27].

It is characteristic that M_c shows a metamagnetic transition at $H_\mathrm{c} = 200\,\mathrm{kOe}$ and reaches M_a at higher fields. In fact, it starts to resemble the M_a-curve with decreasing magnetic field, showing large hysteresis. Upon increasing the field again, M_c is changed into M_a. This metamagnetic transition thus brings about conversion between

Table 2.3: Elastic strain ε_Γ, quadrupole operator O_Γ, and elastic constant C_Γ in the cubic crystal structure.

Elastic strain	Quadruple operator	Elastic constant	Symmetry
$\varepsilon_{\rm B} = \varepsilon_{xx} + \varepsilon_{yy} + \varepsilon_{zz}$	$O_{\rm B} = O_4^0 + 5O_4^4$	$C_{\rm B} = \frac{C_{11}+2C_{12}}{\sqrt{3}}$	Γ_1
$\varepsilon_u = \dfrac{2\varepsilon_{zz} - \varepsilon_{xx} - \varepsilon_{yy}}{\sqrt{3}}$	$O_2^0 = \dfrac{2J_z^2 - J_x^2 - J_y^2}{\sqrt{3}}$	$\dfrac{C_{11} - C_{12}}{2}$	Γ_3
$\varepsilon_v = \varepsilon_{xx} - \varepsilon_{yy}$	$O_2^2 = J_x^2 - J_y^2$		
ε_{xy}	$O_{xy} = J_xJ_y + J_yJ_x$		
ε_{yz}	$O_{yz} = J_yJ_z + J_zJ_y$	C_{44}	Γ_5
ε_{zx}	$O_{zx} = J_zJ_x + J_xJ_z$		

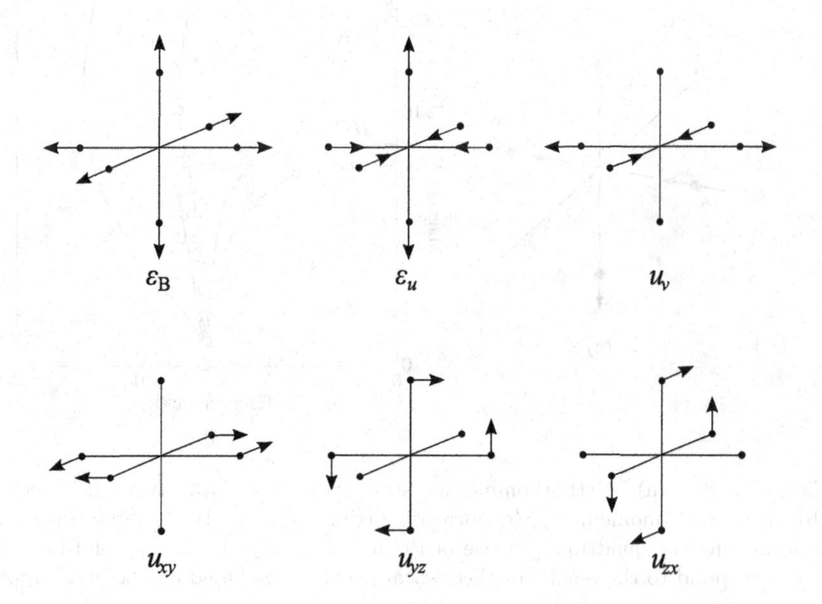

Figure 2.19: Strain of the cubic structure.

the hard c- and easy a-axis magnetizations. To elucidate this metamagnetic transition, we show in Fig. 2.21 the calculated magnetizations (broken lines) based on the CEF effect. We also note that the critical field for this metamagnetic transition increases with increasing temperature, from 80–90 kOe at 1.3–4.2 K to 320 kOe at 50 K, as shown in Fig. 2.22.

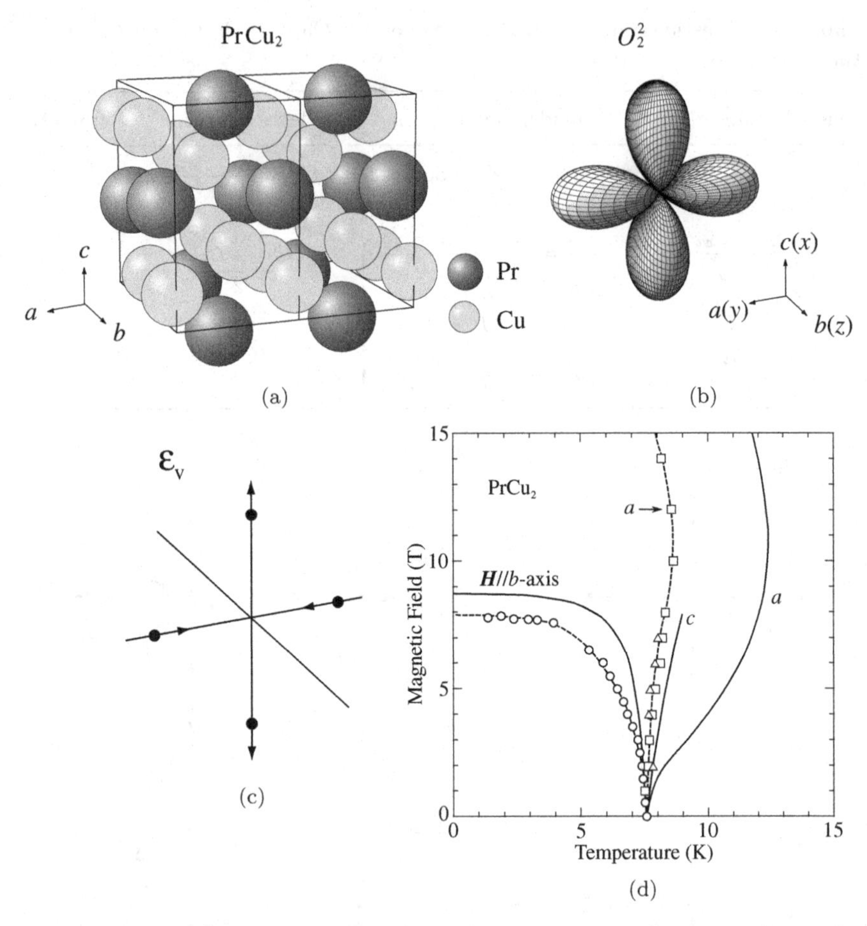

Figure 2.20: (a) Orthorhombic crystal structure with two unit cells, (b) quadrupole moment O_2^2, (c) local distortion ε_v, and (d) the phase diagram showing the ferroquadrupole phase of $PrCu_2$, cited from Ref. [31]. Solid lines in (d) correspond to the results of theoretical calculations based on the quadrupole interactions.

The present metamagnetic transition is also observed in the electrical resistivity, magnetostriction and dHvA experiments [31, 32]. Figure 2.23 shows the corresponding magnetostriction curves. First, the magnetic field was applied along the c-axis, indicated with open circles in Fig. 2.23(a). The magnetostriction ε_c along the c-axis steeply increases at 90 kOe by about 0.5 % and then increases smoothly up to 140 kOe; see the processes (1) and (2) in Fig. 2.23(a).

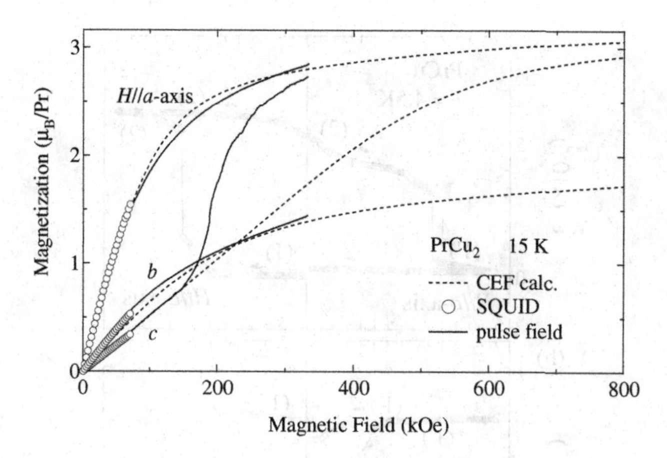

Figure 2.21: Magnetization curves of $PrCu_2$. Thick lines, open circles and dotted lines show the pulse-field experimental curves, SQUID magnetometer data and CEF calculated curves respectively, cited from Ref. [33].

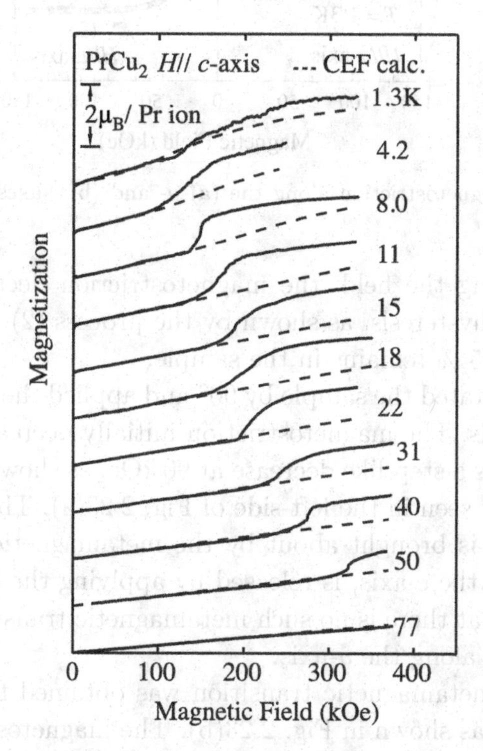

Figure 2.22: Magnetization curves of $PrCu_2$. Thick lines and dotted lines show the experimental and CEF-calculated curves respectively, cited from Ref. [33].

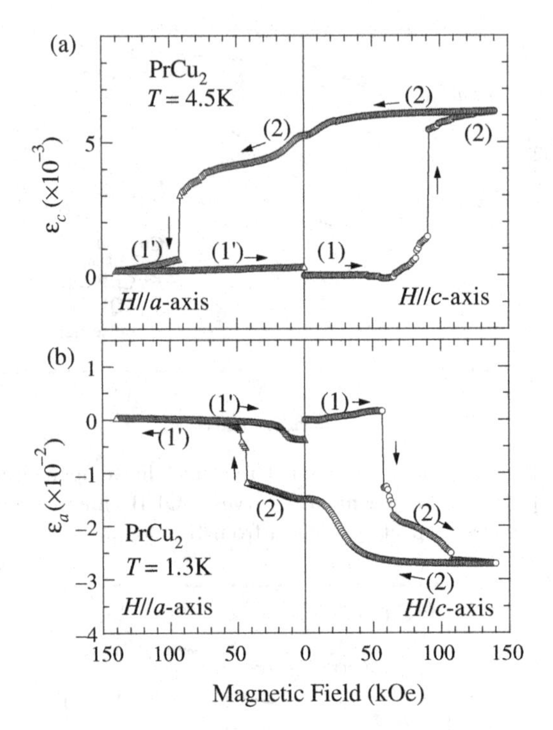

Figure 2.23: Magnetostriction along the (a) a- and (b) c-axes of $PrCu_2$, cited from Ref. [31].

Upon decreasing the field, the magnetostriction decreases slightly, showing large hysteresis, as shown by the process (2). A large residual strain of 0.5 % remains in the sample.

Next, we rotated the sample by 90° and applied the magnetic field along the a-axis. The magnetostriction initially decreases monotonically and shows a step-like decrease at 90 kOe, as shown by processes (2) and (1′), as seen in the left-side of Fig. 2.23(a). The strain in the sample, which is brought about by the metamagnetic transition in the field along the c-axis, is released by applying the field along the a-axis. Note that there is no such metamagnetic transition if we first apply the field along the a-axis.

A similar metamagnetic transition was obtained for the magnetostriction ε_a, as shown in Fig. 2.23(b). The magnetostriction in the field along the c-axis steeply decreases at 60 kOe, as shown by processes (1) and (2) in Fig. 2.23(b). The strain reaches up to −2.7 %

at 140 kOe. The residual strain at 0 kOe is −1.5 %. The strain is released by applying the field along the a-axis, as shown by processes (2) and (1′) in Fig 2.23(b), although the residual strain does not recover completely at 0 kOe, retaining a residual value of −0.4 %.

The present metamagnetic transition is closely related to the distortion in the ac-plane — as the magnetostriction along the c-axis, $\varepsilon_{\rm c}$, is increased, the corresponding $\varepsilon_{\rm a}$ decreases in magnitude. The strain relation of expansion along the c-axis and shrinkage along the a-axis corresponds to $\varepsilon_{\rm v}$ in Fig. 2.20(c). Therefore, the quadrupole operator in this system is O_2^2, as shown in Fig. 2.20(b).

The dHvA effect is a powerful method to detect the metamagnetic transition via electron scattering [32]. Figure 2.24 shows the dHvA oscillations for the field along the a-axis. When we first apply a magnetic field along the a-axis, there is no appreciable change in oscillation, demonstrating a reversible oscillation, as shown in Fig. 2.24(a). On the other hand, the dHvA oscillation is extremely reduced in amplitude after the metamagnetic transition, as shown

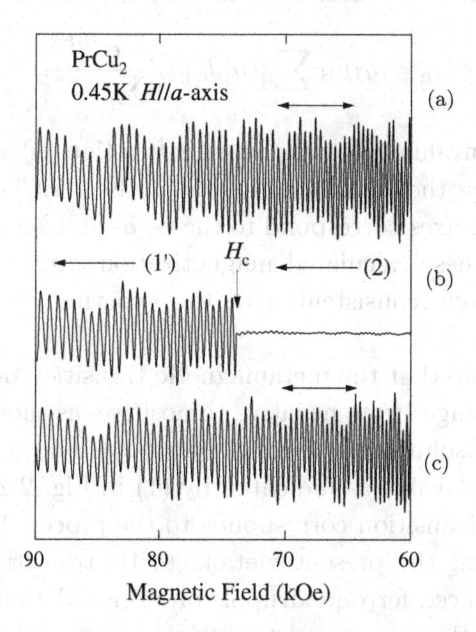

Figure 2.24: dHvA oscillations of $PrCu_2$, cited from Ref. [32]. Note that the magnetic field in the left-hand side is larger than that in the right.

in Fig. 2.24(b), corresponding to process (2) in the left-side of Fig. 2.23(a). As shown in Fig. 2.24(b), the dHvA oscillation is completely recovered after the metamagnetic transition, namely at fields higher than 74 kOe, following the process (1′). The corresponding dHvA oscillation is again reversible against the magnetic field, as shown in Fig. 2.24(c).

We explain the present metamagnetic transition on the basis of the quadrupole interaction $\langle O_2^2 \rangle$. First, we calculated the magnetic susceptibility in the temperature range from about 8 K to room temperature, and determined the CEF parameters and the coupling constant K_{M} in order to reproduce the susceptibility data and the metamagnetic field transition field on the basis of the following Hamiltonian

$$\mathcal{H} = \mathcal{H}_{\mathrm{CEF}} + \mathcal{H}_{\mathrm{Zeeman}} - K_{\mathrm{M}} \langle O_2^2 \rangle O_2^2. \tag{2.69}$$

Figure 2.25(a) shows the calculated magnetization curves at 10 and 30 K, where K_{M} is found to be 0.3. Here, the magnetization M_α ($\alpha = x, y, z$) can be calculated by the following formula

$$M_\alpha = g_J \mu_{\mathrm{B}} \sum_n |\langle n | J_\alpha | n \rangle| \frac{\mathrm{e}^{-\beta E_n}}{\sum_n \mathrm{e}^{-\beta E_n}} \tag{2.70}$$

where the eigenvalue E_n and the eigenfunction $|n\rangle$ are determined by diagonalizing the total Hamiltonian in Eq. (2.69). We note that the x-, y- and z-axes correspond to the c-, a- and b-axes, as shown in Fig. 2.20(b). These calculated magnetization curves in Fig. 2.25(a) are approximately consistent with the experimental data given in Fig. 2.25(b).

Here, we note that the metamagnetic transition occurs when the sign of $\langle O_2^2 \rangle$ changes from negative to positive, as shown in Fig. 2.26. There are three solutions at zero field for $\langle O_2^2 \rangle$. The ground state possesses the lowest value as indicated by (1) in Fig. 2.26. The present metamagnetic transition corresponds to the process from (1) to (2). We can say that the present metamagnetic transition corresponds to the field-induced ferroquadrupole ordering. Metamagnetic transitions based on the same quadrupole interaction are observed in the other compounds, $CeCu_2$, $DyCu_2$ and $TbCu_2$ [34–36].

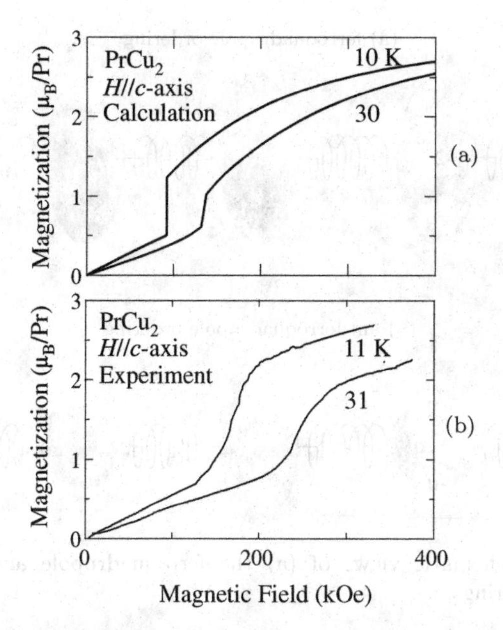

Figure 2.25: (a) Calculated and (b) experimental magnetization curves of $PrCu_2$, cited from Ref. [31].

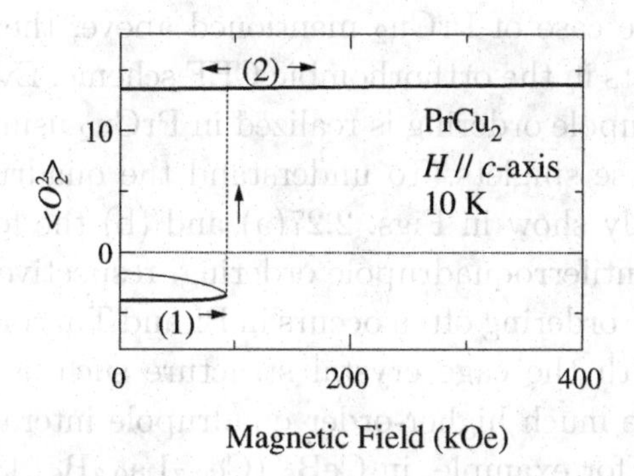

Figure 2.26: Field dependence of the calculated quadrupole moment $\langle O_2^2 \rangle$ in $PrCu_2$, cited from Ref. [31].

There is a possibility that quadrupole ordering occurs in any rare earth compounds — it is often realized in rare earth compounds with the $4f$-CEF quartet ground state such as CeAg and CeB_6. CeAg indicates a ferroquadrupole ordering at $T_Q = 15$ K and a successive ferromagnetic ordering at $T_C = 5.0$ K, as mentioned above. On the

(a) ferroquadrupole ordering

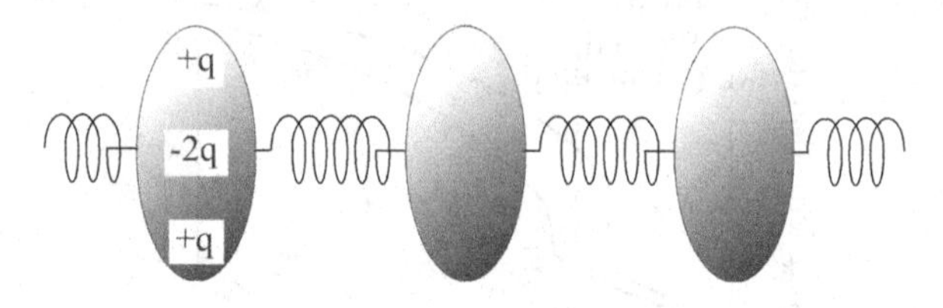

(b) antiferroquadrupole ordering

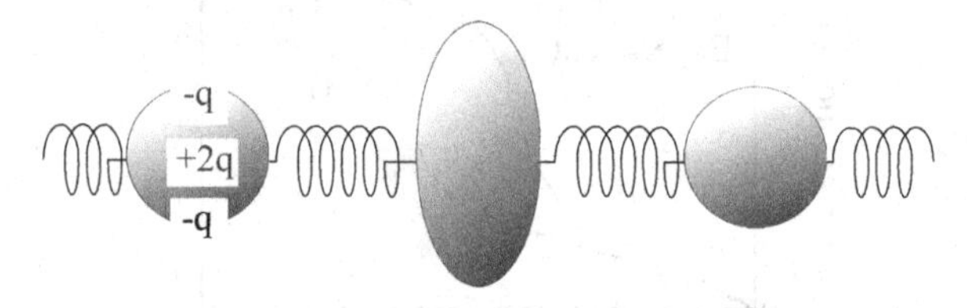

Figure 2.27: Schematic views of (a) the ferroquadrupole and (b) antiferroquadrupole orderings.

other hand, the antiferroquadrupole ordering occurs below T_{Q} = 3.3 K and a successive antiferromagnetic ordering below T_{N} = 2.3 K in $\mathrm{CeB_6}$. In the case of $\mathrm{PrCu_2}$ mentioned above, the $4f$ levels split into nine singlets in the orthorhombic CEF scheme. Even in this case, the ferroquadrupole ordering is realized in $\mathrm{PrCu_2}$ using the magnetic coupling of these singlets. To understand the quadrupole ordering, we schematically show in Figs. 2.27(a) and (b) the ferroquadrupole ordering and antiferroquadrupole ordering, respectively.

Quadrupole ordering often occurs in Pr and Tm compounds, especially those with the cage crystal structure such as $\mathrm{PrFe_4P_{12}}$ and $\mathrm{PrTi_2Al_{12}}$. The much higher-order quadrupole interaction or ordering is realized, for example, in $\mathrm{CeB_6}$ ($\mathrm{Ce_{0.7}La_{0.3}B_6}$) [37, 38].

Chapter 3

Single Crystal Growth and Measuring Methods

High-quality single crystals are grown by various methods such as the Czochralski method, floating zone method, flux method, Bridgman method, and chemical transport method. They are explained in this chapter. For heavy fermion compounds, low-temperature measurements are essential to clarify the electronic properties of the sample because the typical heavy fermion state is formed below 1–10 K. In the experiments, we mainly measure the electrical resistivity, magnetic susceptibility magnetization, specific heat, thermoelectric power, thermal expansion coefficient, thermal conductivity, NQR-NMR, neutron scattering, μSR, magnetoresistance, Hall coefficient, dHvA, and photoelectron spectroscopy, together with the electrical resistivity under pressure. We also describe the methods of measurement in this chapter.

3.1 Single crystal growth

Various techniques for single crystal growth are applied to transition metals, rare earth compounds, and actinide compounds, depending on the melting point and the degree of vapor pressure of the melt. If the vapor pressure is low, the Czochralski pulling method is a powerful method to obtain a single crystal ingot with a large size. An rf-furnace is used in the Czochralski method when the melting point T_m is less than 1500°C. The tungsten crucible can be used for $CeCu_6$ (melting point T_m = 938°C), for example, as shown in Fig. 3.1(a). In this case, the crucible, which is surrounded by an rf-working coil, becomes a heater for the $CeCu_6$ melt. A large ingot of 10 mm in diameter is obtainable in this temperature-stable rf-furnace. A carbon tube acts as a heater when a non-metallic boron nitride (BN) material is used as a crucible.

The Czochralski method is also applied to compounds with high melting points; for example, at a temperature of about 2500°C in the tetra-arc furnace. In such a case, as shown in Fig. 3.1(b), the crucible is unnecessary. The starting materials are set on the water-cooled Cu-hearth plate, and the single crystals are grown by the Czochralski method in the tetra-arc furnace. Typical compounds with high melting points are UPt_3 (T_m = 1700°C), $CeRu_2Si_2$, $CeRh_3B_2$, UB_2 (T_m = 2385°C), and UC (T_m = 2530°C). The melting points of $CeRu_2Si_2$ and $CeRh_3B_2$ are unknown, but are estimated to be 2000–2500°C. These compounds are congruent compounds.

The Czochralski method is also applied to incongruent compounds such as $CeRu_2$, USi_3, and $CeIrSi_3$. For these compounds, the off-stoichiometric starting materials of $CeRu_{1.8}$, $USi_{4.6}$, and $CeIrSi_{3.5}$ are used in the tetra-arc furnace. Large single crystal ingots are shown in Fig. 3.1(c).

The floating zone method is applied to CeB_6 by using a small size working coil in the rf-furnace, which is heated up to 2600°C.

On the other hand, the flux method, the chemical transport method, and the Bridgman method are useful for compounds with high vapor pressure and also for incongruent compounds. Figure 3.2(a) indicates single crystals of $CeRhIn_5$ grown by the In-flux method, where the starting materials and flux were inserted in an alumina crucible, sealed in a quartz tube, heated up to about 1000°C in an electric furnace, and cooled down to lower temperatures. The entire process takes about 20 days. Usually, metals with low-melting points such as Zn, Cd, Al, Ga, In, Sn, Pb, Sb, and Bi are available for flux. In the case of $CeRhIn_5$, excess In itself becomes flux, where the ratio of Ce:Rh:In is 1:1:15.

A single crystal of $CePt_3Si$ is grown by the Bridgman method in a Mo crucible, as shown in Fig. 3.2(b). First, polycrystals of $CePt_3Si$

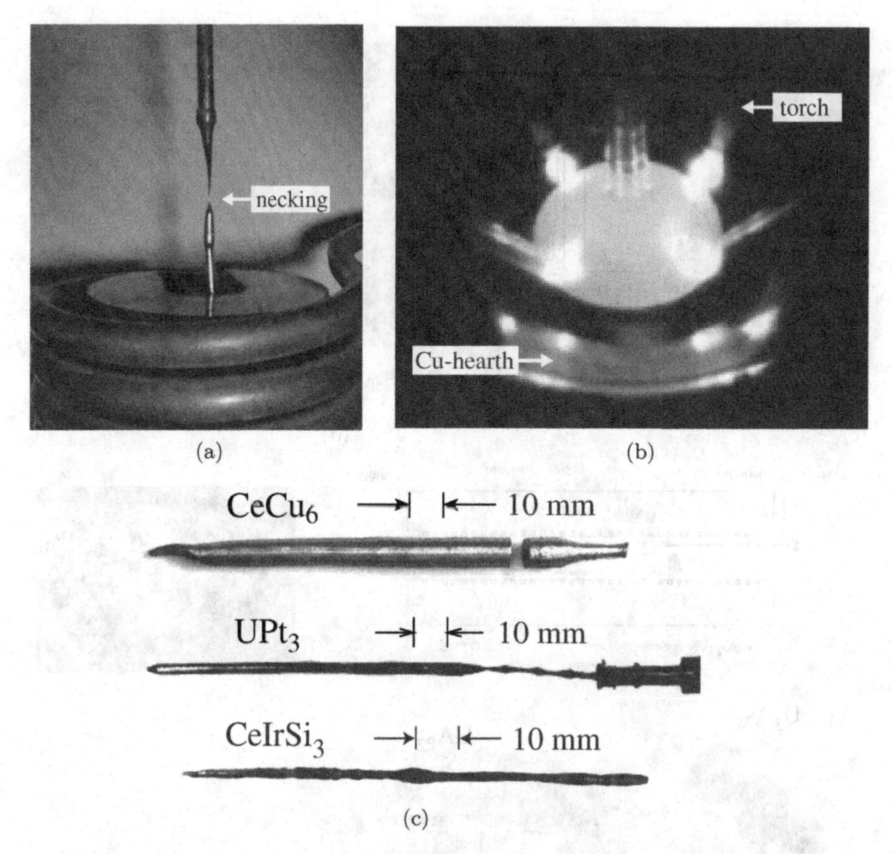

Figure 3.1: (a) $CeCu_6$ ingot in the rf-furnace, (b) pulling process of a UPt_3 ingot in the tetra-arc furnace, and (c) $CeCu_6$, UPt_3, and $CeIrSi_3$ ingots grown by the Czochralski method.

are prepared by arc-melting the starting materials with a ratio of Ce:Pt:Si=1:3:1 and the polycrystals are inserted in a Mo-crucible. A top-plate of the Mo-crucible is sealed using the tetra-arc furnace. The crucible is heated up to 1350°C and cooled down slowly, taking 7 days in total.

Single crystals of U_3As_4(U_3P_4) and UAs_2(UP_2) are grown by the chemical transport method with iodine as the transport agent in an electric furnace with a distinct temperature gradient, as shown in Fig. 3.2(c). Starting materials in powder form are set down at the low-temperature side (830°C) and single crystals are grown at the higher-temperature side (900°C). Note that in transition

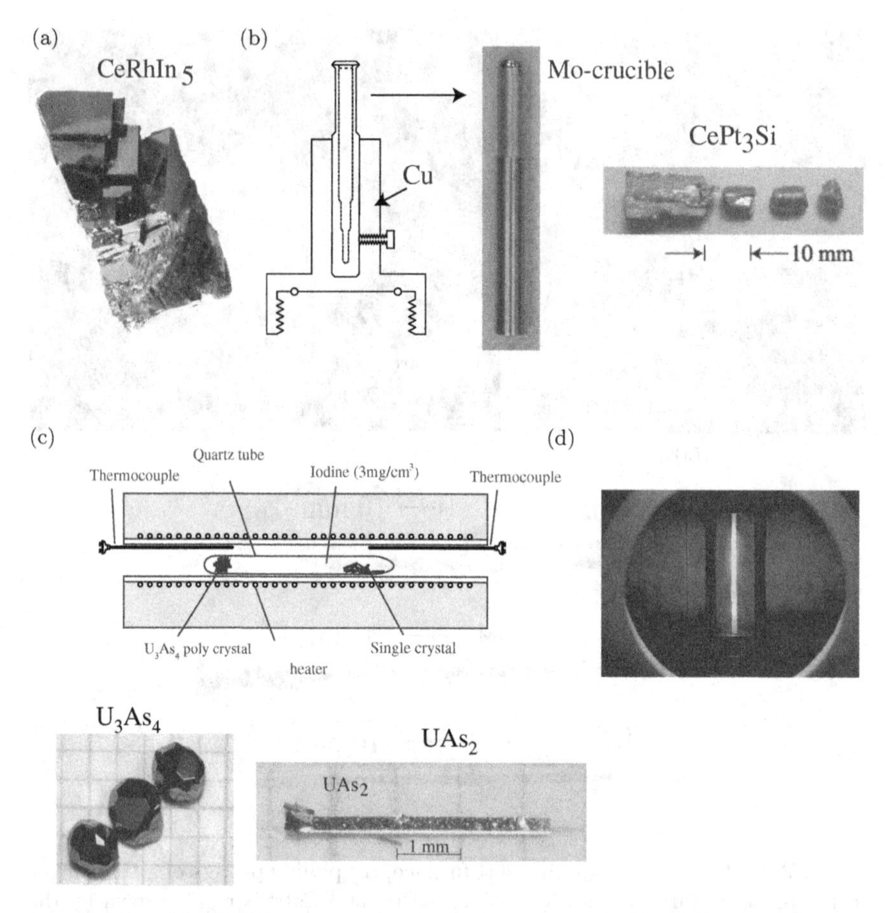

Figure 3.2: (a) $CeRhIn_5$ grown by the In-flux method, (b) sealing set of a top plate of the Mo-crucible using a tetra-arc furnace and a $CePt_3Si$ ingot, (c) an electric furnace with a distinct temperature gradient and U_3As_4 and UAs_2 single crystals, and (d) a solid state electrotransport furnace and a UPt_3 ingot heated up in the furnace.

metal compounds such as CoS_2, their powders are set at the high-temperature side (700°C) while single crystals are grown at the low-temperature side (650°C). $CoBr_2$ powders act as the transport agent as they produce Br_2 gas at high temperatures.

Here, the starting material of a uranium ingot is not of high quality, roughly 99.9% (3N), and needs to be annealed under high vacuum of 10^{-6} Pa via the solid state electrotransport method. Through annealing, an Fe impurity present in quantities of 40 ppm in the starting uranium ingot is reduced to an amount less than 2 ppm, while the Cu impurity is completely removed [39]. Subsequent annealing of the U compound under high vacuum using the electrotransport method drastically improves the quality of the single crystal. This is applied to $CeRu_2$, UPt_3, and URu_2Si_2, as shown in Fig. 3.2(d) for UPt_3. For example, an extremely high-quality single crystal sample of URu_2Si_2 is obtained, where the residual resistivity ratio (RRR) reaches a magnitude greater than 500 [40].

We emphasize that high-quality single crystals are required to accurately determine the characteristic properties in heavy fermion compounds, especially the superconducting property. Cooper pairs are broken by impurities and crystalline defects, and the existence of a residual density of states is crucial in determining the superconducting property from the temperature dependence of nuclear spin-lattice relaxation rate $1/T_1$ below a superconducting temperature T_{sc}. Moreover, the amplitude of the dHvA oscillations, which is proportional to $\exp[-\alpha(m_c^*/H)(T + T_D)]$, is closely related to the sample quality, where $\alpha = 2\pi^2 ck_B/e\hbar$. If the cyclotron effective mass m_c^* is the rest mass of an electron m_0, the temperature of 1 K is usual in the dHvA experiment. A much lower temperature of 0.01 K is needed for $m_c^* = 100m_0$. Even if low temperatures are realized using a dilution refrigerator, the reduction of $\exp[-\alpha(m_c^*/H)T_D]$ is inevitable in the dHvA amplitude. Here, T_D is a so-called Dingle temperature, which is inversely proportional to the scattering lifetime of the carrier τ. Therefore, high sample quality with a large τ value is essential to observe the dHvA oscillations for heavy fermion compounds.

3.2 Measuring methods

3.2.1 Electrical resistivity

The electrical resistivity ρ is measured using a very simple technique, but contains abundant information of the sample under consideration. The electrical voltage V between two lead wires on the sample, with a length l, is measured for a rectangular sample, with a cross-sectional area S in a current I. The electrical resistance R is obtained via Ohm's law, where $R = V/I$. We usually consider the electrical resistivity $\rho = (S/l)R$ instead of R, as it is independent of the sample dimensions and geometry.

Now, we will explain Ohm's law. In the electric field E_x along the x-axis, which is parallel to the direction of the current I, the average velocity of a conduction electron v_x is expressed as

$$m^* \frac{dv_x}{dt} = -eE_x - \frac{m^* v_x}{\tau} \tag{3.1}$$

where $-m^* v_x/\tau$ is the frictional or damping force and τ is the scattering lifetime. In steady state where $dv_x/dt = 0$, we obtain Ohm's law using the expression for a steady current $I = SJ_x = -nev_x S$ (n: carrier density)

$$J_x = \frac{ne^2\tau}{m^*} E_x = \sigma E_x = \frac{1}{\rho} E_x \tag{3.2}$$

$$\rho = \frac{m^*}{ne^2\tau} \tag{3.3}$$

where ρ, as we note earlier, is the electrical resistivity and $\sigma(= 1/\rho)$ is the electrical conductivity. $R = (l/S)\rho$ is obtained from $I = SJ_x$ and $V = lE_x$.

The electrical resistivity $\rho(T)$ consists of four contributions: the electron scattering due to non-magnetic impurities and crystalline defects ρ_0, the electron-phonon scattering $\rho_{\mathrm{ph}}(T)$, the electron-electron scattering $\rho_{\mathrm{e-e}}(T)$, and the magnetic scattering $\rho_{\mathrm{m}}(T)$

$$\rho(T) = \rho_0 + \rho_{\mathrm{ph}}(T) + \rho_{\mathrm{e-e}}(T) + \rho_{\mathrm{m}}(T). \tag{3.4}$$

This yields Mattiessen's rule.

The ρ_0 value is constant with temperature variation. This value is important for one to know the quality of a sample, which is estimated by determining the residual resistivity (RRR $= \rho_{\mathrm{RT}}/\rho_0$, ρ_{RT}: resistivity at room temperature). Note that RRR$\simeq \rho_{\mathrm{ph}}(\simeq \rho$ at RT$)/\rho_0$ if the sample is a simple metal which follows $\rho \simeq \rho_0 + \rho_{\mathrm{ph}} + \rho_{\mathrm{e-e}}$. Of course, a large value of RRR indicates a high-quality sample which corresponds to a low ρ_0 value.

Next, we consider the electrical resistivity due to the electron-phonon scattering $\rho_{\text{ph}}(T)$. When the atom (ion) is displaced by δR_n from the regular position at R_n, the corresponding potential of the lattice $V(\boldsymbol{r})$ is

$$V(\boldsymbol{r}) = \sum_n v\left(\boldsymbol{r} - \boldsymbol{R_n} - \delta\boldsymbol{R_n}\right) \tag{3.5}$$
$$= V_0(\boldsymbol{r}) + \delta V(\boldsymbol{r})$$
$$\delta V(\boldsymbol{r}) = -\sum_n \left(\frac{dv}{dr}\right) \cdot \delta\boldsymbol{R_n}. \tag{3.6}$$

The conduction electron is scattered from this displacement of the ion. Namely, the wave function $\Psi_{\boldsymbol{k}}(\boldsymbol{r})$ of the conduction electron with the wave vector $\boldsymbol{k}$ is scattered into $\Psi_{\boldsymbol{k}'}(\boldsymbol{r})$ with the wave vector $\boldsymbol{k'}$. The scattering probability W is given as follows

$$W_{\boldsymbol{k}\to\boldsymbol{k}'} \propto \left(\int \Psi^*_{\boldsymbol{k}'}(\boldsymbol{r})\delta V(\boldsymbol{r})\Psi_{\boldsymbol{k}}(\boldsymbol{r})d^3\boldsymbol{r}\right)^2 \tag{3.7}$$

under the following momentum and energy conservations

$$\boldsymbol{k} - \boldsymbol{k}' \pm \boldsymbol{q} = \boldsymbol{G} \tag{3.8}$$

$$E_{\boldsymbol{k}'} - E_{\boldsymbol{k}} = \pm\hbar\omega_q \tag{3.9}$$

where $\boldsymbol{q}$ is the momentum of a phonon with an energy $\hbar\omega_q$, $\boldsymbol{G}$ is a reciprocal vector, and the scattering process is called an Umklapp process. In a perfect lattice of $\rho_0 = 0$, the ρ_{ph} term becomes zero at 0 K because $\delta V(\boldsymbol{r})$ is zero at 0 K, meaning $W_{\boldsymbol{k}\to\boldsymbol{k}'} = 0$ and $\tau(\propto 1/W_{\boldsymbol{k}\to\boldsymbol{k}'}) \to \infty$.

The temperature dependence of $\rho_{\text{ph}}(T)$ is based on the well-known Grüneisen's formula. It is proportional to T above the Debye temperature Θ_{D}, and proportional to T^5 far below the Debye temperature. Grüneisen's formula is

$$\rho_{\text{ph}}(T) = \frac{C}{M\Theta_{\text{D}}}\left(\frac{T}{\Theta_{\text{D}}}\right)^5 \int_0^{\Theta_{\text{D}}/T} \frac{x^5}{(1-e^{-x})(e^x-1)}dx \tag{3.10}$$
$$\propto T \left(T \gtrsim \frac{\Theta_{\text{D}}}{2}\right)$$
$$\propto T^5 \ (T \ll \Theta_{\text{D}})$$

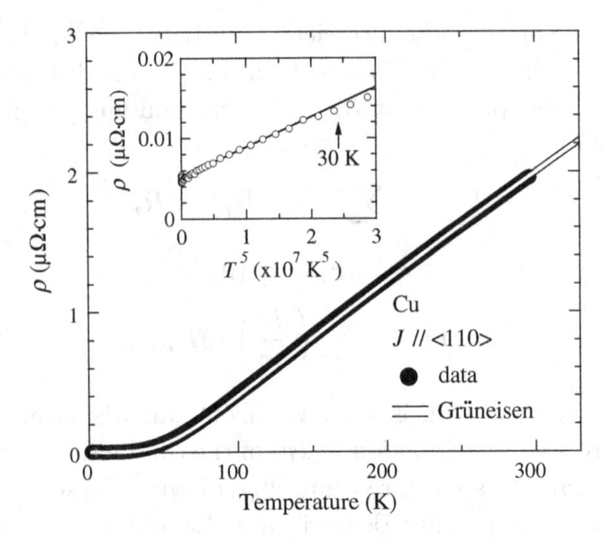

Figure 3.3: Temperature dependence of the electrical resistivity in Cu, where a solid line represents Grüneisen's formula for the Debye temperature $\Theta_{\rm D} = 344$ K. The inset shows the T^5 dependence of the resistivity below 20–30 K.

where C is constant and independent of the compound, and M is the mass of the compound. Figure 3.3 shows the temperature dependence of the electrical resistivity in a single crystal of Cu as an example. Grüneisen's formula for the Debye temperature $\Theta_{\rm D} = 344$ K has been fairly applied to the experimental result, revealing a T^5 dependence of the resistivity below 20–30 K.

The $\rho_{\rm mag}$ term is due to the scattering of the conduction electrons with the spins of magnetic ions. Here, we consider the electron scattering due to the localized $4f$ electrons in rare earth compounds. This is expressed as follows

$$\rho_{\rm mag} \sim J_{\rm cf}^2 (g_J - 1)^2 \sum_{m_s, m_s', i, j} \langle m_s', j \, | \boldsymbol{s} \cdot \boldsymbol{J} | \, m_s, i \rangle \, p_i f_{ij} \tag{3.11}$$

$$p_i = \frac{e^{-E_i/k_{\rm B}T}}{\sum_i e^{-E_i/k_{\rm B}T}}$$

$$f_{ij} = \frac{2}{1 + e^{-(E_i - E_j)/k_{\rm B}T}}$$

where m_s and m'_s correspond to the initial and final spin states of the conduction electron, respectively, and i and j also correspond to the initial and final $4f$-CEF levels. This means that ρ_{mag} is based on the Friedel oscillation or the de Gennes factor and the occupation of $4f$ electrons in the CEF scheme. When the temperature is larger than the overall splitting energy Δ in the CEF scheme, Eq. (3.11) becomes temperature-independent as

$$\rho_{\text{mag}}(T > \Delta) \sim J_{\text{cf}}^2(g_J - 1)^2 J(J+1). \tag{3.12}$$

Below the magnetic ordering temperature T_{mag}, the ρ_{mag} term decreases steeply with decreasing temperature. In the case of a ferromagnet, $\rho_{\text{mag}}(T) \sim T^2$ is observed experimentally, which is explained by electron scattering due to the ferromagnetic spin wave. Contrastingly $\rho_{\text{mag}}(T)$ does not have a simple temperature dependence for an antiferromagnet.

We present here the experimental results of rare earth compounds RCu_2Si_2 with the $ThCr_2Si_2$-type tetragonal structure [41], as shown in Fig. 3.4, where $4f$ electrons are localized. Note that $CeCu_2Si_2$, $EuCu_2Si_2$, and $YbCu_2Si_2$ are excluded in the present discussion. $CeCu_2Si_2$ and $YbCu_2Si_2$ are heavy fermion compounds and $EuCu_2Si_2$ is a nearly trivalent compound. It is found that the Néel temperature, T_N, in RCu_2Si_2 roughly follows the de Gennes factor, as shown in Fig. 3.5(a). That is,

$$T_N \sim (g_J - 1)^2 J(J+1) \tag{3.13}$$

while the Néel temperature of $PrCu_2Si_2$ deviates highly from this scaling. This is due to the quadrupole interaction. The magnetic resistivities ρ_{mag} are obtained by subtracting the resistivity of a non-$4f$ reference compound $LuCu_2Si_2$, [$\rho(LuCu_2Si_2)$-$\rho_0(LuCu_2Si_2)$] from [$\rho(RCu_2Si_2)$-$\rho_0(RCu_2Si_2)$], namely, $\rho_{\text{mag}} = [\rho(RCu_2Si_2) - \rho_0(RCu_2Si_2)] - [\rho(LuCu_2Si_2) - \rho_0(LuCu_2Si_2)]$. The temperature dependence of $\rho_{\text{mag}}(T)$ is shown in Fig. 3.5(b). Note that $\rho_{\text{mag}}(T)$ above 200 K is almost temperature-independent because the temperature above 200 K is larger than the overall CEF splitting

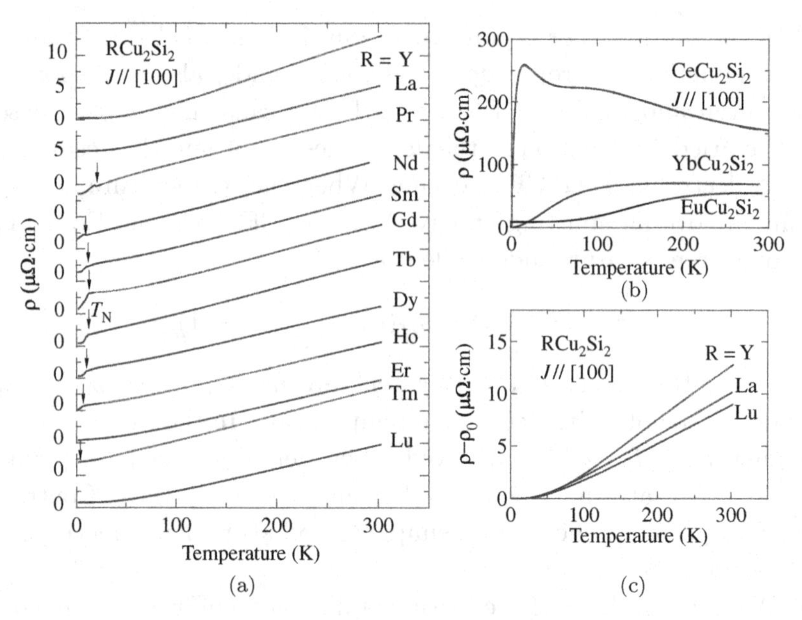

Figure 3.4: Temperature dependence of the electrical resistivities in (a) RCu_2Si_2 (R: Y, La, Pr, Nd, Sm, Gd, Tb, Dy, Ho, Er, Tm, and Lu), (b) $CeCu_2Si_2$, $EuCu_2Si_2$ and $YbCu_2Si_2$, and (c) YCu_2Si_2, $LaCu_2Si_2$, and $LuCu_2Si_2$, where the residual resistivities ρ_0 were subtracted in these compounds, cited from Ref. [41].

energy Δ in RCu_2Si_2. $\rho_{\mathrm{mag}}(200\ \mathrm{K})$ is roughly explained by the de Gennes factor, but is well scaled by T_{N}, as shown in Fig. 3.5(c), meaning that both $\rho_{\mathrm{mag}}(200\ \mathrm{K})$ and T_{N} share the same magnetic origin.

In the heavy fermion state, shown later, the magnetic specific heat C_{mag} is changed into the electronic specific heat C_{e}. This means that the localized $4f$ electrons move in the crystal. The Fermi liquid state with an extremely large effective mass is realized at low temperatures. The magnetic resistivity is correspondingly changed into the resistivity due to electron-electron scattering

$$\rho_{\mathrm{e-e}} = AT^2 \tag{3.14}$$

where A is extremely large; $A = 1.0\ \mu\Omega/\mathrm{K}^2$ in a heavy fermion compound $CeCu_2Si_2$ with $\gamma = 1.1\ \mathrm{J/(K^2{\cdot}mol)}$ and $A = 0.040\ \mu\Omega/\mathrm{K}^2$ in $YbCu_2Si_2$ with $\gamma = 150\ \mathrm{mJ/(K^2{\cdot}mol)}$. Note that the A values in the usual s and p electron systems are extremely small in magnitude,

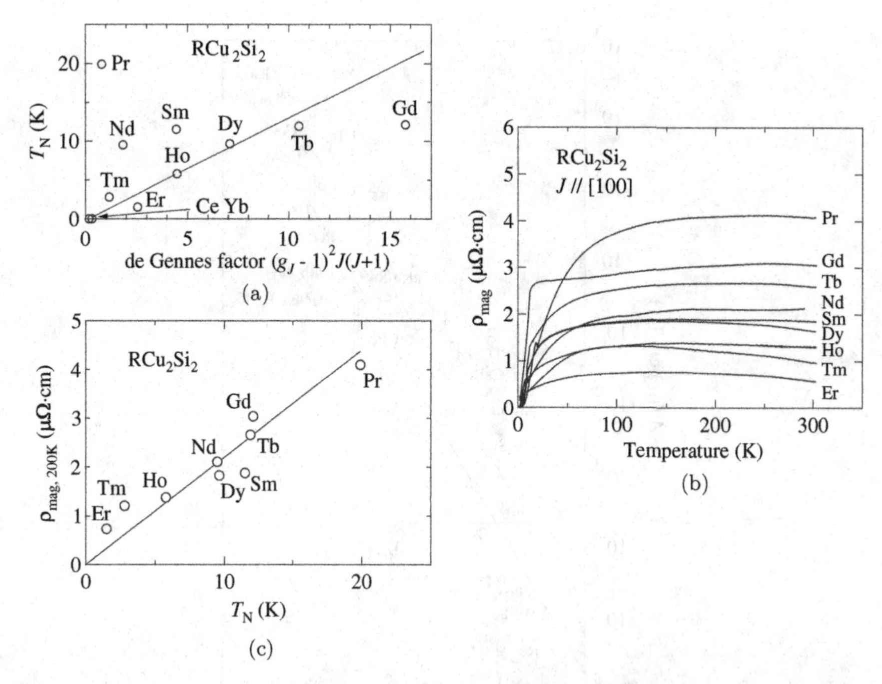

Figure 3.5: (a) Néel temperature T_N vs de Gennes factor, (b) temperature dependence of the magnetic resistivity ρ_{mag}, and (c) ρ_{mag} at 200 K vs T_N in RCu_2Si_2, cited from Ref. [41].

making it difficult to measure the A values accurately, as noted in Sec. 1.3. Figure 3.6 shows a Kadowaki–Woods plot, that is, the relation between the A and γ values [42]. Note that the Kadowaki–Woods plot is expanded experimentally and theoretically [43, 44].

3.2.2 Magnetic susceptibility and magnetization

Nowadays the magnetic susceptibility and magnetization are measured by using the commercial superconducting quantum interference device (SQUID) magnetometer and by an induction method in a long-pulse magnet with pulse duration of 20–50 ms in strong magnetic fields up to 500 kOe or 50 T. At high temperatures, the $4f$ electrons in most rare earth compounds, including the Ce compounds, are localized. The CEF theory is thus well applicable to the magnetic property of the rare earth compounds. For example, the CEF theory is applied to the sixfold-degenerate $4f$ energy levels in

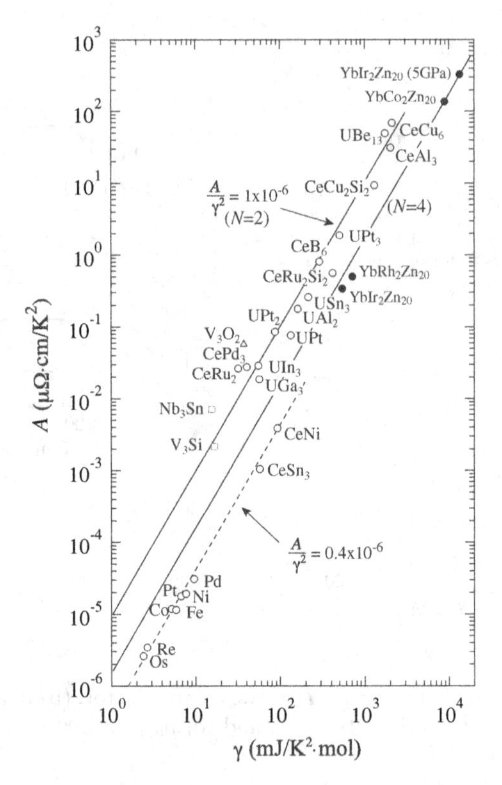

Figure 3.6: Generalized Kadowaki-Woods relation in Ce, Yb, U, and other compounds, cited from Refs. [42–44].

the Ce compound with the non-cubic crystal structure, which split into three Kramers doublets. The Hamiltonian of this system is given by

$$\mathcal{H} = \mathcal{H}_{\rm CEF} + \mathcal{H}_{\rm Zeeman}. \tag{3.15}$$

Here, $\mathcal{H}_{\rm CEF}$ is expressed as

$$\mathcal{H}_{\rm CEF} = B_2^0 O_2^0 + B_4^0 O_4^0 + B_4^4 O_4^4 \tag{3.16}$$

in the tetragonal symmetry and

$$\mathcal{H}_{\rm CEF} = B_2^0 O_2^0 + B_2^2 O_2^2 + B_4^0 O_4^0 + B_4^2 O_4^2 + B_4^4 O_4^4 \tag{3.17}$$

in the orthorhombic symmetry, where B_l^m and O_l^m are the CEF parameters and the Stevens operators respectively. The CEF susceptibility is given by

$$\chi^i_{\text{CEF}} = N(g_J\mu_B)^2\frac{1}{Z}\left(\sum_{m\neq n}|\langle m\,|J_i|\,n\rangle|^2\frac{1-e^{-\frac{\Delta_{m,n}}{k_BT}}}{\Delta_{m,n}}e^{-\frac{E_n}{k_BT}}\right.$$
$$\left.+\frac{1}{k_BT}\sum_n|\langle n\,|J_i|\,n\rangle|^2\,e^{-\frac{E_n}{k_BT}}\right) \tag{3.18}$$

and

$$Z = \sum_n e^{-\frac{E_n}{k_BT}} \tag{3.19}$$

where g_J is the Landé g-factor (6/7 for Ce^{3+}), J_i is a component of the angular momentum and $\Delta_{m,n} = E_n - E_m$. The magnetization can be calculated as

$$M_i = g_j\mu_B\sum_n|\langle n\,|J_i|\,n\rangle|\frac{e^{-\frac{E_n}{k_BT}}}{Z}. \tag{3.20}$$

Thus, the CEF susceptibility is also given by

$$\chi_{\text{CEF}} = \lim_{H\to 0}\frac{dM}{dH}. \tag{3.21}$$

The magnetic susceptibility which includes the molecular field contribution λ_i is given as

$$\chi_i^{-1} = (\chi_{\text{CEF}})^{-1} - \lambda_i \tag{3.22}$$

The eigenvalue E_n and eigenfunction $|n\rangle$ are determined by diagonalizing the total Hamiltonian

$$\mathcal{H} = \mathcal{H}_{\text{CEF}} - g_j\mu_BJ_i\,(H_i + \lambda_iM_i) \tag{3.23}$$

where the second term is the Zeeman term and the third is a contribution from the molecular field described in Sec. 2.2.

Here, we present an example of magnetic susceptibility and magnetization in $NdRhIn_5$ with the tetragonal structure [45]. Figures 3.7(a) and (b) show the temperature dependence of the magnetic susceptibility χ and reciprocal susceptibility $1/\chi$, and magnetization M in $NdRhIn_5$, respectively, revealing the Néel temperature

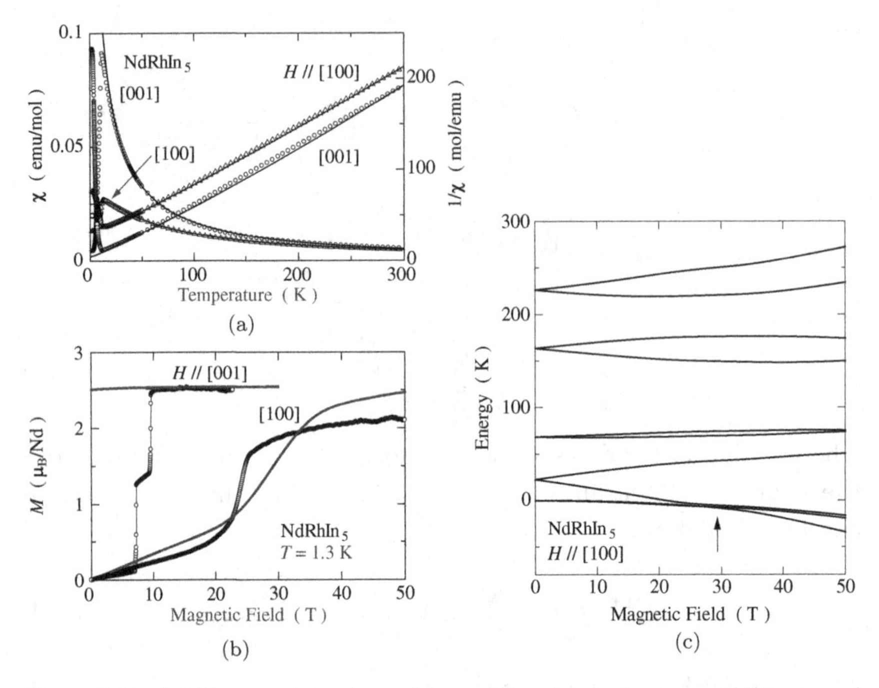

Figure 3.7: (a) Temperature dependence of the magnetic susceptibility χ and reciprocal susceptibility $1/\chi$, (b) magnetization M, and (c) magnetic field dependence of $4f$-CEF levels in NdRhIn$_5$, cited from Refs. [45, 46].

$T_{\rm N}$ = 11.2 K and a saturation moment of $\mu_{\rm s} = 2.5\ \mu_{\rm B}$/Nd for an antiferromagnetic easy-axis of $H \| [001]$. From the slope of $1/\chi$, the effective magnetic moment $\mu_{\rm eff}$ and the paramagnetic Curie temperature $\theta_{\rm p}$ are obtained: $\mu_{\rm eff} = 3.54\ \mu_{\rm B}$/Nd and $\theta_{\rm p} = -32$ K for $H\|[100]$, and $\mu_{\rm eff} = 3.50\ \mu_{\rm B}$/Nd and $\theta_{\rm p} = 5.6$ K for $H\|[001]$. The $\mu_{\rm eff}$ value is close to $\mu_{\rm eff} = g_J\sqrt{J(J+1)} = 3.62\ \mu_{\rm B}$/Nd for $g_J = 8/11$ and $J = 9/2$ ($S = 3/2$ and $L = 6$). Quite evidently, a saturation moment of $2.5\ \mu_{\rm B}$/Nd is smaller than $g_J J = 3.27\ \mu_{\rm B}$/Nd.

In order to understand the anisotropy in the magnetic susceptibility and magnetization, together with the magnetic entropy, (not shown here), the CEF analyses mentioned above are performed. The solid lines in Figs. 3.7(a) and (b) result from CEF calculations. The CEF parameters B_l^m, energy levels and the corresponding wave functions, together with the molecular field contribution

λ_i, are summarized in Table 3.1 below. Note that the two-step magnetization for $H\|[001]$ is not explained by the CEF model. The solid line for $H\|[001]$ just indicates the saturated value of the magnetization.

Note that the hard-axis magnetization for $H\|[100]$ in Fig. 3.7(b) indicates a metamagnetic-like feature at approximately 30T. Although the experimental results are obtained at 1.3 K, in the antiferromagnetic state, the observed metamagnetic-like feature of magnetization for $H\|[100]$ can be reproduced by the calculated magnetization curve based on the present CEF scheme in Table 3.1. The origin of the metamagnetic-like increase of magnetization is a CEF level-crossing effect between the CEF-ground state and the second excited state in magnetic fields, as shown in Fig. 3.7(c). Note that the CEF levels consist of five Kramers doublets at 0 kOe.

The magnetization at 1.3 K indicates a characteristic metamagnetic transition with two steps for $H\|[001]$ [46]. In the first metamagnetic transition at $H_{\rm m1}$, the magnetization becomes half of the saturation moment 2.5 $\mu_{\rm B}$/Nd, and reaches the saturation moment at $H_{\rm m2}$. This magnetization is calculated on the basis of effective spin Hamiltonian under the assumption that the ground-state magnetic structure, which was verified by neutron scattering experiments [47], is a commensurate antiferromagnetic structure with a magnetic wave vector $q = (\frac{1}{2}\,0\,\frac{1}{2})$ below $T_{\rm N}$, as shown in Fig. 3.8.

The effective spin Hamiltonian is

$$\mathcal{H} = -2\sum_{(i,j)} J_{ij}\boldsymbol{S}_i \cdot \boldsymbol{S}_j - g'\mu_{\rm B}\boldsymbol{H} \cdot \sum_i \boldsymbol{S}_i \tag{3.24}$$

where J_{ij} denotes the Ising-type magnetic exchange interaction between spins $\boldsymbol{S}_i$ and $\boldsymbol{S}_j$ at sites i and j, respectively. The saturation moment is expressed as $\mu_{\rm s} = g'\mu_{\rm B}S$. Note that g' is not g_J but depends on the CEF scheme and the magnetic exchange interaction in magnitude. Here, four magnetic sublattices, which consist of $\boldsymbol{S}_1$, $\boldsymbol{S}_2$, $\boldsymbol{S}_3$, and $\boldsymbol{S}_4$, and three exchange interactions, $J_{ij} = J_0$, J_1, and J_2 are taken into account, as shown in Fig. 3.8. J_0 and J_1 are the nearest- and next-nearest-neighbor intra-layer exchange interactions, and J_2 corresponds to the inter-layer interaction.

Table 3.1: CEF parameters, energy levels and the corresponding wave functions in $NdRhIn_5$, cited from Ref. [45].

CEF parameters										
						λ (emu/mol)$^{-1}$			χ_0 (emu/mol)	
B_2^0 (K)	B_4^0 (K)	B_4^4 (K)	B_6^0 (K)	B_6^4 (K)		$\lambda_{x,y} = -7.5$			$\chi_0^{x,y} = -3.0 \times 10^{-4}$	
-1.21	-0.013	0.03	0.0019	-0.003		$\lambda_z = -4.0$			$\chi_0^z = -5.0 \times 10^{-4}$	
Energy levels and wave functions										
E (K)	$\vert +9/2\rangle$	$\vert +7/2\rangle$	$\vert +5/2\rangle$	$\vert +3/2\rangle$	$\vert +1/2\rangle$	$\vert -1/2\rangle$	$\vert -3/2\rangle$	$\vert -5/2\rangle$	$\vert -7/2\rangle$	$\vert -9/2\rangle$
226.2	0	0	-0.930	0	0	0	-0.368	0	0	0
226.2	0	0	0	-0.368	0	0	0	-0.930	0	0
163.6	0	0	-0.368	0	0	0	0.930	0	0	0
163.6	0	0	0	0.930	0	0	0	-0.368	0	0
68.3	0.925	0	0	0	-0.380	0	0	0	-0.015	0
68.3	0	-0.015	0	0	0	-0.380	0	0	0	0.925
22.5	0.380	0	0	0	0.918	0	0	0	0.111	0
22.5	0	0.111	0	0	0	0.918	0	0	0	0.380
0	0	-0.994	0	0	0	0.109	0	0	0	0.028
0	-0.028	0	0	0	-0.109	0	0	0	0.994	0

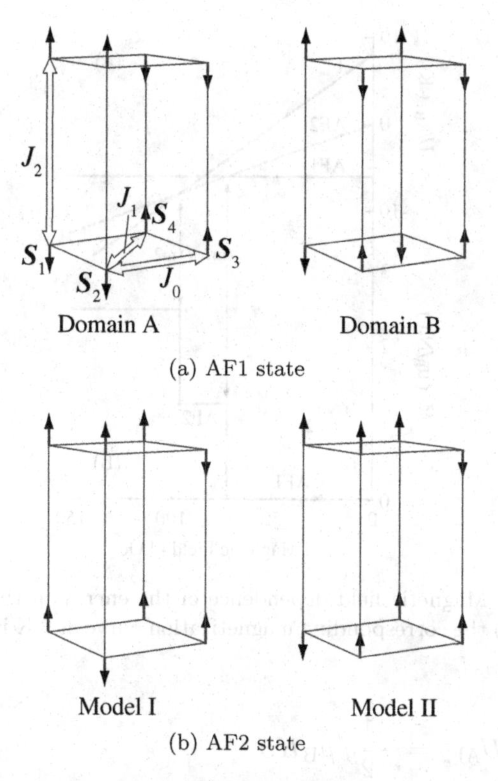

Figure 3.8: Magnetic structures in $NdRhIn_5$ for (a) the antiferromagnetic (AF1) state and (b) the AF2 state, cited from Ref. [46].

The magnetic structure is assumed to change with increasing magnetic fields as follows. Below H_{m1}, the numbers of the up spins and the down spins are equal, and the system is antiferromagnetic — the magnetic state is called the AF1 state. In the intermediate region ($H_{m1} < H < H_{m2}$) or the AF2 state, spins on the three sublattices point upwards and the left sublattice points downwards. The net magnetic moment thus becomes half of the saturation moment. When the magnetic field exceeds H_{m2}, the system becomes field-induced ferromagnetic — the magnetic state is known as the F state.

On the basis of the Hamiltonian (see Eq. (3.24)), the energies per rare earth atom in each phase are written as follows

$$U_{AF1} = (4J_1 + 2J_2)S^2 \tag{3.25}$$

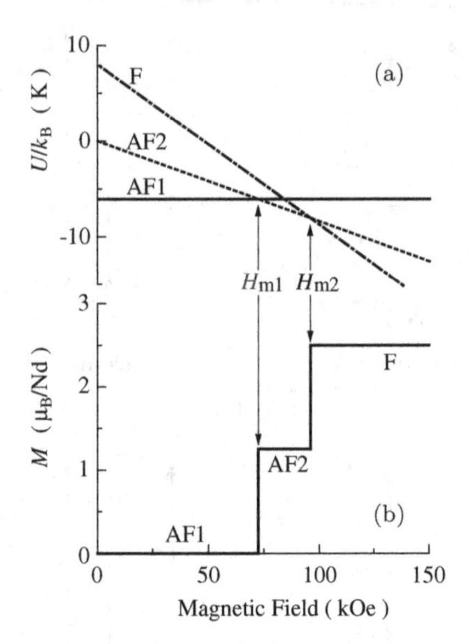

Figure 3.9: (a) Magnetic field dependence of the energy in the AF1, AF2 and F phases and (b) the corresponding magnetization curve at 0 K in $NdRhIn_5$, cited from Ref. [46].

$$U_{\mathrm{AF2}} = -\frac{1}{2}g'\mu_{\mathrm{B}}HS \tag{3.26}$$

$$U_{\mathrm{F}} = (-4J_0 - 4J_1 - 2J_2)S^2 - g'\mu_{\mathrm{B}}HS. \tag{3.27}$$

Figure 3.9 demonstrates the magnetic field dependence of the energy in each state and the corresponding magnetization curve for $NdRhIn_5$. With increasing magnetic fields, the energy of the AF1 state (U_{AF1}) does not change, while the energies of the AF2 state (U_{AF2}) and F state (U_{F}) decrease as a linear function of magnetic field, as shown in Fig. 3.9(a) by the broken and dashed-and-dotted lines, respectively. U_{AF2} crosses U_{AF1} at H_{m1}, and U_{F} crosses U_{AF2} at H_{m2}. Correspondingly, the magnetization shows two steps from 0 to $\mu_{\mathrm{s}}/2$ at H_{m1}, and from $\mu_{\mathrm{s}}/2$ to μ_{s} at H_{m2}. Two metamagnetic transition fields H_{m1} and H_{m2} are thus calculated from Eqs. (3.25)–(3.27) as follows:

$$H_{\mathrm{m1}} = (-8J_1 - 4J_2)S/g'\mu_{\mathrm{B}} \tag{3.28}$$

$$H_{\mathrm{m2}} = (-8J_0 - 8J_1 - 4J_2)S/g'\mu_{\mathrm{B}}. \tag{3.29}$$

Table 3.2: Néel temperature T_N, effective moment μ_{eff}, paramagnetic Curie temperature θ_p, saturation moment μ_s, metamagnetic transition fields H_{m1} and H_{m2} and exchange interactions J_0 and $J_1+J_2/2$, cited from Ref. [46].

	$NdRhIn_5$	$TbRhIn_5$	$DyRhIn_5$	$HoRhIn_5$
T_N (K)	11.6	47.3	28.1	15.8
$\mu_{eff}^{[100]}$ (μ_B)	3.54	10.20	10.80	10.69
$\mu_{eff}^{[001]}$ (μ_B)	3.50	9.72	10.91	10.64
$\theta_p^{[100]}$ (K)	−32	−88	−58	−27
$\theta_p^{[001]}$ (K)	5.6	−9.2	6.4	4.2
μ_s (μ_B)	2.5	5.8	9.7	9.8
H_{m1} (kOe)	70	97	53	34
H_{m2} (kOe)	93	150	82	54
J_0S^2/k_B (K)	−0.49	−2.4	−2.4	−1.6
$(J_1 + J_2/2)S^2/k_B$ (K)	−1.5	−4.7	−4.3	−2.8

From these equations, the following relation is obtained

$$H_{m1} - H_{m2} = -8J_0S/g'\mu_B. \tag{3.30}$$

Since $H_{m1} - H_{m2}$ is positive as determined from the experiments, J_0 should be negative, indicating antiferromagnetic characteristics, and can be derived uniquely from Eq. (3.30). On the other hand, J_1 and J_2 cannot be determined separately from only the present results but are obtained in the form of $J_1 + J_2/2$. The obtained values of J_0 and $J_1 + J_2/2$ are summarized in Table 3.2, together with those of the other similar compounds of $TbRhIn_5$, $DyRhIn_5$, and $HoRhIn_5$. The calculated magnetization curve in $NdRhIn_5$ is shown in Fig 3.9(b).

From these analyses, we propose possible two kinds of magnetic structures for the AF2 state, as shown in Fig. 3.8(b). These structures, though energetically equivalent, reveal individual characteristic features. As mentioned before, the ground-state magnetic structure in $NdRhIn_5$ is antiferromagnetic, with the propagation vector q = (1/2, 0, 1/2). Since $NdRhIn_5$ forms a tetragonal crystal structure, there exist two magnetic domains A and B, as shown in Fig. 3.8(a). The model I structure for the AF2 state in Fig. 3.8(b) consists of ferromagnetic and antiferromagnetic (100) planes so that the two-domain structures may still exist in the AF2 state. On the other hand, for the model II structure, it is possible to make the same magnetic structure for domains A and B, as shown in Fig. 3.8(b), and the system will be a single-domain magnetic structure in the AF2 state.

3.2.3 Specific heat

The specific heat (heat capacity) C is measured by the quasi-adiabatic heat pulse method. We give a heat pulse ΔQ to the sample and measure a change of the temperature ΔT, then

$$C = \frac{\Delta Q}{\Delta T} = \frac{I \cdot V \cdot \Delta t}{\Delta T} \tag{3.31}$$

where I and V are the current and the voltage flowing to the heater, respectively, and Δt is the duration of heating. At low temperatures, the specific heat is written as the sum of electronic, lattice, magnetic and nuclear contributions,

$$\begin{aligned} C &= C_{\mathrm{e}} + C_{\mathrm{ph}} + C_{\mathrm{mag}} + C_{\mathrm{nuc}} \\ &= \gamma T + \beta T^3 + C_{\mathrm{mag}} + \frac{A}{T^2} \end{aligned} \tag{3.32}$$

$$A = \frac{\hbar^2 \gamma_n^2}{3k_{\mathrm{B}}^2} I(I+1) H_{\mathrm{in}}^2 \tag{3.33}$$

where A, γ and β are the constants.

The electronic term C_{e} is linear in T. If we can neglect the magnetic and nuclear contributions, it is convenient to exhibit the experimental values of C as a plot of C/T versus T^2:

$$\frac{C}{T} = \gamma + \beta T^2. \tag{3.34}$$

Then we can estimate the electronic specific heat coefficient γ. Using the density of states $D(E_{\mathrm{F}})$, the coefficient can be expressed as

$$\gamma = \frac{2\pi^2}{3} k_{\mathrm{B}}^2 D(E_{\mathrm{F}}). \tag{3.35}$$

Since $D(E_{\mathrm{F}})$ is proportional to the electron mass, γ possesses an extremely large value in the heavy fermion compound.

Next, we consider $C_{\rm ph}$ based on the Debye T^3 law:

$$C_{\rm ph} = 9Nk_{\rm B}\left(\frac{T}{\Theta_{\rm D}}\right)^3 \int_0^{\Theta_{\rm D}/T} \frac{x^4 e^x}{(e^x - 1)^2} dx$$

$$\simeq \frac{12\pi^4 Nk_{\rm B}}{5}\left(\frac{T}{\Theta_{\rm D}}\right)^3 \tag{3.36}$$

$$\equiv \beta T^3 \left(T < \frac{\Theta_{\rm D}}{50}\right) \tag{3.37}$$

where $\Theta_{\rm D}$ is the Debye temperature and N is the number of atoms. For the actual lattices, the temperatures at which the T^3 approximation holds are quite low. Temperatures below $T = \Theta_{\rm D}/50$ may be required to get a reasonably pure T^3 law, where $\Theta_{\rm D}$ is usually in the range 150–400 K.

If $4f$ levels split into the $4f$-CEF levels i in the paramagnetic state, the energy per magnetic ion is given by

$$E_{\rm CEF} = \langle E \rangle = \frac{\sum_i n_i E_i \exp(-E_i/k_{\rm B}T)}{\sum_i \exp(-E_i/k_{\rm B}T)} \tag{3.38}$$

where E_i and n_i are the energy and the degenerate degree on the $4f$-CEF level i. The magnetic contribution to the specific heat is given by

$$C_{\rm Sch} = \frac{\partial E_{\rm CEF}}{\partial T}. \tag{3.39}$$

This contribution $C_{\rm Sch}$ is called the Schottky term. Here, the magnetic entropy of the $4f$ electron $S_{\rm mag}$ is defined as

$$S_{\rm mag} = \int_0^T \frac{C_{\rm Sch}}{T} dT, \tag{3.40}$$

but may also be described as

$$S_{\rm mag} = R \ln W \tag{3.41}$$

where W is a state number at temperature T. Note $W = 2J + 1$; for example, $W = 6$ for $J = 5/2$ in Ce^{3+}. Therefore we acquire information about the $4f$-CEF levels.

In the magnetically ordered sate, C_{mag} is

$$C_{\text{mag}} \propto T^{3/2} \text{ (ferromagnetic ordering)} \tag{3.42}$$

$$\propto T^{3} \text{ (antiferromagnetic ordering).} \tag{3.43}$$

When the antiferromagnetic magnon is accompanied by an energy gap Δ_{m}, Eq. (3.43) is modified to $C_{\text{mag}} \propto T^3 \exp(-\Delta_{\text{m}}/k_{\text{b}}T)$.

The last term in Eq. (3.32) corresponds to the nuclear contribution to the specific heat, revealing a nuclear Schottky tail of A/T^2. The nuclear specific heat can be calculated as follows: In a magnetic field, a nucleus with a spin I will have $2I + 1$ energy levels due to the Zeeman splitting. The magnetic field refered to here does not directly correspond to the extarnal manetic field, but an internal magnetic field or a hyperfine field and an electric field due to a nuclear quadrupole interaction. Nuclear spin splitting does not need the external magnetic field to occur. Therefore, the average energy of the nuclear system $\langle E_n \rangle$, can be written as

$$\langle E_n \rangle = -\frac{d \ln Z}{d(1/k_{\text{B}}T)} \tag{3.44}$$

where Z is the partition function given by

$$Z = \sum_{i=-I}^{I} \exp(-E_{ni}/k_{\text{B}}T). \tag{3.45}$$

The nuclear specific heat with N nuclei is thus calculated by

$$C_{\text{nuc}} = N\frac{d\langle E_n \rangle}{dT}$$
$$= \frac{R}{(k_{\text{B}}T)^2} \frac{\displaystyle\sum_{i=-I}^{I}\sum_{j=-I}^{I} (E_{ni}^2 - E_{ni}E_{nj}) \exp[-(E_{ni}+E_{nj})/k_{\text{B}}T]}{\displaystyle\sum_{i=-I}^{I}\sum_{j=-I}^{I} \exp[-(E_{ni}+E_{nj})/k_{\text{B}}T]} \tag{3.46}$$

where $R = Nk_B$ is the gas constant. At low temperatures, the nuclear specific heat C_{nuc} increases exponentially, while at high temperatures, it decreases in proportion to T^{-2} with increasing temperature, revealing A/T^2 in Eqs. (3.32) and (3.33)

In order to calculate C_{nuc} from Eq. (3.46), the position of each energy level must be known. The hyperfine interaction between electrons and the nucleus arises from two sources: the magnetic interaction which is proportional to $\boldsymbol{\mu}_n \cdot \boldsymbol{H}_{in}$ and the quadrupole interaction which is proportional to Qq. Here, $\boldsymbol{\mu}_n = \gamma_n \hbar I$ is the nuclear magnetic moment, where γ_n is the nuclear gyromagnetic ratio; H_{in} is the internal magnetic filed or effective magnetic field at the nucleus, Q is the nuclear quadrupole moment, and q is the electric field gradient at the nucleus. Considering these two terms, the energy levels of the system can be expressed as

$$E_{ni}/k_B = -aI_z + P\left[I_z^2 - \frac{1}{3}I(I+1)\right] \quad (3.47)$$

$$(I_z = I, I-1, \cdots, -I)$$

where

$$a = \frac{\gamma_n \hbar H_{in}}{k_B} \quad (3.48)$$

is a magnetic interaction parameter and

$$P = \frac{3e^2qQ}{4k_B I(2I-1)} \frac{(3\cos^2\theta - 1)}{2} \quad (3.49)$$

is the quadrupole coupling constant. Here, θ is the angle between the direction of magnetic field and the principal axis of the electric field gradient tensor (z-axis), and we assume $\theta = 0$. The magnetic interaction would separate the hyperfine energy levels evenly but the quadrupole term distorts this even spacing.

Here, we present the low-temperature specific heat of an antiferromagnet $HoCu_2Si_2$ with a Néel temperature $T_{N1} = 5.8$ K and a successive antiferromagnetic transition at $T_{N2} = 4.7$ K [48]. The temperature dependence of the specific heat below 2.5 K in $HoCu_2Si_2$ is shown in Fig. 3.10(a). The top of the nuclear Schottky peak is

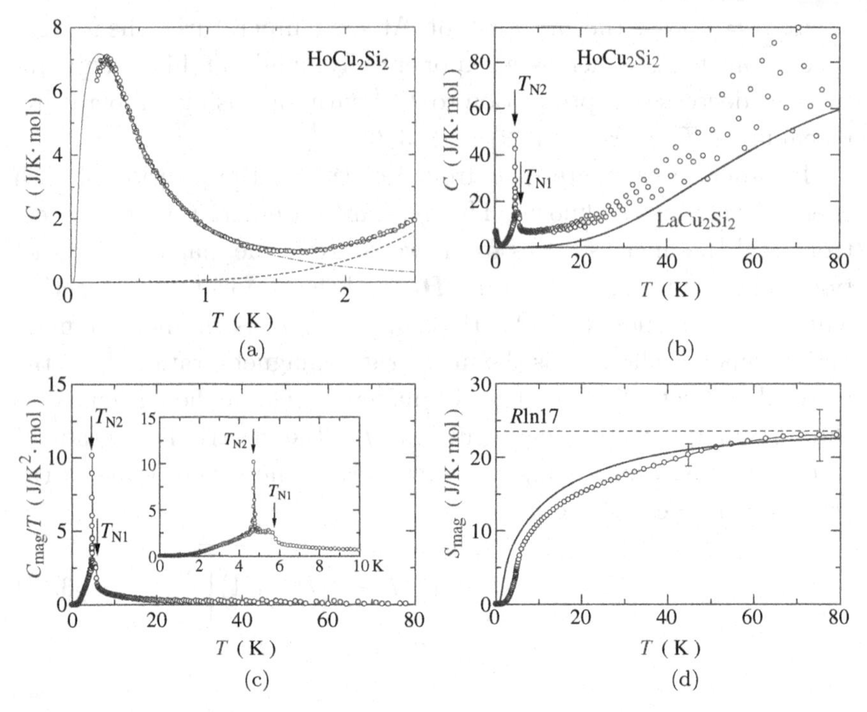

Figure 3.10: (a) Low-temperature specific heat C in $HoCu_2Si_2$. Dashed and dashed-and-dotted lines are $C_e + C_{ph} + C_{mag}$ and C_{nuc}, respectively, and a solid line represents the total specific heat. (b) Higher-temperature specific heat C in $HoCu_2Si_2$ and $LaCu_2Si_2$. Temperature dependence of (c) the magnetic specific heat C_{mag} and (d) the magnetic entropy, where the magnetic entropy calculated on the basis of CEF scheme is shown by a solid line, cited from Ref. [48].

observed around 0.25 K. In order to analyze the nuclear Schottky peak, we assume the high temperature specific heat, $C_e + C_{ph} + C_{mag}$, as $50T^3 + 8.5T^5$ (mJ/K·mol), as shown by a dashed line in Fig. 3.10(a). Note that the low-temperature specific heat C of a non-4f reference compound $LaCu_2Si_2$ is $C(= C_e + C_{ph}) = 3.2 + 0.3T^3$ (mJ/K·mol). The extremely small contribution of C_e to C in $HoCu_2Si_2$ is neglected. C_{ph} is included in a T^3-term of C_{mag}. The parameters $a = 0.215$ K and $P = 0.003$ K are estimated by fitting Eq. (3.46) to the experimental data. The calculated nuclear specific heat C_{nuc} is shown by a dashed-and-dotted line, and the total specific heat is expressed by a solid line.

The internal magnetic field H_{in} and the electric field gradient at the Ho nucleus q are estimated as 500 T and -3.4×10^{25} cm^3, respectively, from the values of a and P that have been determined. The present value of internal magnetic field $H_{\text{in}} = 500$ T is substantially smaller than $H_{\text{in}} = 900$ T calculated for a pure Ho metal by assuming the free ion value of $\langle J_z \rangle = J = 8$ [49]. From the results of the inelastic neutron scattering experiment [50], and the magnetic susceptibility and magnetization [45], the $4f$-CEF ground state of $HoCu_2Si_2$ is a singlet and the first excited state is a doublet at approximately 5 K. It is, therefore, expected that the ordered moment in the antiferromagnetic state is reduced from the free ion value of $g_J J \mu_B = 10\ \mu_B$, where g_J $(= 5/4)$ is the Landé g-factor. In fact, the value of the ordered moment was reported to be 6.5 μ_B from the neutron diffraction measurements [51]. The reduction ratio of the effective magnetic field, $500/900 = 0.65$, is approximately similar to the ratio of the ordered moment at $6.5/10 = 0.65$. Therefore, the relatively small value of the effective magnetic field, $H_{\text{in}} = 500$ T in $HoCu_2Si_2$ is due to the reduced $\langle J_z \rangle$ value of the CEF ground state.

Next, we can obtain experimentally the magnetic specific heat C_{mag} from the specific heat in a wide temperature range from 0 to 80 K. We assume that the C_{ph} of $HoCu_2Si_2$ is the same as the C_{ph} of $LaCu_2Si_2$, which is shown in Fig. 3.10(b) by a solid line. Subtracting C_{ph} of $LaCu_2Si_2$ and C_{nuc} obtained above from the total specific heat C of $HoCu_2Si_2$, we obtain C_{mag}, as shown in Fig. 3.10(c), together with S_{mag} in Fig. 3.10(d). The observed magnetic entropy S_{mag} reaches about $R \ln 3$ at T_{N1}, which is consistent with the CEF scheme determined from the inelastic neutron scattering experiment [50]. Above T_{N1}, S_{mag} increases gradually with increasing temperature and approaches $R \ln(2J+1) = R \ln 17$ for $J = 8$ around 80 K. These features are in very good agreement with the calculated entropy change on the basis of the CEF scheme, as shown in Fig. 3.10(d) by a solid line. Note that the overall $4f$-CEF splitting energy is about 100 K.

Finally, note that a nuclear Schottky peak as in $HoCu_2Si_2$ is not observed experimentally in the other RCu_2Si_2 compounds. This is a rare case. Usually, a nuclear Schottky tail, which is expressed as

A/T^2 in Eq. (3.33), is observed in the low-temperature specific heat experiments.

3.2.4 Thermoelectric power

A temperature gradient in a long and thin rectangular sample should be accompanied by an electric field in the opposite direction. When the temperature difference between two lead wires, ΔT and the potential difference, namely, the voltage ΔV are measured, the thermoelectric power (thermopower) S is defined as

$$S = \frac{\Delta V}{\Delta T}. \tag{3.50}$$

S consists of an electron diffusion term S_{e}, a phonon-drag term S_{ph}, and a spin-drag term S_{mag} in a magnetic material. That is,

$$S = S_{\mathrm{e}} + S_{\mathrm{ph}} + S_{\mathrm{mag}}. \tag{3.51}$$

Here, electrons are diffused from the side with a high temperature, to the side with the lower temperature, which leads to the electron diffusion term S_{e}, given by

$$S_{\mathrm{e}} = -\frac{\pi^2}{3}\frac{k_{\mathrm{B}}^2 T}{|e|}\left[\frac{\partial \ln \sigma(\varepsilon)}{\partial \varepsilon}\right]_{\varepsilon=\varepsilon_{\mathrm{F}}} \tag{3.52}$$

where $\sigma(\varepsilon)$ again refers to the electrical conductivity. A flow of phonons also drags conduction electrons from the higher-temperature side to the lower-temperature side, described by the phonon-drag term. In typical non-magnetic metals, a T-linear dependence of S, due mainly to S_{e}, is changed with decreasing temperature and shows a board peak at around 30–70 K, roughly at $T \simeq 0.2\Theta_{\mathrm{D}}$, which is due to S_{ph}. In magnetic metals, the spin-drag term contributes to S. The Kondo effect is the most prominent contribution.

Figure 3.11 shows the logarithmic scale for temperature dependence of the thermoelectric power in $\mathrm{Ce}_x\mathrm{La}_{1-x}\mathrm{Cu}_6$($x$=0, 0.094, 0.29, 0.50, 0.73, 0.90, 0.99, and 1.0) [52]. Note that the small value of $0 \pm 2\ \mu\mathrm{V/K}$ in LaCu_6 is considered to be usual for the non-magnetic metals. On the other hand, the thermoelectric power of CeCu_6 possesses

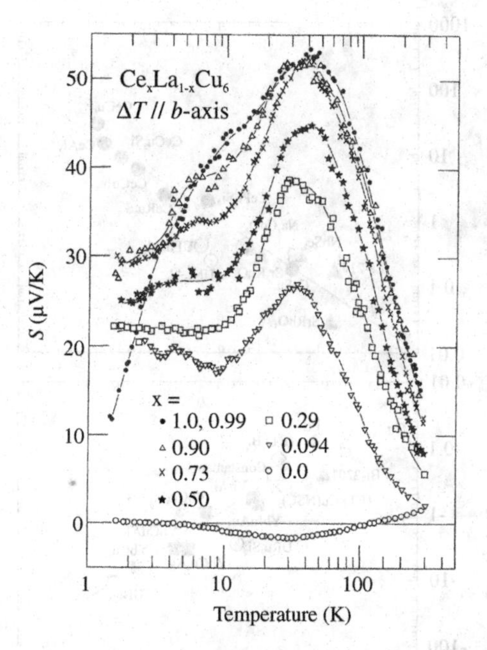

Figure 3.11: Logarithmic scale of temperature dependence of the thermoelectric power S in $Ce_xLa_{1-x}Cu_6$, cited from Ref. [52].

a huge broad peak of 53 μV/K at 40 K and a shoulder-like structure at 5 K. The former peak most likely comes from the combined phenomenon of electron scattering due to the Kondo effect and the CEF effect, while the temperature at which the shoulder-like structure arises approximately corresponds to the Kondo temperature T_K in $CeCu_6$.

The thermoelectric power decreases monotonically with a decrease in the Ce concentration x. Note that the thermoelectric power in $CeCu_6$ decreases steeply with decreasing temperature. It is simply expected that an extremely large value of $S/T\,(T = 0)$ is formed at lower temperatures because of $S = 0$ at 0 K. In fact, a value of S/T at lower temperatures than 0.5 K is obtained as $S/T\,(T = 0) = 20$–$30\ \mu V/K^2$ [53]. This large value of $S/T\,(T = 0)$ correlates with the γ value of 1600 mJ/(K^2·mol) in $CeCu_6$. Figure 3.12 shows a relation between $S/T\,(T = 0)$ and γ values, which is called a Behnia plot [54].

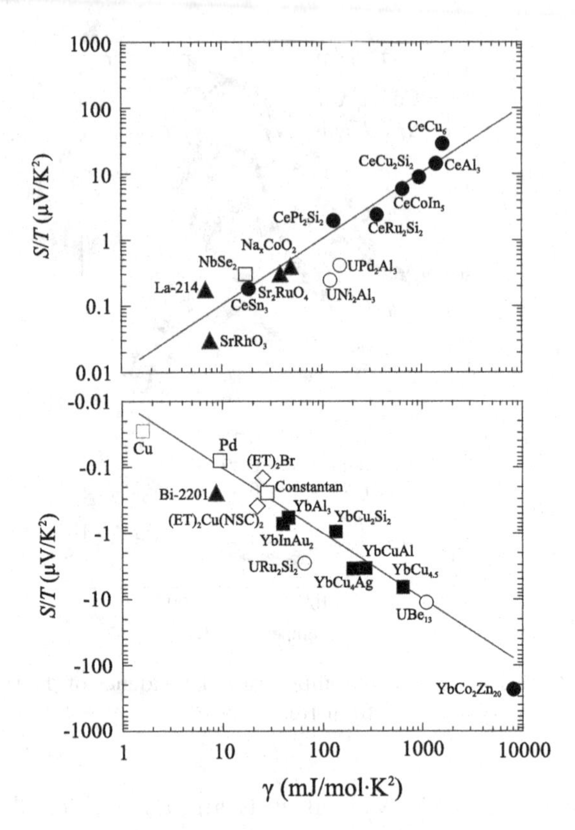

Figure 3.12: Relation between $S/T\,(T=0)$ and γ values, cited from Ref. [54].

3.2.5 Thermal expansion coefficient

The thermal expansion $\Delta\ell/\ell$ is a relative change of the lattice parameter ℓ, which is measured by the capacitance method and/or the strain gauge method. The thermal expansion coefficient α is a temperature derivative of $\Delta\ell/\ell$,

$$\alpha = \frac{d\left(\frac{\Delta\ell}{\ell}\right)}{dT}. \tag{3.53}$$

α is closely related to the specific heat C though the relation $\alpha = (\mathit{\Gamma}/3B_{\mathrm{T}}V)\,C$, where $\mathit{\Gamma}$ is the Grüneisen constant, B_{T} is the bulk modulus, and the inverse of the bulk modulus $1/B_{\mathrm{T}}$ is the compressibility. The thermal expansion coefficient is the sum of an electronic contribution α_e, the phonon contribution α_{ph}, and the magnetic contribution α_{mag} as in the specific heat C, so that

$$\alpha = \alpha_e + \alpha_{ph} + \alpha_{mag} \tag{3.54}$$

$$\alpha_e = \frac{\Gamma_e}{3B_T V} C_e = aT \tag{3.55}$$

$$\alpha_{ph} = b\left(\frac{T}{\Theta_D}\right)^3 \int_0^{\Theta_D/T} \frac{x^4 e^x}{(e^x - 1)^2} dx \tag{3.56}$$

$$\alpha_{mag} = \frac{\Gamma_{mag}}{3B_T V} C_{mag}. \tag{3.57}$$

Here, we present in Figs. 3.13(a) and (b) the temperature dependences of linear thermal expansions along the tetragonal [001] direction $\Delta\ell/\ell \| [001]$ (c-axis) and perpendicular to the [001] direction $\Delta\ell/\ell \perp [001]$, together with the corresponding thermal expansion coefficient $\alpha \| [001]$ and $\alpha \perp [001]$ in $EuNi_2P_2$ with the $ThCr_2Si_2$-type tetragonal structure respectively, revealing that the thermal expansion coefficient is almost constant from room temperature to about 100 K, increases in magnitude below 100 K, possesses a maximum at 40 K, and decreases steeply with decreasing temperature [26].

Most of the Eu compounds are divalent and order magnetically. $EuNi_2P_2$ is, however, a heavy fermion compound without magnetic ordering. The γ value is 93 mJ/(K^2·mol). The electrical resistivity ρ and thermoelectric power S are also shown in Figs. 3.13(c) and (d), respectively.

The electronic specific heat coefficient and Kondo temperature in $CeRu_2Si_2$ are determined to be $\gamma = 350$ mJ/(K^2·mol) and $T_K = 20$ K, respectively [55]. The temperature dependence of α in $CeRu_2Si_2$ shows a peak at about 10 K, which is half of $T_K = 20$ K [55]. A relation between γ and T_K is discussed later, revealing a simple relation of $T_K \sim \gamma^{-1}$ in Eq. (5.16). On the other hand, if the same scaling is applied to $EuNi_2P_2$, $T_K = 75$ K is obtained using $\gamma = 93$ mJ/(K^2·mol). The corresponding maximum is observed at $T_{\alpha max} = 40$ K leading to $T_K = 80$ K in $EuNi_2P_2$. From the coincidence of the estimated Kondo temperatures obtained by using two different methods, the temperature dependence of the thermal

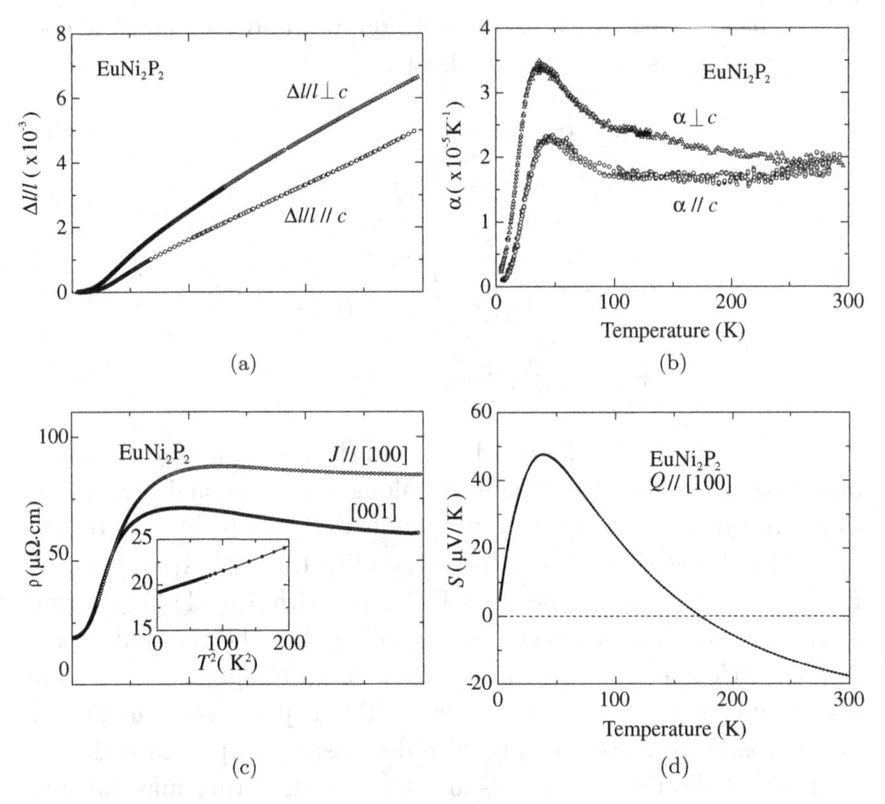

Figure 3.13: Temperature dependence of (a) linear thermal expansion $\Delta\ell/\ell$, (b) the corresponding thermal expansion coefficient α, (c) the electrical resistivity ρ, and (d) thermoelectric power S in $EuNi_2P_2$ with the $ThCr_2Si_2$-type tetragonal structure, cited from Ref. [26].

expansion in $EuNi_2P_2$ is related to and apparently reflects the heavy fermion state.

3.2.6 Thermal conductivity

When a temperature gradient dT/dz exists along a long and thin rectangular sample (parallel to the z-axis), a steady state flow of heat (energy transmitted across unit area per unit time) j_z is given as

$$j_z = -\kappa \frac{dT}{dz}. \tag{3.58}$$

In the case of the electron contribution to κ, the thermal conductivity κ_e becomes

$$\kappa_e = \frac{1}{3}\tau v_F^2 C_e \tag{3.59}$$

or

$$\kappa_e = L_0 \frac{T}{\rho} \quad \left(L_0 = \frac{\pi^2 k_B^2}{3e^2} \right) \tag{3.60}$$

where τ is the scattering lifetime, v_F is the Fermi velocity of conduction electrons, and C_e is the electronic specific heat. L_0 is called the Lorentz number, where equals 2.44×10^{-8} W·Ω/K^2, and a relation of Eq. (3.60) is called the Wiedemann-Franz law.

In typical non-magnetic metals, κ_e becomes approximately temperature-independent at high temperatures ($T > \Theta_D/2$) because the electrical resistivity ρ is proportional to temperature, $\rho \sim T$. However, κ_e is proportional to temperature, $\kappa_e \sim T$ when ρ becomes constant at low temperatures. Note that these explanations, based on Eq. (3.60), are well satisfied if the scattering lifetime τ is the same between κ_e and ρ in the temperature dependence. Nevertheless this is not valid in an intermediate temperature region. The scattering lifetime τ in the temperature gradient becomes short compared with τ in the electric field. The thermal conductivity thus possesses a peak, for example, at 15 K in Au. The peak becomes sharper as the residual resistivity becomes smaller. A similar peak is observed at 20 K in a $4f$-itinerant superconductor $CeRu_2$ with the cubic structure, as shown in Fig. 3.14.

In addition to the thermal conductivity κ_e due to conduction electrons mentioned above, the phonon contribution κ_{ph} is added to the total thermal conductivity κ

$$\kappa = \kappa_e + \kappa_{ph} \tag{3.61}$$

where v_F and C_e in Eq. (3.59) are replaced by the velocity of sound and the specific heat due to phonons C_{ph} for κ_{ph}.

In rare earth compounds, the magnetic thermal conductivity is also added to the total thermal conductivity. A simple case is that

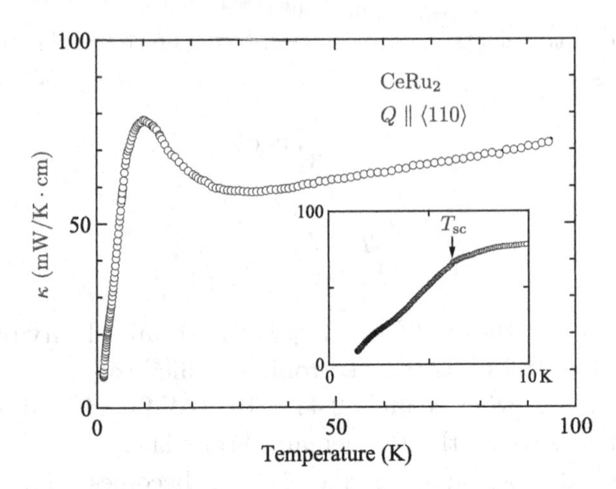

Figure 3.14: Temperature dependence of the thermal conductivity κ in $CeRu_2$ with the cubic structure. Figures courtesy of M. Sera. The inset shows the superconducting transition at $T_{sc} = 6.3$ K.

the magnetic resistivity ρ_{mag} becomes temperature-independent at high temperatures ($T > \Delta$, Δ: CEF splitting energy) and the corresponding thermal conductivity κ_{mag} becomes proportional to temperature, $\kappa_{mag} \sim T$. The total thermal conductivity $\kappa (= \kappa_e + \kappa_{ph} + \kappa_{mag})$ of magnetic metals thus does not have a simple temperature dependence.

3.2.7 Nuclear magnetic resonance

A nucleus has a spin momentum I, for example $I = 3/2$ in ^{63}Cu (natural abundance 69.09%) and ^{65}Cu (30.91%) or $I = 1/2$ in Pt. Note that Ce atoms in $CeCu_6$ and U atoms in UPt_3 possess no nuclear spin moments. The $4f$ electrons in $CeCu_6$ and the $5f$ electrons in UPt_3, however, interact magnetically with nuclear spin moments of Cu atoms and those of Pt atoms, respectively. The electronic states of $CeCu_6$ and UPt_3 are thus studied via the corresponding nuclear spin states, namely, by measuring the spin-lattice relaxation time T_1 and the Knight shift K in nuclear magnetic resonance (NMR) experiments.

First, we consider the NMR. The nucleus with the spin $\boldsymbol{I}$ has a magnetic moment $\boldsymbol{\mu}_n = \gamma_n \hbar \boldsymbol{I}$, where γ_n is the nuclear

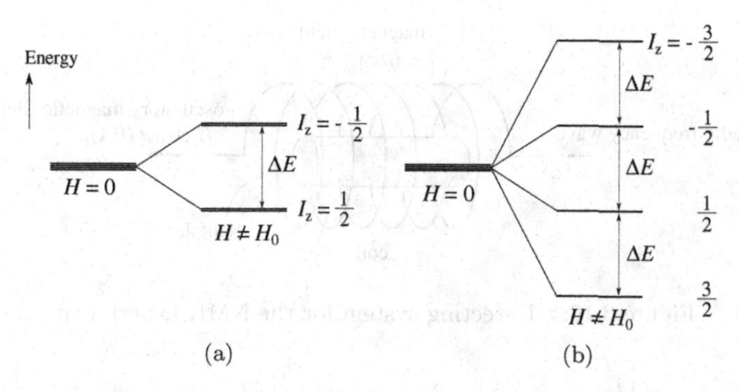

Figure 3.15: Split nuclear spin levels for (a) $I = 1/2$ and (b) $I = 3/2$ in an external magnetic field H_0.

gyromagnetic ratio. As shown in Fig. 3.15, the nuclear spin levels are split into $2I + 1$ levels with an equal energy separation ΔE if the nucleus is placed in an external magnetic field $\boldsymbol{H} = (0, 0, H_0)$. So,

$$\mathcal{H} = -\boldsymbol{\mu}_{\rm n} \cdot \boldsymbol{H} \tag{3.62a}$$

$$E_n = -\gamma_{\rm n} \hbar H_0 I_z \tag{3.62b}$$

$$\Delta E = \gamma_{\rm n} \hbar H_0 \tag{3.63}$$

where $I_z = I, I - 1, I - 2, \cdots, -I$. The nuclear spin system undergoes resonance absorption if the oscillatory magnetic field $H_1 \cos \omega t$ is applied along the x-axis perpendicularly to $\boldsymbol{H} = (0, 0, H_0)$ as in Fig. 3.16, with the condition

$$\omega_0 = \gamma_{\rm n} H_0 \tag{3.64}$$

where the oscillatory frequency is, for example, 42.6 MHz in the case of "the proton" for $H_0 = 10$ kOe.

Here, we define $\boldsymbol{M}(t)$ as the nuclear magnetization averaged over the sample at time t when the magnetic field $\boldsymbol{H}$ is applied along the z-axis. The value of the z-component of $\boldsymbol{M}(t)$, $M_z(t)$, will exponentially recover to the equilibrium value M_0. The equation of motion for $\boldsymbol{M}(t)$ is then described by

$$\frac{dM_z(t)}{dt} = \gamma_{\rm n} \left[\boldsymbol{M}(t) \times \boldsymbol{H}\right]_z + \frac{1}{T_1}\left(M_0 - M_z(t)\right) \tag{3.65}$$

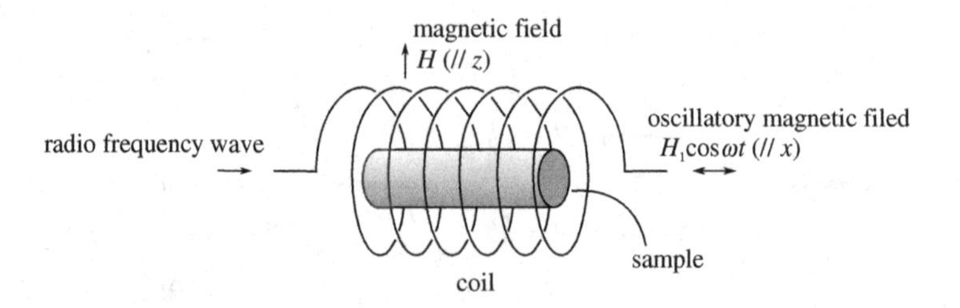

Figure 3.16: Detecting system for the NMR experiment.

$$\frac{dM_{x,y}}{dt} = \gamma_{\rm n} \left[\boldsymbol{M}(t) \times \boldsymbol{H}\right]_{x,y} - \frac{M_{x,y}}{T_2} \tag{3.66}$$

where T_1 is the longitudinal relaxation time and T_2 is the transverse relaxation time. In the case of $H \parallel z$, $\left[\boldsymbol{M}(t) \times \boldsymbol{H}\right]_z = 0$ is realized. Then, Eq. (3.65) becomes

$$\frac{M_0 - M_z(t)}{M_0} = e^{-\frac{t}{T_1}} \quad \text{or} \quad M_z(t) = M_0(1 - e^{-\frac{t}{T_1}}). \tag{3.67}$$

This means that the nuclear spin system arrives at thermal equilibrium with the surrounding lattice. Therefore, T_1 is called the nuclear spin-lattice relaxation time. The absorption energy is transferred from the nuclear spin system to the electron spin system in the lattice. The electron spin system is thus studied by measuring the temperature dependence of the spin-lattice relaxation rate $1/T_1$.

The magnetic field, felt exactly by the nucleus, is not H_0 but $H_0 + H_{\rm in}$, where $H_{\rm in}$ is an internal magnetic field or a hyperfine field based on the polarization of the surrounding electron spin system. The nuclear spin Hamiltonian of the nucleus with the spin $\boldsymbol{I}$ is given by

$$\begin{aligned} \mathcal{H} &= -\gamma_{\rm n}\hbar \boldsymbol{I} \cdot \boldsymbol{H}_0 - \gamma_{\rm n}\hbar \boldsymbol{I} \cdot \boldsymbol{H}_{\rm in} + \frac{h\nu_Q}{6}\left\{3I_z^2 - I(I+1)\right\} \\ &= -\gamma_{\rm n}\hbar I \cdot H_0\,(1+K) + \frac{h\nu_Q}{6}\left\{3I_z^2 - I(I+1)\right\}. \end{aligned} \tag{3.68}$$

The first term corresponds to Eq. (3.62a) or Eq. (3.62b), revealing the Zeeman interaction between the nuclear magnetic moment

$\boldsymbol{\mu}_\mathrm{n} = \gamma_\mathrm{n}\hbar \boldsymbol{I}$ and the external magnetic field $\boldsymbol{H}_0$. The second term indicates the Zeeman interaction between the nuclear magnetic moment and the internal magnetic field $\boldsymbol{H}_\mathrm{in}$, where $K = H_\mathrm{in}/H_0$ is the Knight shift. Here, the internal magnetic field is based on the electron spin $\boldsymbol{S}$, including the f-electron spin fluctuation at the position of the nucleus with spin $\boldsymbol{I}$. The $-\gamma_\mathrm{n}\hbar\boldsymbol{I}\cdot\boldsymbol{H}_\mathrm{in}$ term is defined as $A(\boldsymbol{r})\boldsymbol{S}\cdot\boldsymbol{I}$. The internal field thus fluctuates with time: $\boldsymbol{H}_\mathrm{in}(t) = \langle \boldsymbol{H}_\mathrm{in}\rangle + \delta\boldsymbol{H}(t)$. The average value $\langle \boldsymbol{H}_\mathrm{in}\rangle$ produces an extra field ΔH in the nuclear position and changes the resonance magnetic field. The Knight shift is better defined as $K = \Delta H/H_0$ or $\langle \boldsymbol{H}_\mathrm{in}\rangle/H_0$. The term $\delta\boldsymbol{H}(t)$ is related to the spin-lattice relaxation rate $1/T_1$, which causes relaxation of the nuclear spin, to be shown later.

The third term represents the nuclear quadrupole interaction between the electric field gradient q and the nuclear quadrupole moment Q. Here, ν_Q is the nuclear quadrupole frequency defined as

$$\nu_Q = \frac{3e^2qQ}{2I(2I-1)h} \tag{3.69}$$

$$eq = \left.\frac{\partial^2 V}{\partial z^2}\right|_{r=0}. \tag{3.70}$$

Note that the charge distribution of the nucleus $\rho_\mathrm{n}(\boldsymbol{r})$ is not spherical but ellipsoidal in an electronic potential $V(\boldsymbol{r})$ when the nuclear spin I is larger than 1. The corresponding energy is

$$E = \int \rho_\mathrm{n}(\boldsymbol{r})V(\boldsymbol{r})d^3\boldsymbol{r}. \tag{3.71}$$

The nuclear quadrupole term is obtained from Eq. (3.71)

$$\mathcal{H}_Q = \frac{1}{4}\left.\frac{\partial^2 V}{\partial z^2}\right|_{r=0} \int \left(2z^2 - x^2 - y^2\right)\rho_\mathrm{n}(x,y,z)d^3\boldsymbol{r} \tag{3.72}$$

$$eQ = \int \left(2z^2 - x^2 - y^2\right)\rho_\mathrm{n}(x,y,z)d^3\boldsymbol{r} \tag{3.73}$$

and

$$\mathcal{H}_Q = \frac{e^2qQ}{4I(2I-1)}\left\{3I_z^2 - I(I+1)\right\}. \tag{3.74}$$

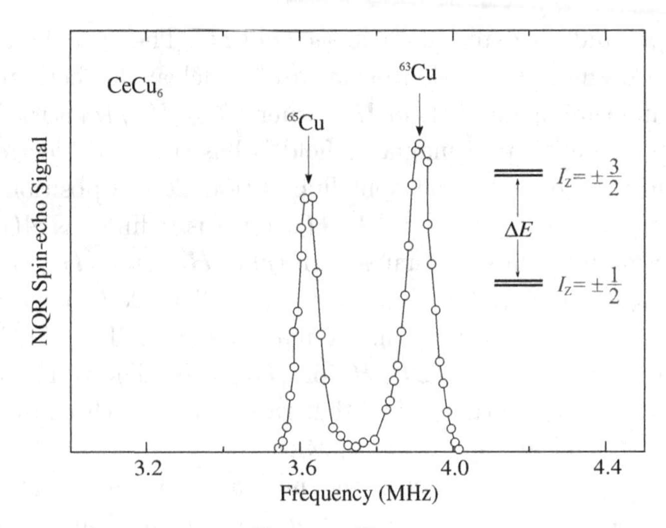

Figure 3.17: NQR spectrum of ^{65}Cu and ^{63}Cu in $CeCu_6$, courtesy of Y. Kitaoka.

The NMR experiment is thus based on Eq. (3.68). The zero-field NMR experiment is realized when the electric field based on Eq. (3.74) splits the nuclear spin levels into several multiplets, and the transition between them causes resonant absorption of the radio frequency. This is called nuclear quadrupole resonance (NQR). A key feature is that the experiment does not need an external magnetic field H_0.

Figure 3.17 represents the NQR spectrum of ^{65}Cu and ^{63}Cu in $CeCu_6$. In the case of Cu($I = 3/2$), the nuclear spin levels are split into two levels of $I_z = \pm 1/2$ and $I_z = \pm 3/2$ based on the nuclear quadrupole interaction. The resonance occurs between two levels, producing one peak for each ^{65}Cu and ^{63}Cu, respectively, where the resonance peak is obtained as a function of the frequency for the spin-echo signal.

Splitting of the nuclear spin levels is also caused by the second term of Eq. (3.68), namely due to the internal magnetic field $H_{\rm in}$ without the external magnetic field H_0. For example, $EuRu_2P_2$ is a ferromagnet of Eu^{2+} ($T_{\rm C} = 29$ K). ^{31}P-NMR is detected under zero external magnetic field due to the internal field based on a large ordered moment of Eu^{2+} below $T_{\rm C}$, namely 7 $\mu_{\rm B}$/Eu at 0 K.

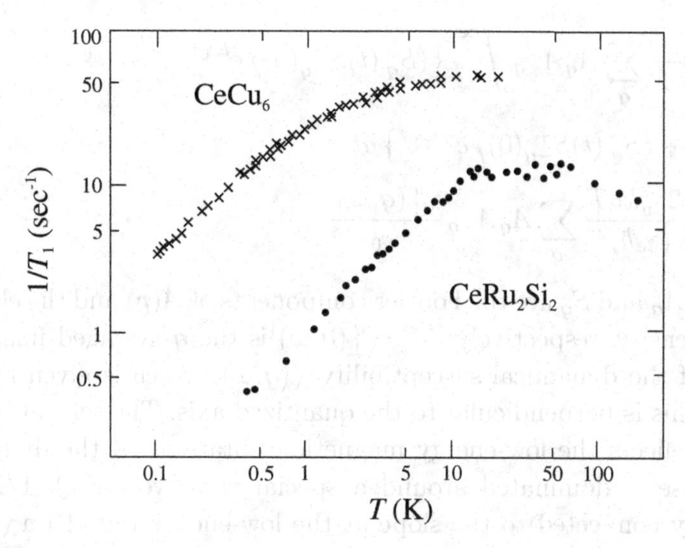

Figure 3.18: Temperature dependence of the nuclear spin-lattice relaxation rate $1/T_1$ for $CeCu_6$ and $CeRu_2Si_2$, cited from Ref. [56].

Figure 3.18 shows the temperature dependence of the nuclear spin-lattice relaxation rate $1/T_1$ for $CeCu_6$ and $CeRu_2Si_2$ [56]. For $CeCu_6$, $1/T_1$ of ^{63}Cu was measured in the NMR experiment, under zero magnetic field as mentioned above. On the other hand, $1/T_1$ of ^{21}Si ($I = 1/2$) in $CeRu_2Si_2$ was measured under the condition that the c-axis ([001] direction) is aligned along the external magnetic field. Note that the sample of $CeRu_2Si_2$ is powdered and the orientation of the sample is set along a magnetization easy-axis, namely for the tetragonal [001] direction, under the magnetic field. As seen in Fig. 3.18, $1/T_1$ is almost temperature-independent above 6 K in $CeCu_6$ and 20 K in $CeRu_2Si_2$. In both compounds, $4f$ electrons are localized at high temperatures.

Here, we consider the spin-lattice relaxation rate $1/T_1$ based on the fluctuating component of the internal field $\delta\boldsymbol{H}(t)$, which is related to the transition probability W from the nuclear spin component $I_z = m$ to $m+1$, namely $W_{m\to m+1}$ and $W_{m+1\to m}$

$$\frac{1}{T_1} = \frac{\gamma_n^2}{4}\int_{-\infty}^{\infty}\left\{\left\langle \delta H^+(t)\delta H^-(0)\right\rangle e^{i\omega_0 t} + \left\langle \delta H^-(t)\delta H^+(0)\right\rangle e^{-i\omega_0 t}\right\} dt$$

$$
\begin{aligned}
&= \frac{\gamma_{\mathrm{n}}^2}{4} \sum_{\boldsymbol{q}} A_{\boldsymbol{q}} A_{-\boldsymbol{q}} \int_{-\infty}^{\infty} \{ \langle S_{\boldsymbol{q}}^{+}(t) S_{-\boldsymbol{q}}^{-}(0) \rangle e^{i\omega_0 t} \\
&\quad + \langle S_{\boldsymbol{q}}^{-}(t) S_{-\boldsymbol{q}}^{+}(0) \rangle e^{-i\omega_0 t} \} \, dt \\
&= \frac{2\gamma_{\mathrm{n}}^2 k_{\mathrm{B}} T}{(\gamma_{\mathrm{e}} \hbar)^2} \sum_{\boldsymbol{q}} A_{\boldsymbol{q}} A_{-\boldsymbol{q}} \frac{\chi_{\perp}''(\boldsymbol{q}, \omega_0)}{\omega_0}
\end{aligned} \tag{3.75}
$$

where $A_{\boldsymbol{q}}$ and $S_{\boldsymbol{q}}$ are the Fourier components of $A(\boldsymbol{r})$ and the electron spin density, respectively [57]. $\chi_{\perp}''(\boldsymbol{q}, \omega)$ is the $\boldsymbol{q}$-averaged imaginary part of the dynamical susceptibility $\chi(\boldsymbol{q}, \omega)$, which is given by $\chi' + i\chi''$. This is perpendicular to the quantized axis. The relaxation rate $1/T_1$ reflects the low-energy magnetic excitations. If the dynamical response is dominated around a special wave vector Q, $1/T_1 T$ is directly connected to the slope at the low-energy tail of $\mathrm{Im}\,\chi(Q, \omega)$ around $\omega \simeq 0$. Here, the relaxation is mainly due to the magnetic moments of localized-f electrons and the conduction electrons.

When the localized-$4f$ electrons of $CeCu_6$ and $CeRu_2Si_2$ fluctuate independently, the dynamical susceptibility is $\boldsymbol{q}$-independent, and the corresponding relaxation rate $1/T_1$ becomes

$$\frac{1}{T_1} \sim T\chi(T) \tag{3.76a}$$

or

$$\frac{1}{T_1} \sim T \frac{C}{T + \theta_{\mathrm{p}}}. \tag{3.76b}$$

Therefore, $1/T_1$ becomes almost temperature-independent because the paramagnetic Curie temperature θ_{p} is small, as shown in Fig. 3.18.

With further decrease in temperature, $1/T_1$ decreases steeply in magnitude and reaches a characteristic behavior showing $1/T_1 \sim T$ or $1/T_1 T =$ constant, as in Fig. 3.18. This means that a Korringa relation is realized [58] and

$$\frac{1}{T_1} = \pi \gamma_{\mathrm{n}}^2 \hbar \langle A_q A_{-q} \rangle D^2(\varepsilon_{\mathrm{F}}) k_{\mathrm{B}} T. \tag{3.77}$$

Equation (3.77) can be obtained by calculating the dynamical spin susceptibility due to the conduction electrons from Eq. (3.75). We can obtain this relation simply. The relaxation rate of the nuclear spins occurs due to the fluctuating internal field of the conduction electrons. This relaxation, which is small in magnitude compared to the Fermi energy, does not change the corresponding energies of conduction electrons. The transition probability W is related to $D(\varepsilon)n(\varepsilon)D(\varepsilon')(1-n(\varepsilon'))$. Here, $D(\varepsilon)n(\varepsilon)$ is the density of states. The transferred density of states from an energy ε to ε', $D(\varepsilon')(1-n(\varepsilon'))$, requires the states to be unoccupied by the electron. The relaxation rate is

$$\begin{aligned}\frac{1}{T_1} &= \pi\gamma_{\mathrm{n}}^2\hbar\,\langle A_q A_{-q}\rangle \int_0^\infty\int_0^\infty D(\varepsilon)n(\varepsilon)D(\varepsilon')\left(1-n(\varepsilon')\right)d\varepsilon d\varepsilon' \\ &= \pi\gamma_{\mathrm{n}}^2\hbar\,\langle A_q A_{-q}\rangle\, D^2(\varepsilon_{\mathrm{F}})k_{\mathrm{B}}T.\end{aligned}$$

Next, we show in Fig. 3.19 a typical ^{195}Pt($I = 1/2$)-NMR spectrum in $\mathrm{UPt_3}$ with the hexagonal structure (No. 164, $a = 5.712$ Å, $c = 4.864$ Å, $Z = 2$) [59]. Note that the contribution of the nuclear quadrupole interaction is absent in Pt because $I = 1/2$. The present NMR spectrum was obtained as a function of magnetic field at 1 K. The full width at half maximum of the spectrum is 9.5 Oe, revealing a very narrow spectrum. This means that the high-quality single crystal was used in the experiment, guaranteeing the precise Knight

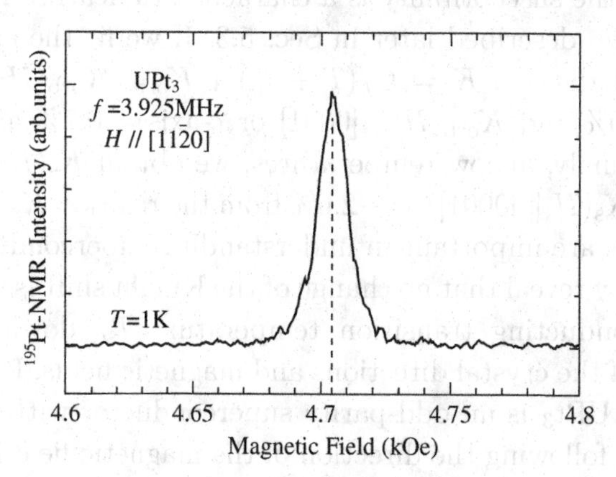

Figure 3.19: ^{195}Pt-NMR spectrum at 1 K for $H \parallel [11\bar{2}0]$ in $\mathrm{UPt_3}$, cited from Ref. [59].

shift measurement for a spin-triplet superconductor in UPt_3 (superconducting transition temperature $T_{sc} = 0.58$ K). Here, the Knight shift consists of K_s, the spin of the electrons, including the $5f$ electrons, which is closely related to the magnetic susceptibility $\chi(T)$, and the temperature-independent orbital part of the electrons, K_{orb} so that

$$K = K_s + K_{orb}, \tag{3.78}$$

which are expressed using the spin susceptibility χ_s, which is closely related to $\chi(T)$ in the $5f$ electrons, and the orbital susceptibility χ_{orb}, as

$$K_s = \frac{H_{in}^{s}}{\mu_B}\chi_s \tag{3.79}$$

$$K_{orb} = \frac{H_{in}^{orb}}{\mu_B}\chi_{orb}. \tag{3.80}$$

The temperature dependence of the Knight shift is scaled by the magnetic susceptibility $\chi(T)$, following the Curie-Weiss law above 30 K, as shown by solid lines in Fig. 3.20. Note that a broad peak at 20 K of the Knight shift for $H \parallel [11\bar{2}0]$, together with the corresponding peak of the susceptibility is a characteristic feature in the heavy fermion state, described later in Sec. 5.3. If we fit the experimental Knight shift data to $K = C/(T + \theta_p) + K_{orb}$, $K_{orb}(H \parallel [11\bar{2}0]$ or a-axis)= 2.0% and $K_{orb}(H \parallel [0001]$ or c-axis)= 0.7% are obtained. Correspondingly, at low temperatures, we obtain $K_s(H \parallel [11\bar{2}0]) = -10\%$ and $K_s(H \parallel [0001]) = -2.6\%$ from the relation $K_s = K - K_{orb}$. These values are important in understanding superconducting properties, as they reveal that no change of the Knight shift is found across the superconducting transition temperature T_{sc} down to 28 mK regardless of the crystal directions and magnetic fields. It is thus suggested that UPt_3 is an odd-parity superconductor with the parallel spin pairing following the direction of the magnetic field in the range 4.4−15.6 kOe without an appreciable pinning of the order parameter to the lattice (see Fig. 6.8(c), shown later).

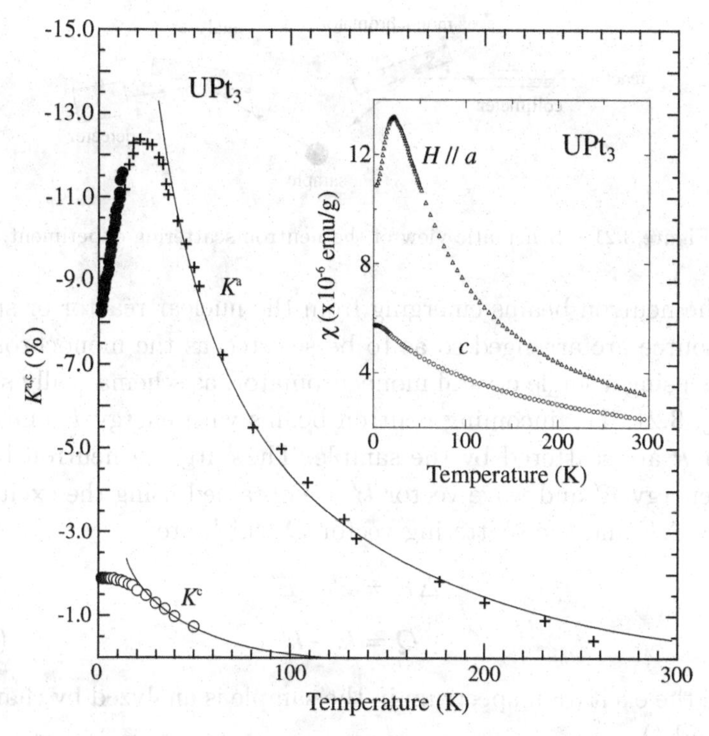

Figure 3.20: Temperature dependence of ^{195}Pt Knight shift for $H \parallel [11\bar{2}0]$(a-axis) and [0001](c-axis) in UPt_3, together with the magnetic susceptibility (inset), cited from Ref. [59].

3.2.8 Neutron scattering

The magnetic structure and magnetic moment in the f-electron systems are found by carrying out neutron scattering experiments. Since the neutron is a neutral particle with a magnetic moment (spin $I = 1/2$, magnetic moment $1.91\mu_{\rm n}$), it interacts only with the magnetic moment of f electrons in rare earth and actinide compounds as well as the nuclei of the underlying lattice. The wavelength of thermal neutrons is within 1–5 Å, which is comparable to the separation of nuclei in crystals. The neutrons have an energy of 5–10 meV (1 meV=11.66 K) which are comparable to the excitation energies of phonons and magnons. We can thus study the crystal and magnetic structures, together with dynamical properties [60].

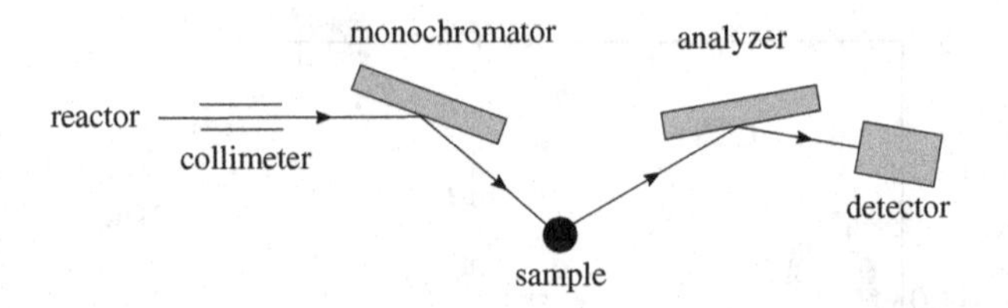

Figure 3.21: Schematic view of the neutron scattering experiment.

The neutron beams emerging from the nuclear reactor or spallation source are arranged so as to be selected as the monochromatic beams using a single crystal monochromator, as schematically shown in Fig. 3.21. The incoming neutron beams with energy E and wave vector $\boldsymbol{k}$ are scattered by the sample. The outgoing neutron beams with energy E' and wave vector $\boldsymbol{k}'$ are obtained using the excitation energy ΔE and the scattering vector $\boldsymbol{Q}$ which are

$$\Delta E = E - E' \tag{3.81}$$

$$\boldsymbol{Q} = \boldsymbol{k} - \boldsymbol{k}' \tag{3.82}$$

where the excitation spectrum in the sample is analyzed by changing ΔE and Q.

We will simply assume the incident neutron as a plane wave

$$\Psi = e^{i\boldsymbol{k}\cdot\boldsymbol{r}}. \tag{3.83}$$

The neutron-nucleus scattering is isotropic and the scattered wave of the neutron is

$$\Psi = -\frac{b}{r}e^{-i\boldsymbol{k}\cdot\boldsymbol{r}}. \tag{3.84}$$

Here, b is the scattering length. The differential cross-section associated with this nuclear interaction is

$$\left(\frac{d\sigma}{d\Omega}\right)_{\text{nuc}} = |b|^2 \tag{3.85}$$

and the total scattering cross-section corresponding to the total number of neutrons scattered per second and per incident flux unit is

$$\sigma = \int \frac{d\sigma}{d\Omega} d\Omega = 4\pi b^2. \tag{3.86}$$

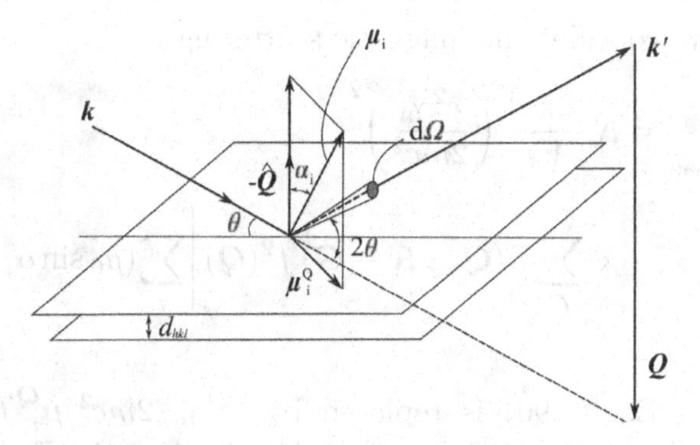

Figure 3.22: Configuration of the nuclear scattering and the magnetic scattering.

For an unpolarized neutron beam, the differential cross-section consists of both the nuclear scattering and the magnetic scattering

$$\frac{d\sigma}{d\Omega} = \left(\frac{d\sigma}{d\Omega}\right)_{\text{nuc}} + \left(\frac{d\sigma}{d\Omega}\right)_{\text{mag}} . \tag{3.87}$$

First, we consider the nuclear scattering, namely the Bragg reflection from the crystal, with the reciprocal lattice vector $\boldsymbol{G} = h\boldsymbol{a}^* + k\boldsymbol{b}^* + l\boldsymbol{c}^* = (h\,k\,l)$. As shown in Fig. 3.22, the neutron wavelength λ, the Bragg angle θ for the incoming beam, and the lattice spacing d_{hkl} must satisfy the Bragg reflection:

$$2d_{hkl}\sin\theta = \lambda. \tag{3.88}$$

This relation is changed using $2\pi/\lambda = k$ and $2\pi/d_{hkl} = Q$ to

$$\frac{4\pi\sin\theta}{\lambda} = Q \quad \text{or} \quad 2k\sin\theta = Q \tag{3.89}$$

which are equivalent to Eq. (3.82). The differential cross-section for the nuclear scattering is given, as in the X-ray scattering, by

$$\left(\frac{d\sigma}{d\Omega}\right)_{\text{nuc}} = N\frac{(2\pi)^3}{V_0}\sum_{G}\delta(\boldsymbol{Q}-\boldsymbol{G})\left|\sum_i b_i e^{i\boldsymbol{Q}\cdot\boldsymbol{r}_i}\right|^2 \tag{3.90}$$

where N is the number of nuclei, V_0 the volume of the unit cell, b_i the scattering length for the i-th nucleus, and $\boldsymbol{r}_i$ the positional vector for the i-th nucleus. From Eq. (3.90), the Bragg reflection peak is obtained at $\boldsymbol{Q} = \boldsymbol{G}$ by changing $Q(= 4\pi\sin\theta/\lambda)$ or θ.

Next, we consider the magnetic scattering

$$\left(\frac{d\sigma}{d\Omega}\right)_{\text{mag}} = N\frac{(2\pi)^3}{V_0}\left(\frac{e^2\gamma_{\text{n}}}{2mc^2}\right)^2 \times \sum_G \delta\left(\boldsymbol{Q} \pm \boldsymbol{K} - \boldsymbol{G}\right) f^2(\boldsymbol{Q}) \left|\sum_i (\mu_0 \sin\alpha_i) e^{i\boldsymbol{Q}\cdot\boldsymbol{r}_i}\right|^2 \tag{3.91}$$

where b_i in Eq. (3.90) is replaced by $(e^2\gamma_{\text{n}}/2mc^2)\boldsymbol{\mu}_i^{\boldsymbol{Q}} f(\boldsymbol{Q})$. Note that m and γ_{n} are the mass and gyromagnetic ratio of the neutron, respectively. $\boldsymbol{\mu}_i^{\boldsymbol{Q}} = \boldsymbol{\mu}_i - \hat{\boldsymbol{Q}}(\hat{\boldsymbol{Q}} \cdot \boldsymbol{\mu}_i)$ corresponds to the component of the magnetic moment perpendicular to the scattering vector $\boldsymbol{Q}$ and $f(\boldsymbol{Q})$ is a magnetic structure factor. Here, $\hat{\boldsymbol{Q}}$ is a unit vector for $\boldsymbol{Q}$. The magnitude of $\boldsymbol{\mu}_i^{\boldsymbol{Q}}$ corresponds to $\mu_0 \sin\alpha_i$ when the magnitude of the magnetic moment is μ_0 and α_i is the angle between $\boldsymbol{\mu}_i$ and $\boldsymbol{Q}$, as shown in Fig. 3.22. The magnetic moment is simply expressed as

$$\boldsymbol{\mu}_i = \boldsymbol{\mu}_0 e^{i\boldsymbol{K}\cdot\boldsymbol{r}_i} \tag{3.92}$$

where $\boldsymbol{K} = 0$ corresponds to a ferromagnet and $\boldsymbol{K} = 0.5c^*$ corresponds to an antiferromagnet, for example, where the magnetic unit cell is double compared with the unit cell along the tetragonal c-axis. Even if the f electrons are localized, the space distribution of the magnetic moment, the magnetic density $M(\boldsymbol{r})$, depends on $\boldsymbol{r}$. The corresponding magnetic factor $f(\boldsymbol{Q})$ is expressed as

$$f(\boldsymbol{Q}) = \int M(\boldsymbol{r}) e^{-i\boldsymbol{Q}\cdot\boldsymbol{r}} d^3\boldsymbol{r} \tag{3.93}$$

which represents the Fourier transform of $M(\boldsymbol{r})$. Then $f(\boldsymbol{Q})$ decreases steeply as a function of $Q = 4\pi\sin\theta/\lambda$. The magnetic Bragg reflection peak is thus obtained from Eq. (3.91) at $\boldsymbol{Q} = \boldsymbol{G} \pm \boldsymbol{K}$ by changing $Q(= 4\pi\sin\theta/\lambda)$ or θ.

Here, we show in Fig. 3.23 the magnetic scans as a function of Q along the $[1\,1\,\ell]$ direction in an antiferromagnet UPd_2Si_2 with

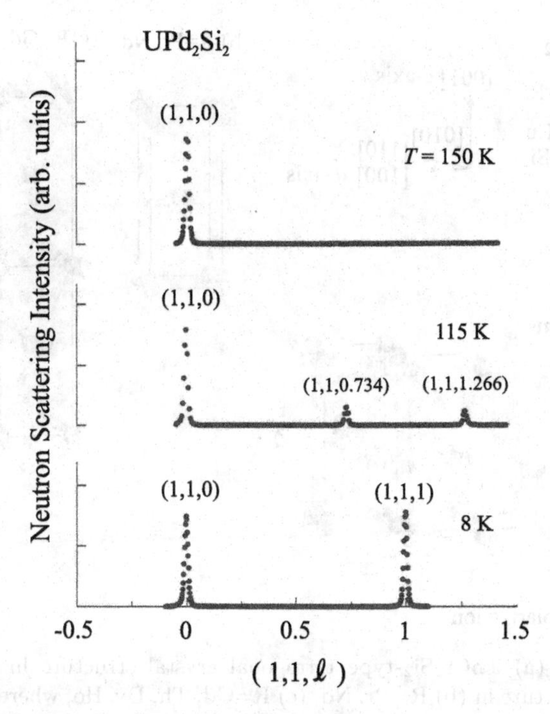

Figure 3.23: Magnetic scans as a function of Q along the $[11\ell]$ direction in an antiferromagnet UPd_2Si_2 (T_N = 133 K and T'_N = 110 K) with the $ThCr_2Si_2$-type tetragonal structure, cited from Ref. [61].

the $ThCr_2Si_2$-type tetragonal structure (No. 139, a = 4.089 Å, c = 10.051 Å, Z = 2) [61]. This compound indicates a strong uniaxial anisotropy along the c-axis ([001] direction) with the Néel temperature T_N = 135 K and a successive magnetic transition T'_N = 108 K. In the paramagnetic state at 150 K, only a nuclear Bragg reflection (110) peak is observed. On the other hand, two new peaks are found at Q = (1, 1, 0.73) and (1, 1, 1.27) at 115 K in the temperature region from T_N to T'_N = 108 K, which are reducible to a propagation vector $K = 0.73c^*$. Below T'_N = 108 K, namely at 8 K, the two peaks disappear completely and a new reflection peak is strongly observed at Q = (1, 1, 1), revealing $K = 1c^*$. Moreover, the magnetic reflection peak is very weak along $[0\,0\,\ell]$. The magnetic structure below

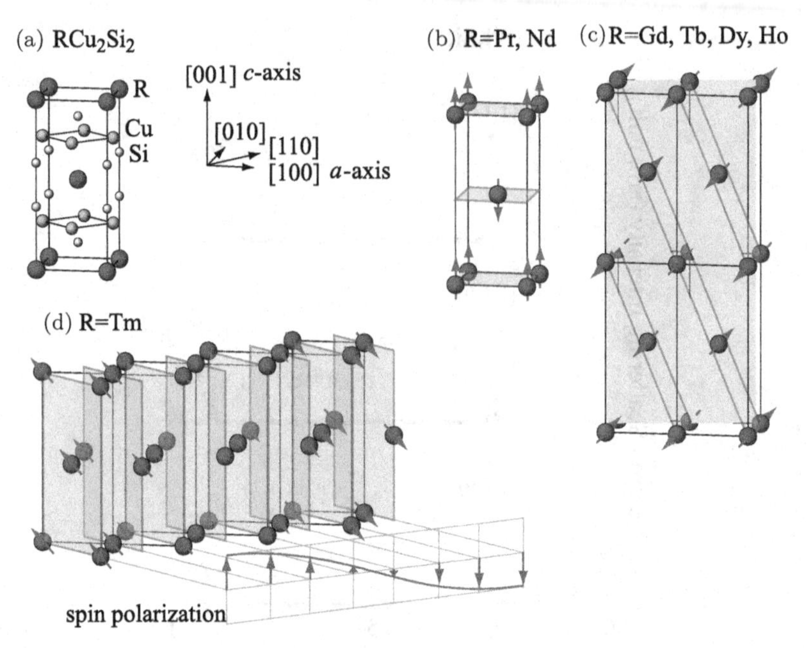

Figure 3.24: (a) $ThCr_2Si_2$-type tetragonal crystal structure in RCu_2Si_2 and magnetic structure in (b) R=Pr, Nd, (c) R=Gd, Th, Dy, Ho, where the magnetic moments in $GdCu_2Si_2$ are oriented along the [0 1 0] direction but the orientation of moments is different in the other RCu_2Si_2 (R=Th, Dy, Ho) and R=Tm, cited from Ref. [41].

$T'_{\rm N} = 108$ K is thus concluded to correspond to the magnetic structure labelled as type-I in Fig. 3.24(b).

The various magnetic structures in RCu_2Si_2 (R: rare earth) are presented in Fig. 3.24 [41]. The magnetic structure of $TmCu_2Si_2$ is sinusoidally modulated along the [1 1 0] directions [41]. In the case of UPd_2Si_2, the magnetic structure at $T'_{\rm N} < T < T_{\rm N}$ is an incommensurate-longitudinal-sine-wave modulation.

The observed intensities are

$$I = RC(Q)\frac{d\sigma}{d\Omega} \tag{3.94}$$

where R is an instrumental factor and $C(Q)$ is a function which contains the Lorentz factor, absorption and excitation corrections, and other factors. The ordered moment $\mu_0 = 1.9\,\mu_B/U$ is determined by calculating I_{mag} (110)/I_{nuc} (110) at 8 K. These are based on the elastic neutron scattering, $\Delta E = 0$ in Eq. (3.81).

Next, we consider the inelastic neutron scattering ($\Delta E \neq 0$). The corresponding differential cross-section is

$$\frac{d^2\sigma}{d\Omega d\omega} = \left(\frac{e^2\gamma_n}{2mc^2}\right)^2 g^2 f^2(\boldsymbol{Q})\frac{|\boldsymbol{k}'|}{|\boldsymbol{k}|}\sum_{\alpha,\beta}\left(\delta_{\alpha\beta} - \hat{\boldsymbol{Q}}_\alpha\hat{\boldsymbol{Q}}_\beta\right) S_{\alpha\beta}(\boldsymbol{Q},\omega) \tag{3.95}$$

$$S_{\alpha\beta}(\boldsymbol{Q},\omega) = \frac{1}{2\pi}\int_{-\infty}^{\infty} \langle S_\alpha(-\boldsymbol{Q},0)S_\beta(\boldsymbol{Q},t)\rangle\, e^{-i\omega t}dt \tag{3.96}$$

where $\langle\ \rangle$ indicates the thermal average. When we compare Eqs. (3.95) and (3.96) with Eq. (3.91), the inelastic scattering requires $|\boldsymbol{k}'| \neq |\boldsymbol{k}|$ in Eq. (3.95). μ_0 in Eq. (3.91) is represented by gS which depends on $\boldsymbol{Q}$ and ω. The term $(\delta_{\alpha\beta} - \hat{\boldsymbol{Q}}_\alpha\hat{\boldsymbol{Q}}_\beta)$ is a directional factor, perpendicular to $\boldsymbol{Q}$. The spin moment $\boldsymbol{S}$ is a time-dependent operator:

$$\boldsymbol{S}(t) = e^{i\mathcal{H}t/\hbar}\boldsymbol{S}e^{-i\mathcal{H}t/\hbar}. \tag{3.97}$$

The response function $S_{\alpha\beta}(\boldsymbol{Q},\omega)$ is directly related to the generalized susceptibility $\chi(\boldsymbol{Q},\omega)$:

$$S_{\alpha\beta}(\boldsymbol{Q},\omega) = \frac{N\hbar}{\pi}\frac{1}{1-e^{-\hbar\omega/k_BT}}\text{Im}\,\chi_{\alpha\beta}(\boldsymbol{Q},\omega). \tag{3.98}$$

Note that the usual magnetic susceptibility is integrated over $\boldsymbol{Q}$ or ω,

$$\chi = \frac{1}{\pi}\int\frac{\text{Im}\,\chi(Q=0,\omega)}{\omega}d\omega. \tag{3.99}$$

By using inelastic neutron scattering and changing $\boldsymbol{Q}$ and ω, we can study characteristic properties such as phonons, magnons, and the CEF scheme, especially the magnetic correlations in heavy fermion systems. The $4f$-CEF peaks are observed as a function of ω, for example. The corresponding differential cross-section is

$$\frac{d^2\sigma}{d\Omega d\omega} = N\left(\frac{e^2\gamma_{\rm n}}{2mc^2}\right)^2 g^2 f^2(\boldsymbol{Q})\frac{|\boldsymbol{k}'|}{|\boldsymbol{k}|}\sum_{i,j} p_i \left|\langle j\,|S_\perp|\,i\rangle\right|^2 \delta(E_i - E_j + \hbar\omega) \tag{3.100}$$

where $S_\perp$ represents the component of $\boldsymbol{S}$ perpendicular to the scattering vector $\boldsymbol{Q}$, and $\langle j|S_\perp|i\rangle$ is a transition probability between the states $|i\rangle$ and $|j\rangle$, where the corresponding $4f$ energies are E_i and E_j, respectively. The factor p_i is an occupation probability of the state $|i\rangle$ or E_i:

$$p_i = \frac{e^{-E_i/k_{\rm B}T}}{\sum\limits_i e^{-E_i/k_{\rm B}T}}. \tag{3.101}$$

The δ function of Eq. (3.100) becomes infinite at

$$E_j - E_i = \hbar\omega. \tag{3.102}$$

In the exact spectrum, a finite $4f$-CEF peak is observed in this condition. Usually, the spectrum contains a phonon peak, which is subtracted by measuring a similar spectrum in the corresponding non-$4f$ La compound.

Figure 3.25 shows the scattering function derived from the inelastic neutron scattering $d^2\sigma/d\Omega d\omega$ for $PrIr_2Zn_{20}$ at $Q = 3.26$ Å^{-1} [62]. Distinct excitation peaks are seen at 2.5 and 5.6 meV, and their intensities vary slightly with temperature. $PrIr_2Zn_{20}$ crystallizes in the $CeCr_2Al_{20}$-type cubic structure, as shown later in Fig. 5.25(b). From the magnetization and specific heat measurements, the CEF ground state is determined to be the non-Kramers doublet Γ_3, where the CEF scheme consists of Γ_1(singlet), Γ_3(doublet), Γ_4(triplet) and Γ_5(triplet). The present excitation peaks correspond to $4f$-CEF-level intervals between the Γ_3 ground state and the excited states. Based on the CEF Hamiltonian [63], the CEF scheme is determined to be Γ_3 (0 meV) – Γ_4 (2.36 meV = 28 K) – Γ_1 (5.67 meV = 66 K)–Γ_5 (5.80 meV = 67 K). It is noted that the CEF peaks are not observed in heavy fermion compounds such as $CeCu_6$ and $CeRu_2Si_2$. However, new magnetic excitations are observed in these compounds, as explained in Chap. 5.

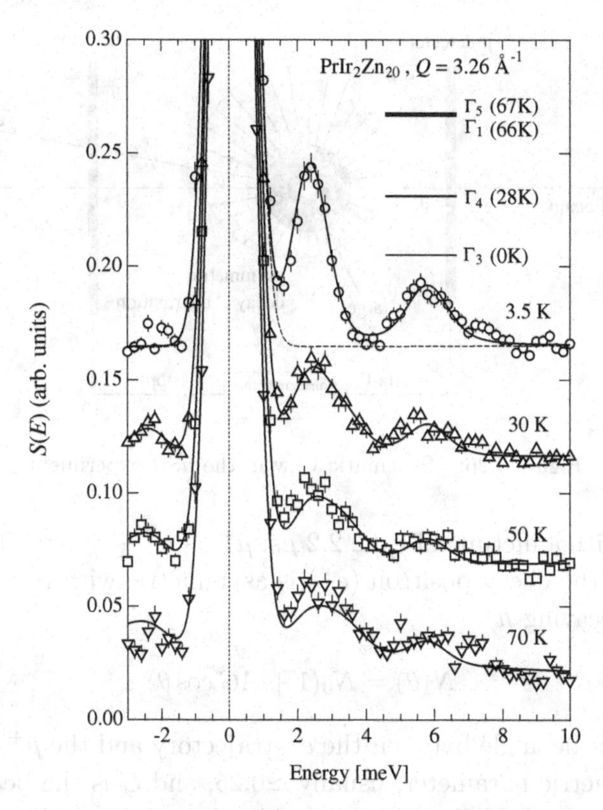

Figure 3.25: Scattering function $S(E)$ evaluated from the inelastic neutron scattering spectra of $PrIr_2Zn_{20}$ with the cubic structure. Each set is shown, with a relative shift of 0.05. The solid lines which connect data joints have been calculated for best fit. Broken lines are due to incoherent scattering intensities and background in the fitting analyses, cited from Ref. [62]. The inset shows the CEF scheme.

3.2.9 μSR

The muon spin rotation, relaxation and resonance (μSR) is a microscopic technique similar to NMR [64]. The muon (μ^+) with a positive charge ($+e$) is usually used in the experiment and possesses a spin $I = 1/2$, resulting in no quadrupole interaction, as discussed in the NMR experiments. The muon beam is characteristically spin ($\boldsymbol{S}_\mu$)-polarized along the flight direction, even in the sample. The spin-polarized muon beam will stop rather homogeneously throughout the sample and decays to a positron (e^+) and two kinds of neutrinos

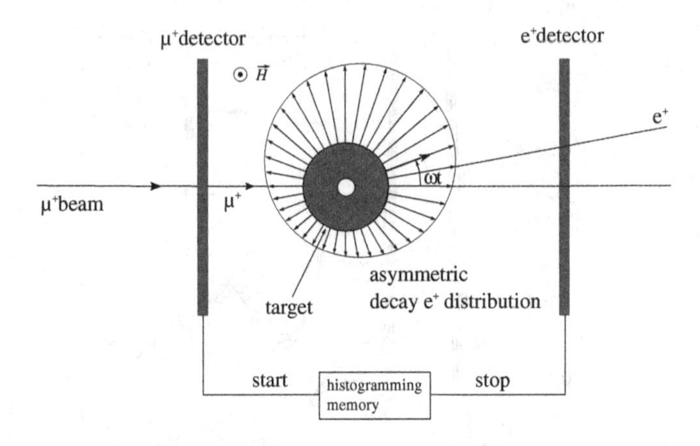

Figure 3.26: Schematic view in the μSR experiment.

$(\bar{\nu_\mu}, \nu_e)$ with a lifetime of $\tau_\mu \simeq 2.2$ μs: $\mu^+ \to e^+ + \bar{\nu_\mu} + \nu_e$. The distribution of the decay positron (e^+) is asymmetric with respect to the spin of decaying μ^+

$$N(\theta) = N_0(1 + AG\cos\theta) \tag{3.103}$$

where θ is the angle between the e^+-trajectory and the μ^+-spin, A is an asymmetric parameter, usually $\simeq$0.25, and G is the beam polarization. Figure 3.26 shows a schematic view of the μSR experiment. The forward and backward e^+ detectors count the number of e^+ as in

$$N(\theta, t) = N_0 e^{-t/\tau_\mu}[1 + AG(t)\cos\theta]. \tag{3.104}$$

Note that the muon decays exponentially, with lifetime τ_μ. Their decay count numbers are as follows

$$N_{\rm F} = N(0, t) = N_0 e^{-t/\tau_\mu}[1 + AG(t)] \tag{3.105}$$

$$N_{\rm B} = N(\pi, t) = N_0 e^{-t/\tau_\mu}[1 - AG(t)]. \tag{3.106}$$

The asymmetric part is thus obtained as follows

$$AG(t) = \frac{N_{\rm F} - N_{\rm B}}{N_{\rm F} + N_{\rm B}}. \tag{3.107}$$

An external magnetic field, which is perpendicular to the muon spin direction $\boldsymbol{S}_\mu$, is applied to the sample. The stopped μ^+ will start to perform a Larmor precession of $\omega_\mu = \gamma_\mu H$ (13.55 kHz/Oe or 135.5 MHz/T). The corresponding distribution is

$$N_{\mathrm{F,B}}(t) = N_0 e^{-t/\tau_\mu}[1 \pm AG_{\mathrm{x}}(t)\cos(\omega_\mu t)]. \tag{3.108}$$

This technique is called the transverse field (TF)-μSR. Here, $G_{\mathrm{x}}(t)$ depends on the nature of the local magnetic field at the muon site, namely the static and fluctuating (dynamic) aspects of magnetism. Note that no precession occurs when the external magnetic field is parallel to the μ^+-spin.

Figure 3.27 shows the μSR spectra at zero magnetic field for $CeAl_3$ with the hexagonal structure (No. 194, a = 6.541 Å, c = 4.610Å, Z = 2), indicating a spontaneous muon spin-precession at 0.5 and 0.05 K [65]. This result was surprising because $CeAl_3$ was

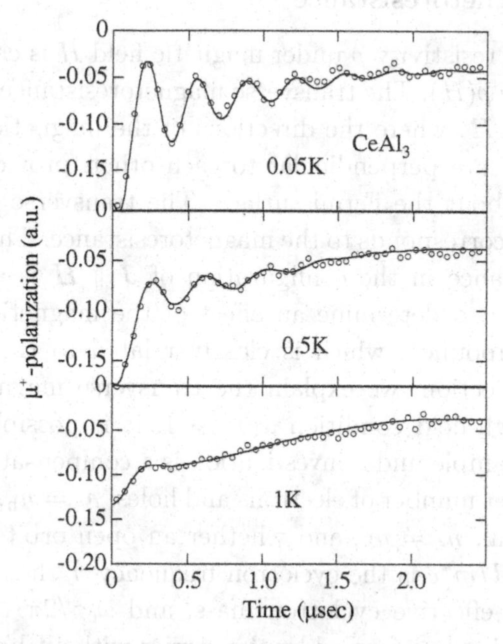

Figure 3.27: μSR spectra obtained with application of zero magnetic field in $CeAl_3$, indicating a spontaneous muon spin-precession at 0.5 and 0.05 K, cited from Ref. [65].

considered to be a heavy-fermion compound without magnetic ordering. The μ^+ in $CeAl_3$ experiences two different magnetic environments. One is associated with a nonzero magnetic field and the other is associated with zero average magnetic field but a nonzero field spreads around this value, leading to the Kubo-Tayabe signal

$$G_z(t) = \frac{1}{3} + \frac{2}{3}\left(1 - \sigma_{\mathrm{KT}}^2 t^2\right) \exp\left(-\frac{1}{2}\sigma_{\mathrm{KT}}^2 t^2\right). \tag{3.109}$$

The Larmor precession frequency reflects an average local magnetic field of 220 Oe at the muon site, implying magnetic ordering with a very small ordered moment of 0.05 μ_{B}/Ce. Later, it was also confirmed from ^{27}Al-NMR spectra that $CeAl_3$ orders antiferromagnetically below 1.2 K, with an ordered moment of 0.3 μ_{B}/Ce at maximum [66].

3.2.10 Magnetoresistance

The electrical resistivity ρ under magnetic field H is called the magnetoresistance $\rho(H)$. The transverse magnetoresistance in the configuration of $\boldsymbol{J} \perp \boldsymbol{H}$, where the directions of the magnetic field $\boldsymbol{H}$ and the current $\boldsymbol{J}$ are perpendicular to each other, provides important information about the Fermi surface. The transverse magnetoresistance usually corresponds to the magnetoresistance. The longitudinal magnetoresistance in the configuration of $\boldsymbol{J} \parallel \boldsymbol{H}$ is also measured experimentally to determine an effect of the magnetic field on the magnetic compounds, which is closely related to their magnetization. In this section, we explain the transverse magnetoresistance. Under the high-field condition $\omega_{\mathrm{c}}\tau \gg 1$, it is possible to find out whether the sample under investigation is a compensated metal with an equal carrier number of electrons and holes, $n_{\mathrm{e}} = n_{\mathrm{h}}$, or an uncompensated metal, $n_{\mathrm{e}} \neq n_{\mathrm{h}}$, and whether an open orbit exists or not. Here, $\omega_{\mathrm{c}} = qH/m_{\mathrm{c}}^* c$ is the cyclotron frequency, τ the scattering lifetime, m_{c}^* the effective cyclotron mass, and $\omega_{\mathrm{c}}\tau/2\pi$ the number of cyclotron motions performed by the carrier without being scattered.

First, we consider a non-magnetic metal with one kind of carrier. In the configuration of $\boldsymbol{J} \parallel x$ and $\boldsymbol{H} \parallel z$, we get the following equation for the carriers:

$$m^*\frac{d\boldsymbol{v}}{dt} = q\left(\boldsymbol{E} + \frac{1}{c}\boldsymbol{v}\times\boldsymbol{H}\right) - \frac{m^*\boldsymbol{v}}{\tau} \tag{3.110}$$

where $\boldsymbol{v} = (v_x,\, v_y,\, v_z)$, $\boldsymbol{E} = (E_x,\, E_y,\, E_z)$ and $\boldsymbol{H} = (0,\, 0,\, H)$. In the steady state $d\boldsymbol{v}/dt = 0$

$$v_x = \frac{\mu}{1+\alpha^2}(E_x + \alpha E_y) \tag{3.111a}$$

$$v_y = \frac{\mu}{1+\alpha^2}(E_y - \alpha E_x) \tag{3.111b}$$

$$v_z = \mu E_z \tag{3.111c}$$

where $\mu(= q\tau/m^*)$ is the mobility and $\alpha = \mu H/c = \omega_c\tau$. Note that $E_y = \alpha E_x$ and $E_z = 0$ because $J_x \neq 0$ and $J_y = J_z = 0$ or $v_y = v_z = 0$. Equation (3.111a) is thus

$$J_x = nqv_x = nq\mu E_x = \sigma E_x = \rho^{-1}E_x. \tag{3.112}$$

This means no magnetoresistance for the corresponding metal because ρ does not depend on the magnetic field $\boldsymbol{H}$. This is because the Lorentz force $(q/c)\boldsymbol{v}\times\boldsymbol{H}$ or αE_x is compensated by the Hall field E_y. The Hall coefficient is defined as

$$R_{\mathrm{H}} = \frac{E_y}{HJ_x} = \frac{1}{nqc} \tag{3.113}$$

where R_{H} becomes negative for an electron carrier with $q = -e$, and becomes positive for a hole carrier with $q = +e$.

Next, we consider two kinds of carriers such as two kinds of electron (or hole) Fermi surfaces or electron and hole Fermi surfaces. The following equations are obtained:

$$\left.\begin{aligned} v_{1x} &= \frac{\mu_1}{1+\alpha_1^2}\left(E_x + \alpha_1 E_y\right),\, v_{1y} = \frac{\mu_1}{1+\alpha_1^2}\left(E_y - \alpha_1 E_x\right) \\ v_{2x} &= \frac{\mu_2}{1+\alpha_2^2}\left(E_x + \alpha_2 E_y\right),\, v_{2y} = \frac{\mu_2}{1+\alpha_2^2}\left(E_y - \alpha_2 E_x\right) \end{aligned}\right\} \tag{3.114}$$

$$J_x = n_1q_1v_{1x} + n_2q_2v_{2x} \tag{3.115}$$

$$J_y = n_1q_1v_{1y} + n_2q_2v_{2y} = 0. \tag{3.116}$$

By putting v_{1y} and v_{2y} of Eqs. (3.114) in Eq. (3.116), the following equation is obtained:

$$E_y = \frac{\dfrac{n_1 q_1 \mu_1 \alpha_1}{1+\alpha_1^2} + \dfrac{n_2 q_2 \mu_2 \alpha_2}{1+\alpha_2^2}}{\dfrac{n_1 q_1 \mu_1}{1+\alpha_1^2} + \dfrac{n_2 q_2 \mu_2}{1+\alpha_2^2}} E_x. \tag{3.117}$$

When $\alpha^2 \ll 1$ (low field), Eq. (3.117) is

$$E_y = \frac{n_1 q_1 \mu_1^2 + n_2 q_2 \mu_2^2}{n_1 q_1 \mu_1 + n_2 q_2 \mu_2} \frac{H}{c} E_x \tag{3.118}$$

and R_{H} is obtained to be

$$R_{\mathrm{H}} = \frac{E_y}{H J_x} = \frac{n_1 q_1 \mu_1^2 + n_2 q_2 \mu_2^2}{(n_1 q_1 \mu_1 + n_2 q_2 \mu_2)^2} \frac{1}{c}. \tag{3.119}$$

The magnetoresistance is also obtained by putting Eqs. (3.114) in Eq. (3.115)

$$\rho(H) = \frac{\dfrac{\sigma_1}{1+\alpha_1^2} + \dfrac{\sigma_2}{1+\alpha_2^2}}{\left(\dfrac{\sigma_1}{1+\alpha_1^2} + \dfrac{\sigma_2}{1+\alpha_2^2}\right)^2 + \left(\dfrac{\sigma_1 \alpha_1}{1+\alpha_1^2} + \dfrac{\sigma_2 \alpha_2}{1+\alpha_2^2}\right)^2}. \tag{3.120}$$

Equation (3.120) is not simplified even if $\alpha^2 \ll 1$, but increases with increasing magnetic field, $\rho(H) \sim H^2$.

When $\alpha^2 \gg 1$ (high field condition), Eq. (3.117) becomes

$$E_y = \frac{n_1 q_1 + n_2 q_2}{\dfrac{n_1 q_1}{\mu_1} + \dfrac{n_2 q_2}{\mu_2}} \frac{H}{c} E_x. \tag{3.121}$$

Equation (3.115) becomes

$$J_x = \left\{ \left(\frac{n_1 q_1}{\mu_1} + \frac{n_2 q_2}{\mu_2} \right) \left(\frac{c}{H} \right)^2 + \frac{(n_1 q_1 + n_2 q_2)^2}{\dfrac{n_1 q_1}{\mu_1} + \dfrac{n_2 q_2}{\mu_2}} \right\} E_x. \tag{3.122}$$

From Eq. (3.122), the magnetoresistance becomes

$$\rho(H) = \frac{(n_1 q_1 + n_2 q_2)^2}{\dfrac{n_1 q_1}{\mu_1} + \dfrac{n_2 q_2}{\mu_2}} \quad \text{when } n_1 q_1 + n_2 q_2 \neq 0 \tag{3.123}$$

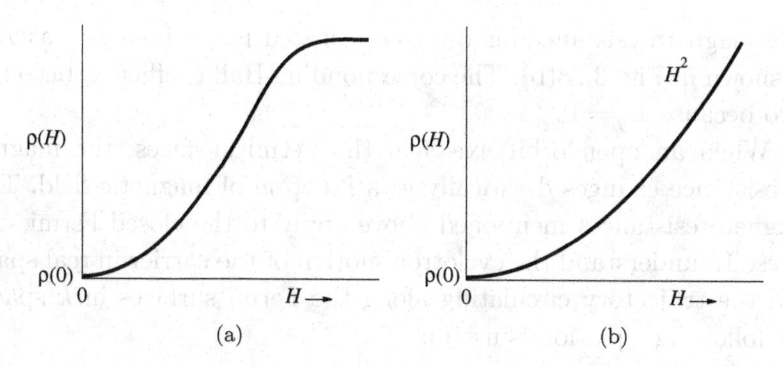

Figure 3.28: Magnetoresistance in the case of (a) an uncompensated metal and (b) compensated metal.

and the Hall coefficient is

$$R_{\mathrm{H}} = \frac{E_y}{HJ_x} = \frac{1}{n_1 q_1 + n_2 q_2}\frac{1}{c}. \tag{3.124}$$

Note that the first term in Eq. (3.122) is neglected because $(c/H^2) \to 0$. The magnetoresistance for an uncompensated metal starts from the value of $\rho(0) = 1/(n_1 q_1 \mu_1 + n_2 q_2 \mu_2)$, increases as $\rho(H) \sim H^2$ at low fields, and finally saturates into Eq. (3.123), as schematically shown in Fig. 3.28(a).

In the case of a compensated metal of $n_1 q_1 + n_2 q_2 = 0$, namely $q_1 = +e$, $q_2 = -e$, and $n_1 = n_2 = n$

$$\rho(H) = \frac{1}{\dfrac{n_1 q_1}{\mu_1} + \dfrac{n_2 q_2}{\mu_2}}\left(\frac{H}{c}\right)^2 \tag{3.125}$$

or

$$\begin{aligned}\frac{\rho(H) - \rho(0)}{\rho(0)} &= \frac{\Delta\rho}{\rho(0)} \\ &= \frac{n_1 q_1 \mu_1 + n_2 q_2 \mu_2}{\dfrac{n_1 q_1}{\mu_1} + \dfrac{n_2 q_2}{\mu_2}}\left(\frac{H}{c}\right)^2 \\ &= |\mu_1||\mu_2|\left(\frac{H}{c}\right)^2 \\ &= |\omega_{c1}\tau_1||\omega_{c2}\tau_2|. \end{aligned} \tag{3.126}$$

The magnetoresistance for the compensated metal increases as H^2, as shown in Fig. 3.28(b). The corresponding Hall coefficient becomes zero because $E_y = 0$.

When an open orbit exists in the Fermi surfaces, the magnetoresistance changes drastically as a function of magnetic field. The magnetoresistances mentioned above apply to the closed Fermi surfaces. To understand the cyclotron motion of the carrier in real space and the trajectory circulating along the Fermi surfaces in k space, the following equation is useful:

$$\frac{d(\hbar \boldsymbol{k})}{dt} = \frac{q}{c}\left(\frac{d\boldsymbol{r}}{dt} \times \boldsymbol{H}\right). \tag{3.127}$$

When the carrier is an electron ($q = -e$), the electron moves by $d\boldsymbol{k}$ along the k_y direction in a period dt under condition $H \parallel k_z$ or $H \parallel z$, the electron moves by $d\boldsymbol{r}$ along the x direction, as shown in Figs. 3.29(a) and (b). In Fig. 3.29(c), we show several orbits of an ellipsoidal electron-Fermi surface in k space and the corresponding cyclotron motions in real space. Orbit ③ is termed "stationary", of which the cross-sectional area is determined in the dHvA experiment. On the other hand, the other orbits drift in real space along the direction parallel or antiparallel to the field. Note that the rotation of an orbit in the hole-Fermi surface is counter-clockwise compared with that in the electron-Fermi surface because $q = +e$.

The open-orbit Fermi surface is illustrated simply in Fig. 3.30(b). In the configuration $\boldsymbol{J} \parallel x$ and $\boldsymbol{H} \parallel z$, the magnetoresistance saturates at high fields because the carrier due to the open orbit moves along the current direction, that is, along the x direction. In the configuration of $\boldsymbol{J} \parallel y$ and $\boldsymbol{H} \parallel z$, the magnetoresistance instead indicates a H^2 dependence at high fields because the carrier due to the open-orbit moves along the x direction, perpendicular to the current. Note that the situation is drastically changed when the magnetic breakdown (magnetic breakthrough) effect is realized in high fields. Namely, the trajectory circulating along the Fermi surface in Fig. 3.30(b) is changed into a large closed one as shown in Fig. 3.30(a).

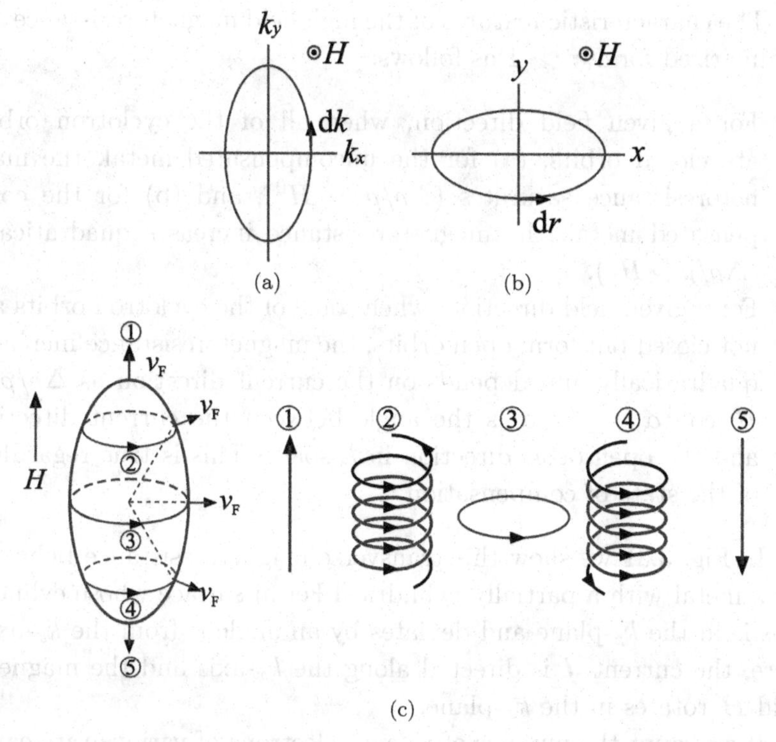

Figure 3.29: (a) Trajectory circulating along the electron-Fermi surface in k space when the magnetic field is applied along the k_z direction, and (b) the cyclotron motion in real space for an electron, together with (c) several orbits of an ellipsoidal electron-Fermi surface in k space and real space when the magnetic field is applied along the ellipsoidal axis.

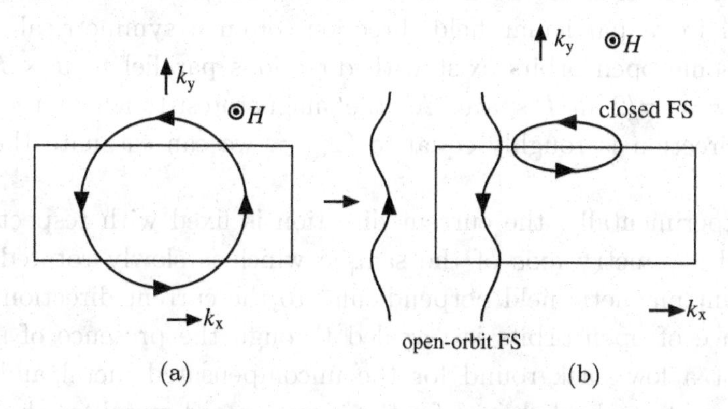

Figure 3.30: (a) Large orbit in a rectangular Brillouin zone, which corresponds to (b), an open-orbit Fermi surface (FS) and a small closed Fermi surface.

The characteristic features of the high field magnetoresistance are summarized for $\omega_c\tau \gg 1$ as follows:

(1) For a given field direction, when all of the cyclotron orbits are closed orbits, (a) for the uncompensated metal, the magnetoresistance saturates ($\Delta\rho/\rho \sim H^0$), and (b) for the compensated metal, the magnetoresistance increases quadratically ($\Delta\rho/\rho \sim H^2$).
(2) For a given field direction, when some of the cyclotron orbits are not closed but form open orbits, the magnetoresistance increases quadratically and depends on the current direction as $\Delta\rho/\rho \sim H^2\cos^2\alpha$, where α is the angle between the current direction and the open orbit direction in $\boldsymbol{k}$-space. This is true regardless of the state of compensation.

In Fig. 3.31 we show this transverse magnetoresistance behavior for a metal with a partially cylindrical Fermi surface whose cylinder axis is in the k_z-plane and deviates by an angle α from the k_x-axis. Here, the current $\boldsymbol{J}$ is directed along the k_x-axis and the magnetic field $\boldsymbol{H}$ rotates in the k_x-plane.

If we count the number of valence electrons of various rare earth and U compounds in the unit cell, most of them are even in number, meaning that they are compensated metals. In this case, the transverse magnetoresistance increases as H^n $(1 < n \leq 2)$ for the general direction of the field. When the magnetoresistance is saturated for a particular field direction, often a symmetrical direction, some open orbits exist with directions parallel to $\boldsymbol{J} \times \boldsymbol{H}$, so that $\alpha = \pi/2$ in $\boldsymbol{k}$-space. As the magnetoresistance in the general direction is roughly equal to $(\omega_c\tau)^2$, we can estimate the $\omega_c\tau$ value.

Experimentally, the current direction is fixed with respect to a crystal symmetry axis of the sample which is slowly rotated in a constant magnetic field perpendicular to the current direction. The presence of open orbits is revealed through the presence of spikes against a low background for the uncompensated metal and dips against a large background for the compensated metal, as shown in Figs. 3.31(c) and (d), respectively.

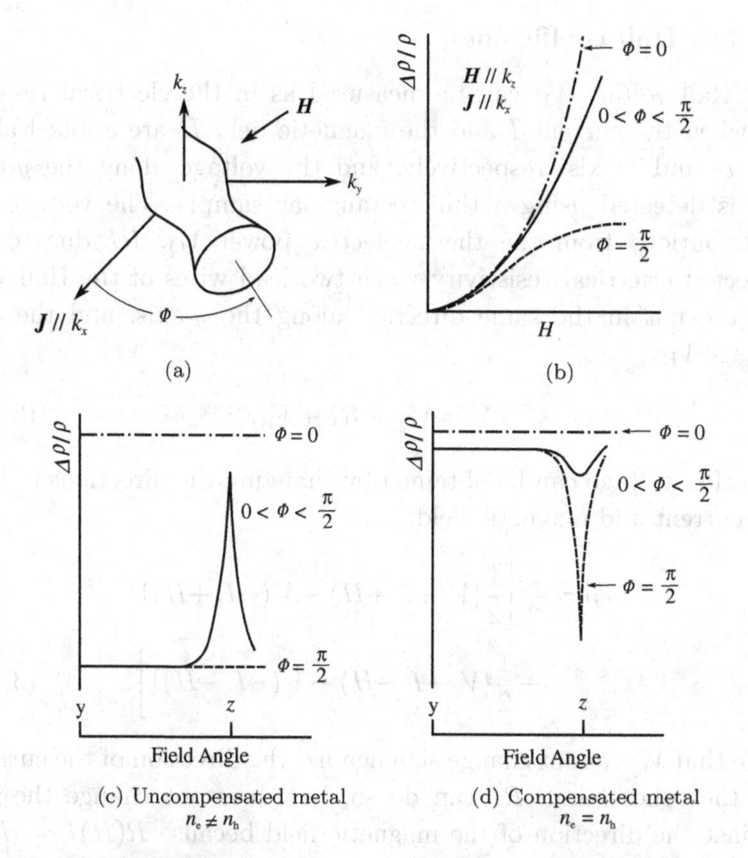

Figure 3.31: Schematic picture of the transverse magnetoresistances in uncompensated and compensated metals when a partially cylindrical Fermi surface exists. The magnetic field H rotates in the k_x-plane.

Magnetic breakdown, sometimes observed in the magnetoresistance, is able to change closed orbits to open orbits, and vice versa, as mentioned above, giving rise to a change in the field dependence of the magnetoresistance between low and high fields.

Lastly we note a characteristic oscillatory behavior in the angular dependence of the magnetoresistance. This occurs for a nearly cylindrical Fermi surface when all the cross-sectional areas of the Fermi surface perpendicular to the field direction become the same. This is called "the Yamaji effect" [67]. See Refs. [68, 69] in a spin-triplet superconductor $SrRuO_4$, for an example.

3.2.11 Hall coefficient

The Hall voltage V_{H} can be measured as in the electrical resistivity when the current I and the magnetic field H are applied along the x- and z-axis, respectively, and the voltage along the y-axis, V_{H}, is detected using a thin rectangular sample. The voltage has contributions from the thermoelectric power V_{Q}, RI due to the expected electrical resistivity when two lead wires of the Hall voltage are not in the same direction along the y-axis, and the Hall voltage V_{H}:

$$V = V_{\mathrm{Q}} + RI + V_{\mathrm{H}}. \tag{3.128}$$

The Hall voltage can be obtained by changing the directions of both the current and magnetic field

$$V_{\mathrm{H}} = \frac{1}{2}\left[\frac{1}{2}\{V(+I,+H) - V(-I,+H)\} - \frac{1}{2}\{V(+I,-H) - V(-I,-H)\}\right]. \tag{3.129}$$

Note that V_{Q} cannot change sign against the direction of the current. On the other hand, RI can do so, but does not change the sign against the direction of the magnetic field because $R(H)I \sim H^2$ in low magnetic fields. The Hall voltage $V_{\mathrm{H}} = R_{\mathrm{H}}HI/a$ is based on the current and magnetic field, where a is the thickness of the sample and R_{H} the Hall coefficient. The Hall voltage is usually smaller than R by two or three orders of magnitude. A thin sample is thus needed in the measurement.

The Hall effect is a good experimental method to detect f electron behavior. The Hall effect of a typical metal is due to the Lorentz force, $(q/c)\boldsymbol{v} \times \boldsymbol{H}$. The Hall coefficient R_0 is independent of temperature and given as $-1/nec$ for one carrier, where n refer to the carrier concentration. This corresponds to Eq. (3.113).

On the other hand, the Hall effect of magnetic compounds is generally considered to be a sum of the ordinary Hall effect mentioned above and an anomalous part dependent on the magnetization M. The Hall resistivity or Hall coefficient in the paramagnetic state is phenomenologically expressed as

$$\rho_{\rm H} = R_0 H + 4\pi R_{\rm s} M \tag{3.130a}$$

or

$$R_{\rm H} = R_0 + 4\pi R_{\rm s} \chi \tag{3.130b}$$

and

$$\chi = C/(T + \theta_{\rm p}) \tag{3.131}$$

where $R_{\rm s}$ is the temperature-independent anomalous Hall coefficient, χ the magnetic susceptibility, C the Curie constant and $\theta_{\rm p}$ the paramagnetic Curie temperature.

Theoretically, the skew (asymmetric) scattering is the main mechanism of the anomalous Hall coefficient. The asymmetry comes from the interaction of conduction electrons with localized d or f electrons which possess orbital angular momenta. In this case, the transition probability of the conduction electrons, scattered from a state $\boldsymbol{k}$ to another state $\boldsymbol{k}'$ due to localized d or f electrons, is not equal to the one scattered from $\boldsymbol{k}'$ to $\boldsymbol{k}$. This leads to the anomalous Hall coefficient. Experimentally, the sign of $R_{\rm s}$ is positive in the ferromagnetic metal, while in the antiferromagnetic metal, it is dependent on the compound.

The Hall effect in the impurity Kondo system was calculated by Coleman *et al.* [70] and by Fert *et al.* [71] on the basis of skew scattering by cerium impurities

$$R_{\rm H} = R_0 + \gamma \tilde{\chi} \rho_{\rm m} \tag{3.132}$$

where γ is $-(15/7) g \mu_{\rm B} k_{\rm B}^{-1} \sin\delta_2 \cos\delta_2$, $\tilde{\chi} = \chi/C$ and $\rho_{\rm m}$ the magnetic resistivity. The skew scattering in this case is due to the d-wave ($l = 2$) and f-wave ($l = 3$) scattering. The phase shift δ_2 is associated with spherical scattering in the channel $l = 2$. The prominent scattering is due to the Kondo scattering in the channel $l = 3$. Therefore, the susceptibility originates from the Ce impurity. $\rho_{\rm m}$ is extremely dominant in the heavy fermion compounds.

Now, we show in Fig. 3.32(a) the temperature dependence of the Hall coefficient in $\mathrm{Ce}_x\mathrm{La}_{1-x}\mathrm{Cu}_6$ [72]. The Hall coefficient becomes

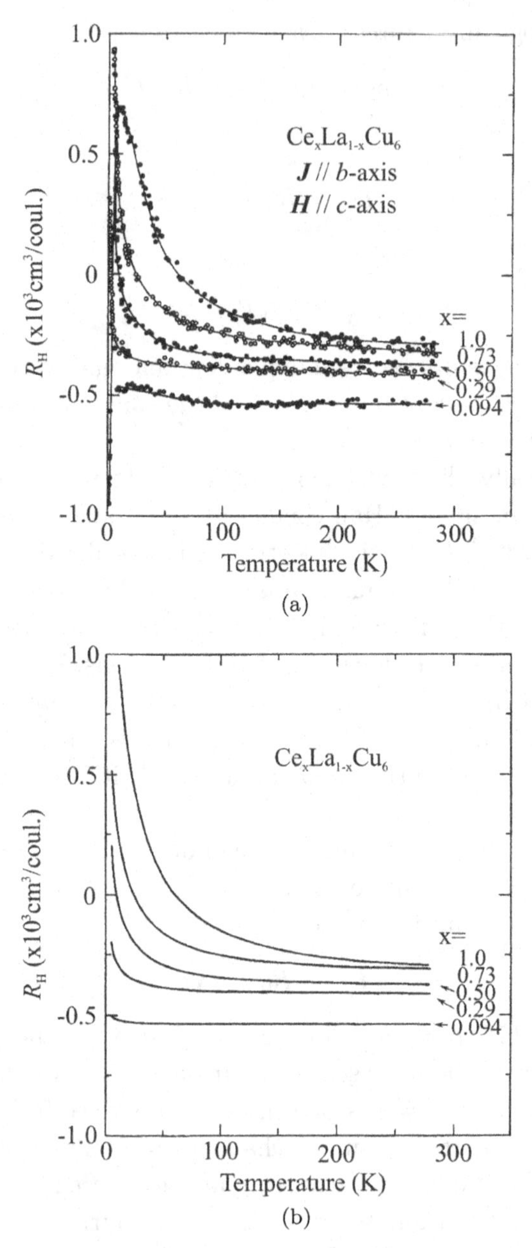

Figure 3.32: (a) Temperature dependence of Hall coefficient in $Ce_xLa_{1-x}Cu_6$ and (b) theoretical curves, cited from Ref. [72].

large in magnitude with increasing concentration of Ce. We tried to fit the data of the susceptibility $\tilde{\chi} = \chi/C$ and the magnetic resistivity $\rho_{\rm m} = \rho_{\rm Ce_xLa_{1-x}Cu_6} - \rho_{\rm LaCu_6}$ for the same sample with Eq. (3.132), proposed by Fert *et al.* To estimate the parameter R_0 and the phase shift δ_2, we plotted the $R_{\rm H}$ vs $\rho_{\rm m}$ for each $\rm Ce_xLa_{1-x}Cu_6$. The data are fitted by a straight line. By using the values of R_0 and δ_2 thus obtained, we show in Fig. 3.32(b) the theoretical curves following Eq. (3.132). The data in Fig. 3.32(a) are in good agreement with the theoretical curves in Fig. 3.32(b).

The Hall coefficient in $\rm Ce_xLa_{1-x}Cu_6$ deviates gradually from the present theoretical curve with decreasing temperature. The system enters the coherent Kondo state, an f-driven heavy fermion state. We show in Fig. 3.33 the Hall coefficient with the logarithmic scale of temperature. The Hall coefficient of $\rm CeCu_6$ is maximum around 10 K, decreases steeply with decreasing temperature and changes sign from positive to negative. Other data at lower temperatures show a gradual decrease, and become constant [73]. The low-temperature

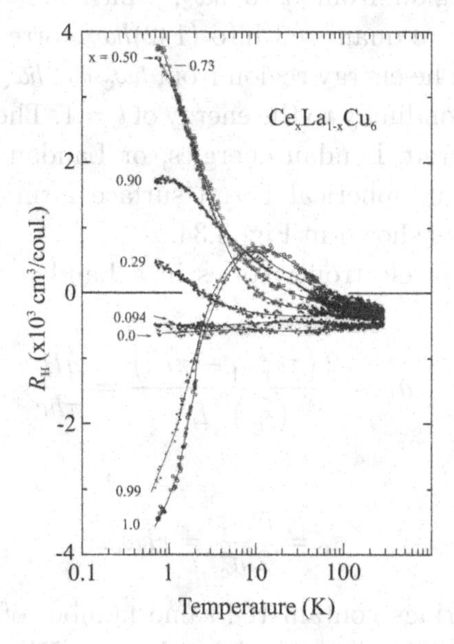

Figure 3.33: Hall coefficient of $\rm Ce_xLa_{1-x}Cu_6$, in the logarithmic scale of temperature, cited from Ref. [72].

Hall coefficient in $CeCu_6$ mainly depends on the carriers with heavy masses.

3.2.12 de Haas-van Alphen effect

We will consider a simple spherical Fermi surface of which the kinetic energy ε is

$$\varepsilon = \frac{\hbar^2 \boldsymbol{k}^2}{2m^*} = \frac{\hbar^2}{2m^*}(k_x^2 + k_y^2 + k_z^2). \tag{3.133}$$

Under a strong magnetic field H along the z-axis, this energy is quantized and forms Landau levels so that

$$\varepsilon = \left(l + \frac{1}{2}\right)\hbar\omega_{\mathrm{c}} + \frac{\hbar^2 k_z^2}{2m^*} \tag{3.134}$$

where the cyclotron frequency $\omega_{\mathrm{c}} = eH/m_{\mathrm{c}}^* c$ and $l = 0, 1, 2, \cdots$. A maximum (or minimum) cross-sectional area is detected in the dHvA experiment. Therefore, the conduction electrons with the $k_z = 0$ plane contribute to the dHvA oscillations. The conduction electrons in the energy region from 0 to $\hbar\omega_{\mathrm{c}}$, which are occupied continuously in energy, are quantized into $(1/2)\hbar\omega_{\mathrm{c}}$ correspondingly to the energy of $l = 0$. The energy region from $\hbar\omega_{\mathrm{c}}$ to $2\hbar\omega_{\mathrm{c}}$ is quantized into $(3/2)\hbar\omega_{\mathrm{c}}$ correspondingly to the energy of $l = 1$. The continuous energies become discrete Landau energies, or Landau rings in k-space. The corresponding spherical Fermi surface form Landau tubes in magnetic fields, as shown in Fig. 3.34.

The number of electronic states in a Landau ring of $l = r$ per unit area is

$$n_r = \frac{2\left(\pi k_{r+1}^2 - \pi k_r^2\right)}{\left(\frac{2\pi}{L}\right)^2 L^2} = \frac{qH}{\pi\hbar c} \tag{3.135}$$

where

$$\varepsilon_r = \frac{\hbar^2 k_r^2}{2m_{\mathrm{c}}^*} = r\hbar\omega_{\mathrm{c}}. \tag{3.136}$$

All the Landau rings contain the same number of electronic states because Eq. (3.135) does not depend on r. We will consider the Landau ring in real space. The conduction electrons with $l = r$

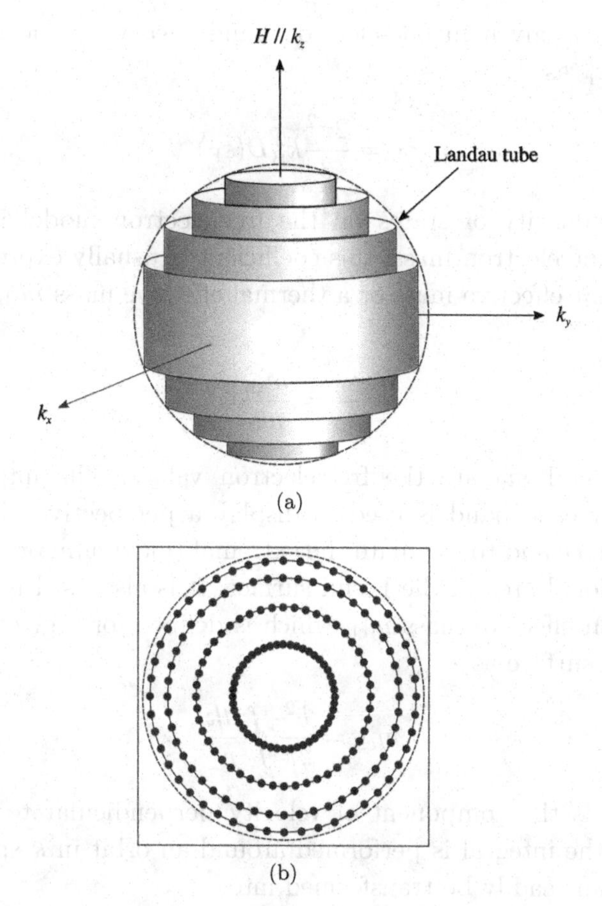

Figure 3.34: (a) Landau tubes and (b) Landau rings (at the cross-section of $k_z = 0$) in magnetic fields.

present the same orbital motion or the same cyclotron motion in real space. An area difference between areas of the cyclotron motion $l = r + 1$ and of $l = r$ in real space, called the flux Φ, is obtained using a relation of $k_r = (q/\hbar c)r_r H$ to be

$$\Phi = H\left(\pi r_{r+1}^2 - \pi r_r^2\right) = \frac{ch}{q} \tag{3.137}$$

which corresponds to a quantum fluxoid in the normal state.

The electronic contribution to the specific heat of a metal at low temperature varies linearly with temperature, $C_{\mathrm{el}} = \gamma_{\mathrm{b}} T$, with the coefficient $\gamma_{\rm b}$ given in one-electron band theory by the density of states at $\varepsilon_{\rm F}$ as

$$\gamma_{\rm b} = \frac{2\pi^2}{3} k_{\rm B}^2 D(\varepsilon_{\rm F}). \tag{3.138}$$

Since the density of states on the free-electron model is proportional to the electron mass, this coefficient is usually expressed by a specific-heat effective mass or a thermal effective mass $m_{\rm tb}$, which is defined by

$$\frac{\gamma_{\rm b}}{\gamma_0} = \frac{m_{\rm tb}}{m_0} \tag{3.139}$$

where γ_0 and m_0 are the free-electron values. The interpolation scheme for each band is used to display a perspective view of the Fermi surface and to calculate the extremal (maximum or minimum) cross-sectional area of the Fermi surface. It is also used to calculate a cyclotron effective mass $m_{\rm cb}$ which is defined for a given orbit on the Fermi surface as

$$m_{\rm cb} = \frac{\hbar^2}{2\pi} \oint \frac{d\boldsymbol{k}}{v_\perp} \tag{3.140}$$

where $v_\perp$ is the component of velocity perpendicular to magnetic field and the integral is performed around an orbit in k space. This formula can readily be transformed into

$$m_{\rm cb} = \frac{\hbar^2}{2\pi} \frac{\partial S(k_{\rm H})}{\partial \varepsilon}\bigg|_{\varepsilon_{\rm F}} \tag{3.141}$$

where $S(k_{\rm H})$ is the cross-sectional area of the Fermi surface which is perpendicular to the field, and $k_{\rm H}$ is the wave number along the field direction. In the dHvA experiment, we can determine the cyclotron mass $m_{\rm c}^*$ at the extremal cross-sectional area, which is compared with the band mass $m_{\rm cb}$ 'or simply, $m_{\rm b}$'.

The experimental values for the low-temperature electronic specific heat coefficient γ and equivalently the thermal effective mass $m_{\rm t}^*$ are usually larger than the theoretical values of $\gamma_{\rm b}$ and $m_{\rm b}$, the latter of which are determined by energy band calculations, as mentioned above. The mass enhancement factor λ is defined by the ratio

$$\frac{\gamma}{\gamma_{\mathrm{b}}} = \frac{m_{\mathrm{t}}^*}{m_{\mathrm{b}}} = 1 + \lambda. \tag{3.142}$$

The existence of λ is ascribed to the many-body effects which cannot be taken into account in the usual band theory. The electron-phonon interaction and the magnetic interaction are considered as the most probable origins of these many-body effects, and their contributions are denoted by λ_{p} and λ_{m}, respectively. Therefore, λ is expressed as a sum of two contributions

$$\lambda = \lambda_{\mathrm{p}} + \lambda_{\mathrm{m}}. \tag{3.143}$$

The experimental cyclotron effective mass m_{c}^* is also usually larger than the theoretical value m_{b} defined in Eq. (3.141). Therefore, the enhancement factor for the cyclotron effective mass can be defined in the same way

$$\frac{m_{\mathrm{c}}^*}{m_{\mathrm{b}}} = 1 + \lambda. \tag{3.144}$$

It should be noted, however, that the magnitude of this enhancement factor may differ from orbit to orbit on the Fermi surface. In Chaps. 5 and 6, we present values of $\gamma/\gamma_{\mathrm{b}}$ and $m_{\mathrm{c}}^*/m_{\mathrm{b}}$ for many compounds. The magnetic contribution λ_{m} can take a huge value in heavy fermion compounds. For example, it amounts to more than 100 in $CeCu_6$.

Under a high magnetic field, the orbital motion of the conduction electron is quantized and forms Landau levels [13, 74]. Therefore, various physical qualities show a periodic variation with H^{-1} since increasing the field strength H causes a sharp change in the free energy of the electron system when the Landau level crosses the Fermi energy. In a three-dimensional system, this sharp structure is observed at extremal areas in $\boldsymbol{k}$-space, perpendicular to the field direction and enclosed by the Fermi energy because the density of states also becomes extremal (maximum or minimum). From the field and temperature dependences of various physical quantities, we can obtain the extremal area of the Fermi surface S_{F}, the cyclotron mass m_{c}^* and the scattering lifetime τ for this cyclotron orbit. The magnetization or the magnetic susceptibility is the most common of these physical quantities, and its periodic character is the de Haas–van Alphen (dHvA) effect. It provides one of the best tools for the investigation of Fermi surfaces of metals.

The theoretical expression for the oscillatory component of magnetization $M_{\rm osc}$ due to the conduction electrons was given by Lifshitz and Kosevich as follows [74]:

$$M_{\rm osc} = \sum_r \sum_i \frac{(-1)^r}{r^{3/2}} A_i \sin\left(\frac{2\pi r F_i}{H} + \beta_i\right) \tag{3.145a}$$

$$A_i \propto F H^{1/2} \left|\frac{\partial^2 S_i}{\partial {k_{\rm H}}^2}\right|^{-1/2} R_{\rm T} R_{\rm D} R_{\rm S} \tag{3.145b}$$

$$R_{\rm T} = \frac{\alpha r m^*_{{\rm c}i} T/H}{\sinh(\alpha r m^*_{{\rm c}i} T/H)} \tag{3.145c}$$

$$R_{\rm D} = \exp(-\alpha r m^*_{{\rm c}i} T_{\rm D}/H) \tag{3.145d}$$

$$R_{\rm S} = \cos(\pi g_i r m^*_{{\rm c}i}/2m_0) \tag{3.145e}$$

$$\alpha = \frac{2\pi^2 k_{\rm B}}{e\hbar}. \tag{3.145f}$$

Here, the magnetization is periodic on $1/H$ and has a dHvA frequency F_i

$$\begin{aligned} F_i &= \frac{\hbar}{2\pi e} S_i \\ &= 1.05 \times 10^{-12}\ [\mathrm{T} \cdot \mathrm{cm}^2] \cdot S_i \\ &= 1.05 \times 10^{-8}\ [\mathrm{Oe} \cdot \mathrm{cm}^2] \cdot S_i \end{aligned} \tag{3.146}$$

which is directly proportional to the i-th extremal cross-sectional area S_i ($i = 1, \ldots, n$). The extremal area indicates a gray plane in Fig. 3.35, where there is one extremal area in a spherical Fermi surface. The factor $R_{\rm T}$ in the amplitude A_i in Eq. (3.145b) is related to the thermal damping at a finite temperature T. The factor $R_{\rm D}$ is also related to the Landau level broadening $k_{\rm B}T_{\rm D}$. Here, $T_{\rm D}$ is due to both the lifetime broadening and inhomogeneous broadening caused by impurities, crystalline imperfections or strains. The factor $T_{\rm D}$ is

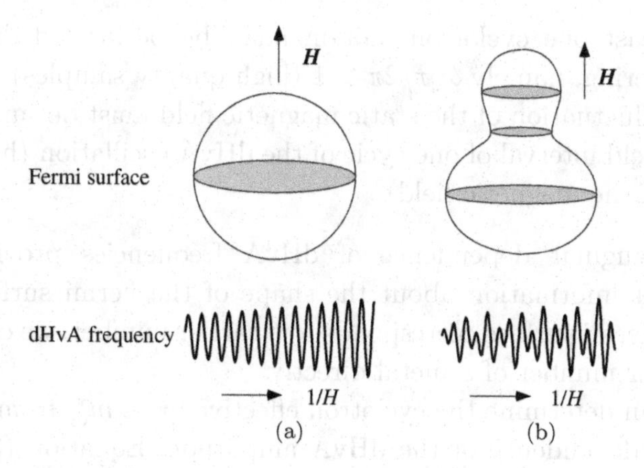

Figure 3.35: Simulations of the cross-sectional area and its dHvA signal for a simple Fermi surface. There is one dHvA frequency in (a), while there are three different frequencies in (b).

called the Dingle temperature and is given by

$$\begin{aligned} T_{\mathrm{D}} &= \frac{\hbar}{2\pi k_{\mathrm{B}}}\tau^{-1} \\ &= 1.22 \times 10^{-12}\ [\mathrm{K}\cdot\mathrm{sec}]\cdot\tau^{-1}. \end{aligned} \tag{3.147}$$

The factor R_{S} is called the spin factor and is related to the difference of phase between the Landau levels due to Zeeman splitting. When $g_i = 2$ (a free electron value) and $m_{\mathrm{c}}^* = 0.5\ m_0$ (m_0: rest mass of an electron), this term becomes zero for $r = 1$. The fundamental oscillation vanishes for all values of the field. This is called the zero spin splitting situation in which the up and down spin contributions to the oscillation are canceled out. This can be useful for determining the value of g_i. Note that in the second harmonics for $r = 2$, the dHvA oscillation should show a full amplitude. The quantity $|\partial^2 S/\partial {k_{\mathrm{H}}}^2|^{-1/2}$ is called the curvature factor. The rapid change of cross-sectional area around the extremal area along the field direction diminishes the dHvA amplitude for this extremal area.

The detectable conditions of dHvA effect are as follows:

(1) The distance between the Landau levels $\hbar\omega_{\mathrm{c}}$ must be larger than the thermal broadening width $k_{\mathrm{B}}T$: $\hbar\omega_{\mathrm{c}} \gg k_{\mathrm{B}}T$ ($\omega_{\mathrm{c}} = eH/m_{\mathrm{c}}^* c$, high fields, low temperatures).

(2) At least one cyclotron motion must be performed during the scattering, namely $\omega_c\tau/2\pi > 1$ (high quality samples).
(3) The fluctuation of the static magnetic field must be smaller than the field interval of one cycle of the dHvA oscillation (homogeneity of the magnetic field).

The angular dependence of dHvA frequencies provides very important information about the shape of the Fermi surface. As a value of Fermi surface corresponds to a carrier number, we can obtain the carrier number of a metal directly.

We can determine the cyclotron effective mass m^*_{ci} from the temperature dependence of the dHvA amplitude. Equation (3.145c) is transformed into

$$\begin{aligned}\log\left\{A_i\left[1-\exp\left(\frac{-2\alpha m^*_{ci}T}{H}\right)\right]/T\right\} \\ = \frac{-\alpha m^*_{ci}}{H}T + \text{const.}\end{aligned} \tag{3.148}$$

Therefore, from the slope of a plot of $\log\{A_i[1-\exp(-2\lambda m^*_{ci}T/H)]/T\}$ vs T at constant field H (what is called a mass plot), the cyclotron effective mass can be obtained, which is later shown in Fig. 5.21(a).

Let us consider the relation between the cyclotron mass and the electronic specific heat coefficient γ. Using density of states $D(\varepsilon_F)$, γ is written as

$$\gamma = \frac{2\pi^2}{3}{k_B}^2 D(\varepsilon_F). \tag{3.149}$$

In the spherical Fermi surface, using $\varepsilon_F = \hbar^2 {k_F}^2/2m^*_c$, we obtain

$$\begin{aligned}\gamma &= \frac{\pi^2}{3}{k_B}^2\frac{V}{2\pi^2}\left(\frac{2m^*_c}{\hbar^2}\right)^{3/2}{\varepsilon_F}^{1/2} \\ &= \frac{{k_B}^2 V}{3\hbar^2}m^*_c k_F\end{aligned} \tag{3.150}$$

where V is the molar volume and $k_F = (S_F/\pi)^{1/2}$. We obtain from Eq. (3.146)

$$\begin{aligned}\gamma &= \frac{k_B{}^2 m_0}{3\hbar^2}\left(\frac{2e}{\hbar}\right)^{1/2} V\frac{m_c^*}{m_0}F^{1/2} \\ &= 2.87 \times 10^{-4}\ [(\mathrm{mJ/K^2 \cdot mol})(\mathrm{mol/cm^3})\mathrm{T}^{-1/2}] \cdot V\frac{m_c^*}{m_0}F^{1/2} \\ &= 2.87 \times 10^{-2}\ [(\mathrm{mJ/K^2 \cdot mol})(\mathrm{mol/cm^3})\mathrm{Oe}^{-1/2}] \cdot V\frac{m_c^*}{m_0}F^{1/2}.\end{aligned} \tag{3.151}$$

In the case of the cylindrical Fermi surface

$$\begin{aligned}\gamma &= \frac{\pi^2}{3}k_B{}^2\frac{V}{2\pi^2\hbar^2}m_c^*k_z \\ &= \frac{k_B{}^2 V}{6\hbar^2}m_c^*k_z\end{aligned} \tag{3.152}$$

where the Fermi wave number k_z is parallel to an axial direction of the cylinder. If we simply regard the Fermi surfaces approximately as a sphere, an ellipse or a cylinder, we can calculate their γ values.

We can determine the Dingle temperature T_D from measuring the field dependence of a dHvA amplitude. Equations (3.145b)–(3.145d) yield

$$\begin{aligned}&\log\left\{A_i H^{1/2}\left[1-\exp\left(\frac{-2\lambda m_{ci}^* T}{H}\right)\right]\right\} \\ &\quad = -\lambda m_{ci}^*(T+T_D)\frac{1}{H} + \text{const.}\end{aligned} \tag{3.153}$$

From the slope of a plot of $\log\{A_i H^{1/2}[1-\exp(-2\lambda m_{ci}^* T/H)]\}$ vs $1/H$ at constant T (called a Dingle plot), T_D can be obtained. Here, the cyclotron effective mass must have been already obtained, which is later shown in Fig. 5.21(b).

We can estimate the mean free path l or the scattering lifetime τ from the Dingle temperature. The relation between an effective mass and lifetime takes the form

$$\hbar k_{\mathrm{F}} = m^* v_{\mathrm{F}} \tag{3.154}$$

$$l = v_{\mathrm{F}} \tau. \tag{3.155}$$

Then Eq. (3.147) is transformed into

$$l = \frac{\hbar^2 k_{\mathrm{F}}}{2\pi k_{\mathrm{B}} m_{\mathrm{c}}^* T_{\mathrm{D}}}. \tag{3.156}$$

When the extremal area can be regarded as an approximate circle, using Eq. (3.146), the mean free path is expressed as

$$\begin{aligned} l &= \frac{\hbar^2}{2\pi k_{\mathrm{B}} m_0} \left(\frac{2e}{\hbar c}\right)^{1/2} F^{1/2} \left(\frac{m_{\mathrm{c}}^*}{m_0}\right)^{-1} T_{\mathrm{D}}{}^{-1} \\ &= 77.6\ [\text{Å} \cdot \mathrm{T}^{-1/2} \cdot \mathrm{K}] \cdot F^{1/2} \left(\frac{m_{\mathrm{c}}^*}{m_0}\right)^{-1} T_{\mathrm{D}}{}^{-1} \\ &= 77.6 \times 10^2\ [\text{Å} \cdot \mathrm{Oe}^{-1/2} \cdot \mathrm{K}] \cdot F^{1/2} \left(\frac{m_{\mathrm{c}}^*}{m_0}\right)^{-1} T_{\mathrm{D}}{}^{-1}. \end{aligned} \tag{3.157}$$

The experiment of the dHvA effect is constructed by using the usual AC-susceptibility field modulation method. Now, we give an outline of the field modulation method with a pick-up coil dHvA system.

A small AC-field, $h_0 \cos \omega t$, is varied on an external field H_0 ($H_0 \gg h_0$) in order to obtain the periodic variation of the magnetic moment M_{osc}. The sample is set up into a pair of balanced coils (pick up and compensation coils), as shown in Fig. 3.36. An induced emf (electromotive force) V_{osc} will be proportional to $\mathrm{d}M_{\mathrm{osc}}/\mathrm{d}t$:

$$\begin{aligned} V_{\mathrm{osc}} &= c \frac{\mathrm{d}M_{\mathrm{osc}}}{\mathrm{d}t} \\ &= c \frac{\mathrm{d}M_{\mathrm{osc}}}{\mathrm{d}H} \frac{\mathrm{d}H}{\mathrm{d}t} \\ &= -c h_0 \omega \sin \omega t \sum_{k=1}^{\infty} \frac{h_0^k}{2^{k-1}(k-1)!} \left(\frac{\mathrm{d}^k M_{\mathrm{osc}}}{\mathrm{d}H^k}\right)_{H_0} \sin k\omega t \end{aligned} \tag{3.158}$$

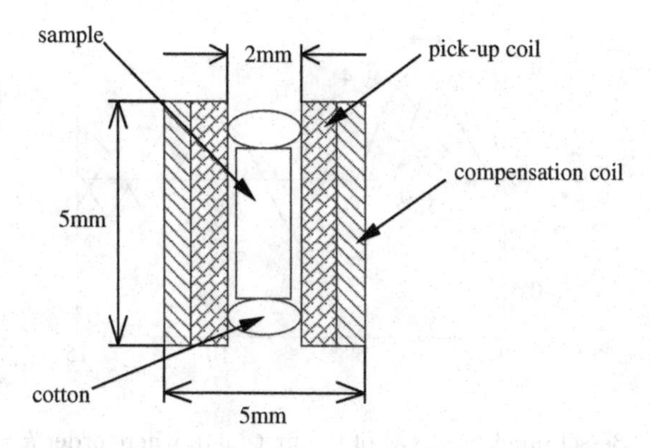

Figure 3.36: Detecting coil and the sample location.

where c is constant and is fixed by variables such as the number of turns in the coil and so on. The higher differential terms of the coefficient of $\sin k\omega t$ are neglected. Calculating $\mathrm{d}^k M/\mathrm{d}H^k$, it becomes

$$V_{\rm osc} = -c\omega A \sum_{k=1}^{\infty} \frac{1}{2^{k-1}(k-1)!} \left(\frac{2\pi h_0}{\Delta H}\right)^k \times \sin\left(\frac{2\pi F}{H} + \beta - \frac{k\pi}{2}\right) \sin k\omega t. \tag{3.159}$$

Here, $\Delta H = H^2/F$. Considering ${h_0}^2 \ll {H_0}^2$, the time dependence of magnetization $M(t)$ is given by

$$M_{\rm osc}(t) = A\left[J_0(\lambda)\sin\left(\frac{2\pi F}{H_0} + \beta\right) + 2\sum_{k=1}^{\infty} kJ_k(\lambda)\cos k\omega t \, \sin\left(\frac{2\pi F}{H_0} + \beta - \frac{k\pi}{2}\right)\right] \tag{3.160}$$

where

$$\lambda = \frac{2\pi F h_0}{{H_0}^2}. \tag{3.161}$$

Here, J_k is k-th Bessel function. Figure 3.37 shows the Bessel function of the first kind for various orders of k. Finally, we can obtain the

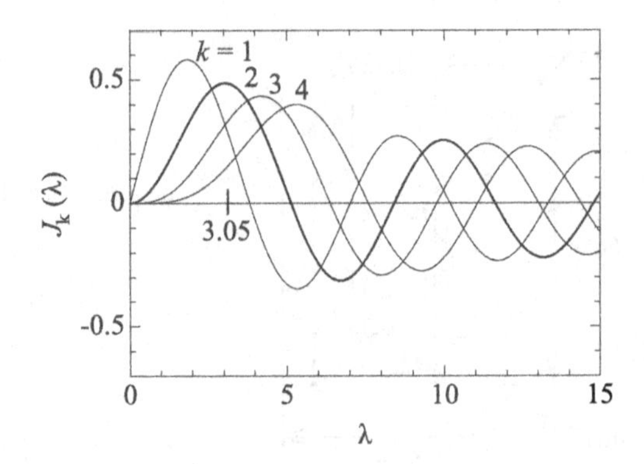

Figure 3.37: Bessel function $J_k(\lambda)$ of the first kind, where order $k = 2$ is usually used.

output emf as

$$V_{\text{osc}} = c\left(\frac{\mathrm{d}M}{\mathrm{d}t}\right) = -2c\omega A\sum_{k=1}^{\infty} kJ_k(\lambda)\sin\left(\frac{2\pi F}{H_0} + \beta - \frac{k\pi}{2}\right)\sin k\omega t. \tag{3.162}$$

The signal is usually detected at the second harmonic of the modulation frequency 2ω using a lock-in amplifier, since this condition may cut off the offset magnetization and then detect the component of the quantum oscillation only. Thus, Eq. (3.162) becomes

$$V_{\text{osc}} = -4c\omega AJ_2(\lambda)\sin\left(\frac{2\pi F}{H_0} + \beta\right)\sin 2\omega t. \tag{3.163}$$

Here, we summarize Eq. (3.163) as follows:

$$V_{\text{osc}} = A\left|\frac{\partial^2 S_{\text{F}}(k_z)}{\partial k_z^2}\right|^{-1/2} R_{\text{T}}R_{\text{D}}R_{\text{S}}\sin\left(\frac{2\pi F}{H} + \beta\right) \tag{3.164}$$

$$A \propto \omega J_2(x)H^{1/2} \tag{3.165}$$

$$R_{\text{T}} = \frac{2\alpha m_{\text{c}}^* T/H}{\sinh(2\alpha m_{\text{c}}^* T/H)}$$

$$R_{\text{D}} = \exp(-\alpha m_{\text{c}}^* T_{\text{D}}/H)$$

$$R_{\text{S}} = \cos(\pi m_{\text{c}}^* g/2m_0)$$

$$\alpha = 2\pi^2 ck_{\mathrm{B}}/e\hbar.$$

$$x = \frac{2\pi F h}{H^2}.$$

The value of A in Eq. (3.165) depends on the detection method. We usually choose the modulation field h_0 to maximize the value of $J_2(\lambda)$ so that $\lambda = 3.05$, as shown in Fig. 3.37.

The Fermi surface was observed by the dHvA experiment in magnetic fields up to 190 kOe, at low temperatures which go down to 20 mK in $CeRu_2Si_2$ with the $ThCr_2Si_2$-type tetragonal structure (No. 139, a = 4.185 Å, c = 9.794 Å, Z = 2) [75, 76], as shown in Fig. 3.38. Figure 3.39 shows the dHvA oscillations for $H \parallel [100]$ and the corresponding fast Fourier transformation (FFT) spectrum. Figures 3.40(a) and (b) show the angular dependence of the dHvA frequency, which was obtained below H_{m}, and the results of the energy band calculation under the assumption that $4f$-electrons in

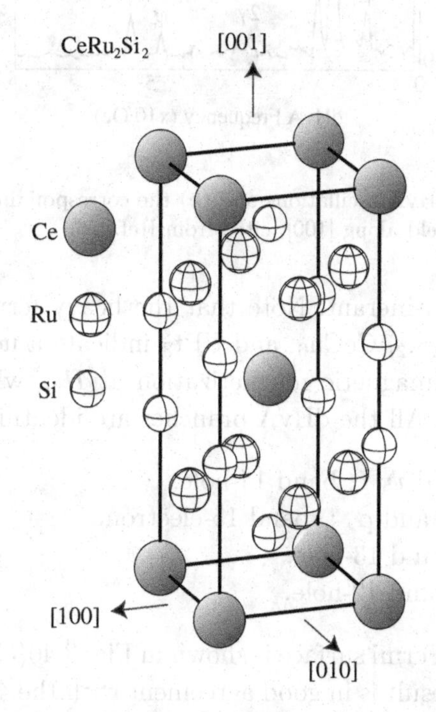

Figure 3.38: $ThCr_2Si_2$-type tetragonal crystal structure of $CeRu_2Si_2$.

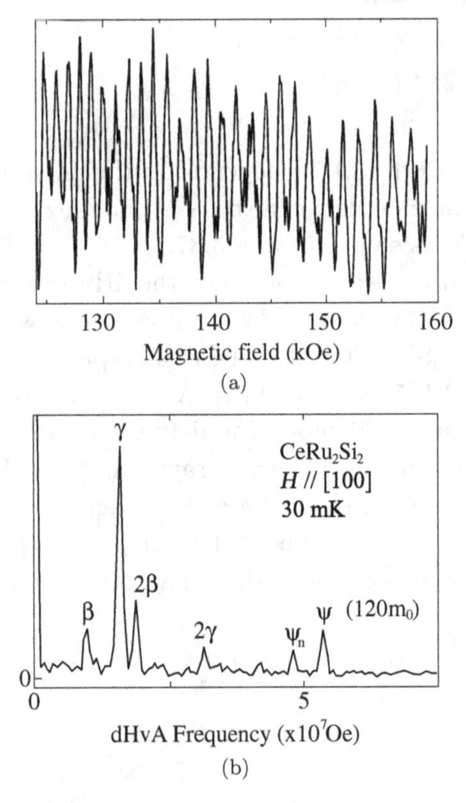

Figure 3.39: (a) dHvA oscillations and (b) the corresponding FFT spectrum of $CeRu_2Si_2$ for the field along [100], cited from Ref. [76].

the Ce ions are itinerant. Note that the heavy fermion compounds including $CeRu_2Si_2$, $CeCu_6$, and UPt_3 indicate a non-linear magnetization or metamagnetic magnetization at H_{m}, which is described later in Sec. 5.3. All the dHvA branches are identified as follows:

branches ψ and ψ_{n} : band 14-hole.
branches κ, ε and α : band 15-electron.
branch γ : band 13-hole.
branch β : band 12-hole.

The theoretical Fermi surface is shown in Fig. 3.40(c) [75]. The experimental dHvA result is in good agreement with the $4f$-itinerant band model, although the calculated Fermi surfaces have a slightly larger

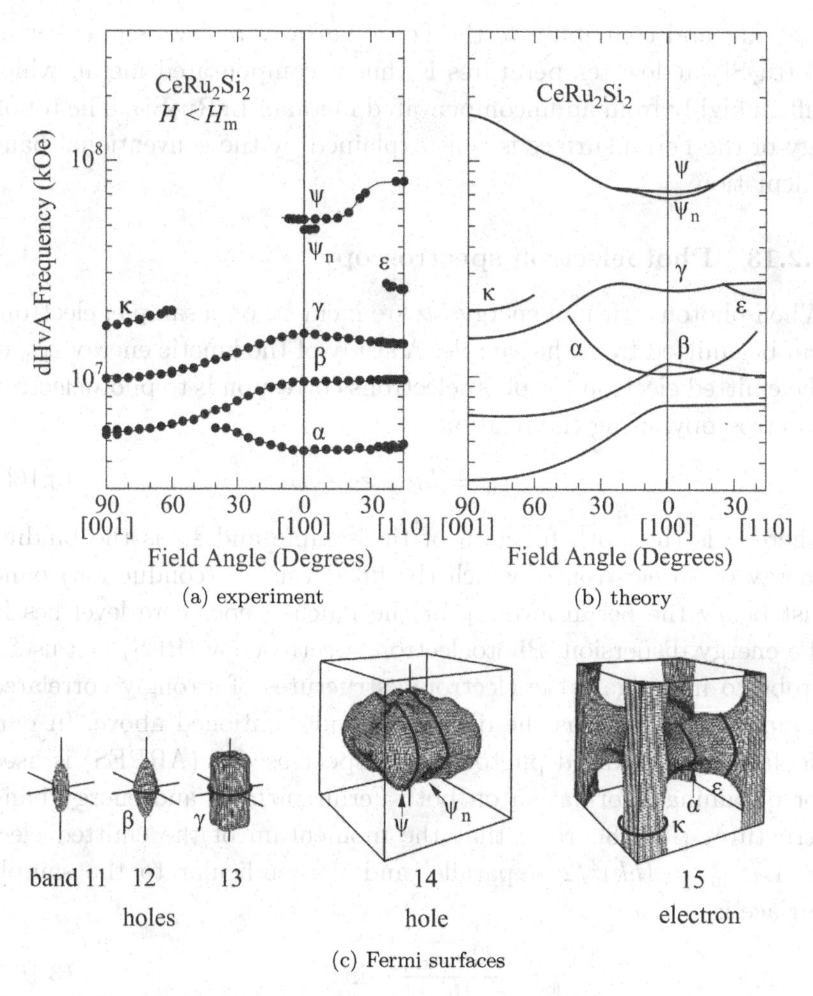

Figure 3.40: (a) Angular dependence of the dHvA frequency below $H_{\rm m}$ and (b) the result of the $4f$-itinerant band calculation of $CeRu_2Si_2$, together with (c) theoretical Fermi surfaces, cited from Refs. [75, 76].

volume than the experimental surfaces. The branches ψ and $\psi_{\rm n}$ have an extremely large cyclotron mass of $120\, m_0$ (m_0: rest mass of an electron). The cyclotron mass is about 60 times larger than the corresponding band mass because the many-body Kondo effect is not included in the usual band calculations. The experimental results are in good agreement with theory indicating that the $4f$ electrons are itinerant and contribute to the Fermi surfaces at low temperatures. $CeRu_2Si_2$ at low temperatures is thus a compensated metal, which differs highly from an uncompensated metal of $LaRu_2Si_2$. The topology of the Fermi surface is thus explained by the conventional band calculation.

3.2.13 Photoelectron spectroscopy

When photons with an energy $h\nu$ are incident on a sample, electrons can be emitted from the sample. A study of the kinetic energy ε_{kin} of the emitted electrons or photoelectrons corresponds to photoelectron spectroscopy, using the relation

$$\varepsilon_{\text{kin}} = h\nu - \varepsilon_{\text{B}} - \phi \tag{3.166}$$

where ϕ is the work function of the sample and ε_{B} is the binding energy of an electron of which the filled valence (conduction) band just below the Fermi level ε_{F} or the much deeper core level lies in the energy dispersion. Photoelectron spectroscopy (PES) is a useful probe to investigate the electronic structures of strongly correlated compounds as well as the dHvA method mentioned above. In particular, angle-resolved photoelectron spectroscopy (ARPES) is used for obtaining information on both Fermi surfaces and energy band structures [77, 78]. Note that the momentum of the emitted electron $[\varepsilon_{\text{kin}} = (\hbar k)^2/2m]$ parallel and perpendicular to the sample surface is

$$k_{\parallel} = \frac{\sqrt{2m\varepsilon_{\text{kin}}}}{\hbar} \cdot \sin\theta \tag{3.167}$$

$$k_{\perp} = \sqrt{\frac{2m}{\hbar^2}\left(\varepsilon_{\text{kin}}\cos^2\theta + V_0\right)} \tag{3.168}$$

where V_0 is the inner potential, for example $V_0 = 12\,\text{eV}$. In the ARPES experiment, a wide area of $k_{\parallel}$ is covered in the cross-section of the Brillouin zone by changing the kinetic energy of the photoelectrons or the incident photon energy $h\nu$ when the detection angle θ is fixed.

Here, incident light sources are classified into laboratory-based light sources and synchrotron radiations. Among laboratory-based

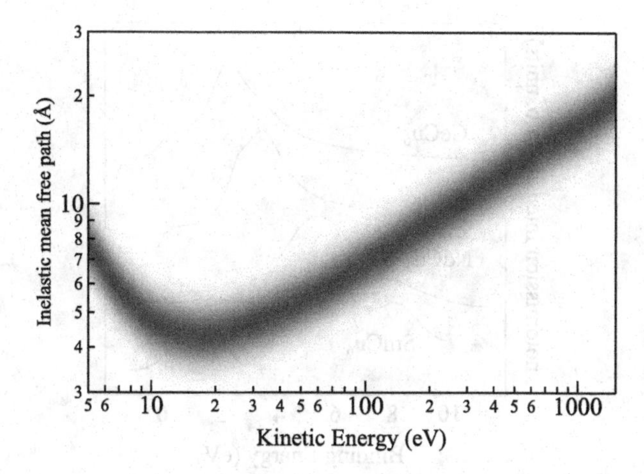

Figure 3.41: Inelastic mean free path of electrons, courtesy of S. Fujimori, cited from Refs. [78, 79].

light sources, He I ($h\nu = 21.2\,\text{eV}$) from a He discharge lamp has been widely used, together with He II ($h\nu = 40.8\,\text{eV}$), a Xe discharge lamp ($h\nu = 8.4\,\text{eV}$), and a laser-based system ($h\nu = 7\,\text{eV}$) of which the resolution is less than 100 μeV. On the other hand, synchrotrons are very intense sources of radiation over an extremely wide energy range. Soft X-rays ($h\nu = 400{-}1200\,\text{eV}$) are usually employed as incident photons, with a resolution of 100 meV.

PES experiments are essentially surface-sensitive. This is because the electrons collected in PES experiments mostly come from a 5–20 Å surface region of the sample [79], as shown in Fig. 3.41. One needs to consider the bulk properties of the electronic structure, as well as contributions from surface layers of the sample appropriately. Usually, clean sample surfaces are used, achieved by clearing the sample in situ under an ultra high vacuum condition of less than 10^{-10} torr.

Figure 3.42 shows the difference spectra at room temperature of $CeCu_6$, $PrCu_6$, $NdCu_6$, and $SmCu_6$ using a technique involving $4d$–$4f$ resonance photoemission [80]. Ce-$4f$ electrons can be directly excited as photoelectrons, but another technique is used. If the incident photon energy is tuned to the Ce-$4d$ absorption energy, a transition from Ce-$4d$ to Ce-$4f$ orbitals, followed by Auger-like

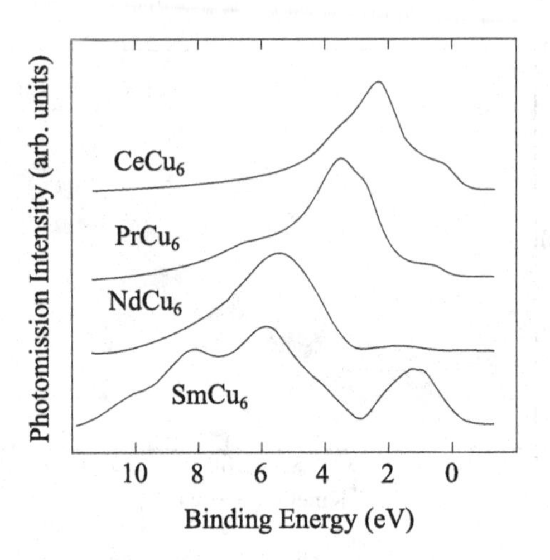

Figure 3.42: Difference spectra of $CeCu_6$, $PrCu_6$, $NdCu_6$, and $SmCu_6$ at room temperature, cited from Ref. [80].

decay, is realized. Namely, as the incident photon energy approaches the Ce-4d absorption edge ($h\nu = 122\,\text{eV}$), the contributions from Ce-4f state are significantly enhanced. The difference between the on-resonance spectrum at $h\nu = 122$ eV and the off-resonance spectrum at $h\nu = 101\,\text{eV}$ gives an energy distribution of the 4f partial density of states in $CeCu_6$, as shown in Fig. 3.42. The 4f spectrum has a main peak at 2.3 eV with weak features on both sides of it. The feature on the low binding energy side indicates that an appreciable amount of the 4f character exists near the Fermi level, consistent with the heavy fermion nature of $CeCu_6$. The feature on the high binding energy side indicates that an appreciable mixing between the Cu-3d states and the Ce-4f states occurs in $CeCu_6$.

The energy distributions of the 4f partial densities of states are also obtained for the other RCu_6 compounds using the difference spectra between the on-resonance spectra and the off-resonance spectra, as shown in Fig. 3.42. The main peak of the R-4f spectrum is shifted to a higher binding energy successively as R is changed from Ce to Sm in RCu_6. This is consistent with the general trend of the binding energy of the 4f level in the light lanthanide. The 4f-partial density of states near the Fermi level decreases successively in the order $CeCu_6$, $PrCu_6$, and $NdCu_6$. In $NdCu_6$, the 4f component does

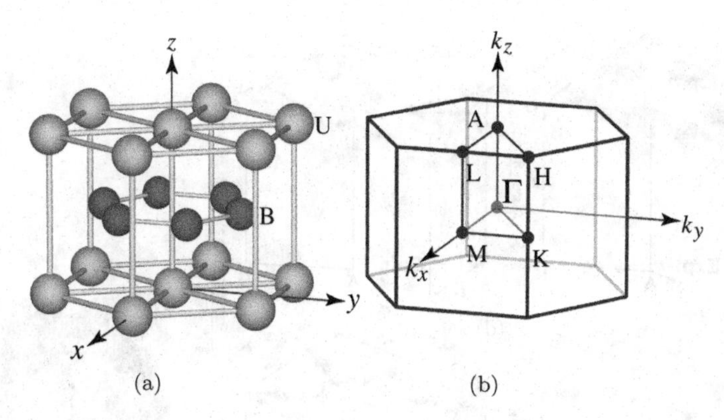

Figure 3.43: (a) Hexagonal crystal structure and (b) the corresponding Brillouin zone of a Pauli paramagnet UB_2, courtesy of S. Fujimori.

not exist at the Fermi edge, meaning that the $4f$ electrons are completely localized. In the $4f$ spectrum of $SmCu_6$, a feature with a weak fine structure is found near the Fermi edge, which is ascribed to the $4f$-contribution of Sm^{2+}. Note that a main peak around 6–8 eV is due to Sm^{3+} in $SmCu_6$. The apparent concentration of Sm^{2+} ions is large, roughly estimated to be 22% in $SmCu_6$. There is a possibility that the Sm^{2+} ions are enhanced in number in the surface region.

Next, we show in Figs. 3.43(a) and (b) the hexagonal crystal structure and the corresponding Brillouin zone in a Pauli paramagnet UB_2. The energy band structure and the Fermi surface are obtained in the ARPES experiment [78]. The spectra, measured at $h\nu = 500$ eV within the A-H-L high-symmetry plane, are shown in Fig. 3.44(a). The spectra show clear energy dispersions along all the high-symmetry lines. The relatively weakly dispersive bands in the vicinity of E_F correspond to the quasi-particle bands with a large contribution of the U-$5f$ states. On the other hand, the strongly dispersive bands at high binding energies ($\varepsilon_b \geq 1$ eV) correspond to bands with large contributions of the B-$2s$ and -$2p$ states. The dispersive nature of the U-$5f$ states is clearly recognized, particularly in the vicinity of ε_F, indicating that the U-$5f$ electrons have a highly itinerant character. Some characteristic features exist in the experimental Fermi surfaces. Within the A-H-L high-symmetry plane, a triangular feature centered at the H point is observed, as shown in Fig. 3.44(a)

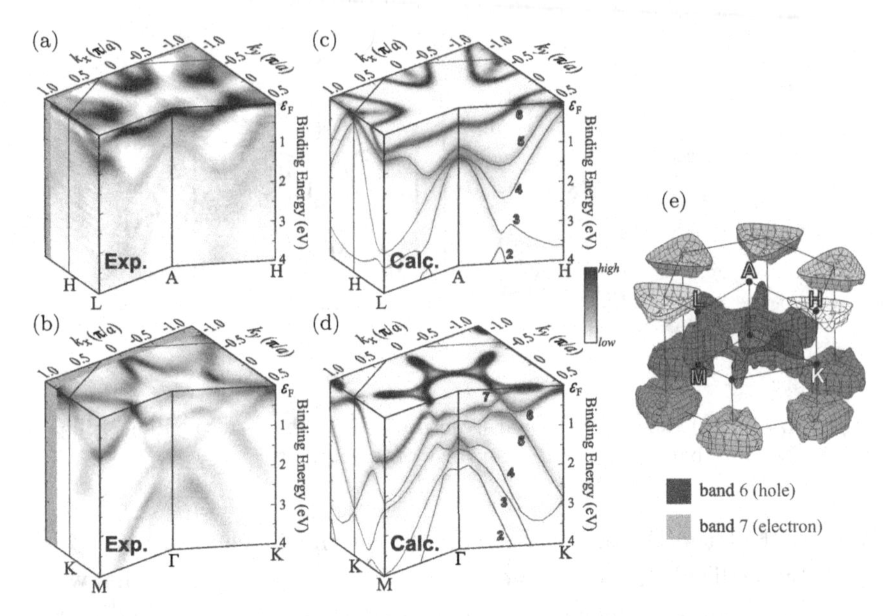

Figure 3.44: Experimental and calculated Fermi surfaces and band structures of UB_2 with the hexagonal sturucture. (a) Experimental Fermi surface mapping and band structure within the A-H-L high-symmetry plane measured at $h\nu = 500\,\text{eV}$. (b) Same representation within the Γ-K-M high-symmetry plane measured at $h\nu = 450\,\text{eV}$. (c) and (d) Results of the band structure calculation for the ARPES spectral functions. (e) Theoretical Fermi surfaces, courtesy of S. Fujimori, cited from Ref. [78].

which has a hole-like dispersion, as seen in the band structure along the A-H high symmetry line. Within the Γ-K-M plane, a star-like feature is realized around the Γ point, as seen in Fig. 3.44(b). These Fermi surfaces and band structures are well explained by the results of the relativistic LAPW band calculations, treating all the U-5f electrons as itinerant, as shown in Figs. 3.44(c) and (d). UB_2 is a compensated metal, where the band 6 forms the hole Fermi surface centered at the H point, and the band 7 forms the star-like electron Fermi surface, as shown in Fig. 3.44(e). The sizes of the Fermi surfaces are slightly different between experiment and theory, but it could be established that the Fermi surface and band structure can be obtained by the ARPES experiment. In the case of heavy fermion compounds, the results of ARPES experiment are roughly or scarcely explained by the $5f$-itinerant band calculations, which originate from the electron correlations.

3.2.14 Pressure experiment

The Kondo temperature T_K is an energy scale in the f electron systems in the range 0–100 K. Therefore, the electronic state is changed by temperature, magnetic field, and pressure. When pressure is introduced to the sample of a Ce compound, for example, an antiferromagnet $CeRhIn_5$, this compound becomes non-magnetic at about 2 GPa, revealing superconductivity. Pressure is thus very useful in controlling the electronic state.

The pressure P is proportional to F/S, where F is the loading force to the sample space and S the corresponding cross-section. There are three kinds of pressure techniques: (1) the piston cylinder type, (2) the multi-anvil type, and (3) the facing anvil type. The first type is convenient, with ample sample space, which is used for all the measurement methods. Here, we describe the electrical resistivity and dHvA measurements using the piston cylinder cell at pressures less than 3 GPa.

One of the main difficulties in the higher-pressure experiments is to obtain sufficiently uniform high pressures. We cannot obtain purely hydrostatic pressures at low temperatures because all the pressure transmitting media, including Dephne oil, Fluorinert, and liquid argon, freeze under pressure when they are cooled to low temperatures. The cubic anvil cell of the second type of technique is designed to attain quasi-hydrostatic pressures up to about 10 GPa [81]. The third type corresponds to indenter-type [82], Bridgman-anvil [83], and diamond-anvil [84] pressure techniques, reaching pressures of 4.5, 7, and 10–100 GPa, respectively. The electrical resistivity, AC-specific heat, and thermoelectric power measurements are characteristically carried out in the modified Bridgman-anvil cell in Ref. [83]. We will explain four kinds of pressure cells in detail.

1) Piston cylinder cell

A cross-sectional view of the clamp-type piston cylinder cell for the electrical resistivity and dHvA measurements is shown

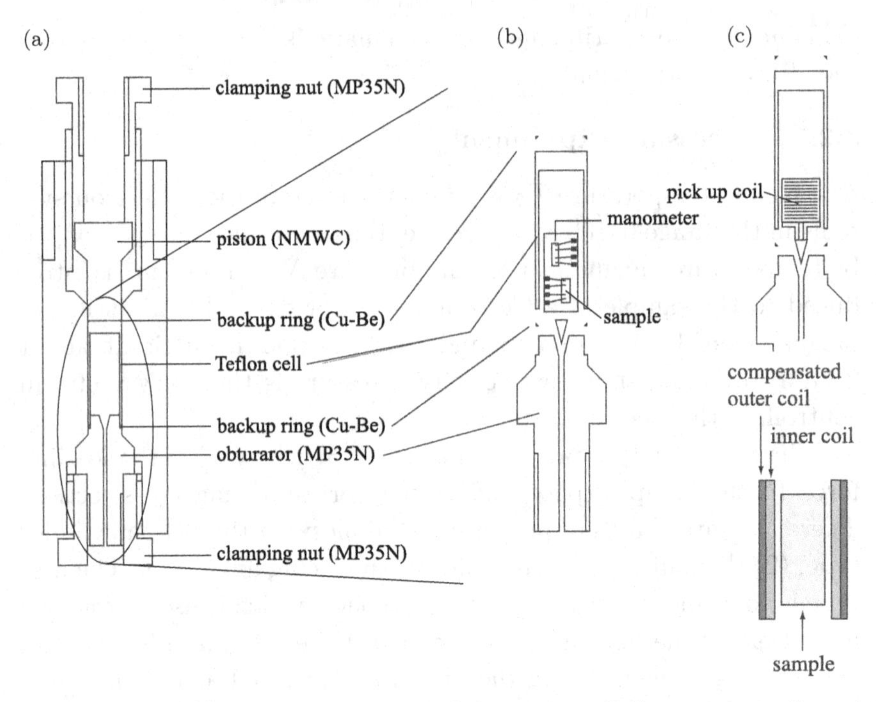

Figure 3.45: (a) Cross-sectional view of the piston-cylinder pressure cell, and setups of (b) the electrical resistivity and (c) dHvA measurements.

in Fig. 3.45(a). The cylinder body, clamping nut, and obturator are made of a nonmagnetic MP35N (nickel-cobalt-chromium-molybdenum) alloy [85]. The inner piston is made of a non-magnetic tungsten carbide (WC). Samples are placed inside a Teflon cell with an inner diameter of 4.2 mm and height of 15 mm, for example. The relatively large sample chamber accommodates a plate for the electrical resistivity measurement or additionally a pick-up coil for the dHvA experiment, as shown in Figs. 3.45(b) and (c), respectively. The electrical leads are introduced into the sample chamber through a hole of the obturator. The hole is sealed by epoxy resin (stycast 2850 FT). The sample chamber is also sealed by the backup rings made of a copper-beryllium (Cu-Be) alloy. A 1 : 1 mixture of Daphne oil (7373) and petroleum ether is used as a pressure-transmitting medium. Pressure inside the cell at low temperatures is calibrated by the superconducting transition temperature of Sn [85]. High pressures up to 3 GPa can be realized using this piston cylinder cell.

Note that the temperature is heated up due to an eddy current of the metallic pressure cell when the modulation field of 80–100 Oe is applied in the dHvA experiment. In the usual dHvA experiment, the modulation frequency 200 Hz is used in the ^{3}He or ^{4}He equipment, but the frequency is reduced to 11 Hz in the dilution refrigerator, and further reduced to 3.5 Hz when the metallic pressure cell is installed. The lowest temperature attained in the dilution refrigerator is increased from 30 mK to 130 mK using the Cu-Be pressure cell and 80 mK in the MP35N pressure cell.

2) Cubic anvil cell

Next, we describe the cubic anvil cell. High pressures up to 8 GPa and 16 GPa are realized using the WC anvil cell and the sintered-diamond anvil cell, respectively, in the temperature range 2–300 K [81]. A characteristic feature in the cubic anvil cell is the maintenance of a stabilized quasi-hydrostatic pressure, which can be realized by applying pressure for a cubic pyrophyllite gasket using six WC anvils, as shown in Fig. 3.46(a). Note that pressure is applied under a constant load. The sample is set in the gasket. A sample with $0.3 \times 0.3 \times 0.8\,\text{mm}^3$ is prepared. Gold wires with 0.02 mm in diameter for the current and voltage leads are contacted with the sample using silver paste (4922N, Dupont). The sample is then set in a Teflon capsule (Teflon cell) of the cubic pyrophyllite gasket, where the pressure medium of Fluorinert is inserted, and the gasket is closed. Note that the four leads of the sample are contacted by gold foil, as shown in Fig. 3.46(b), and the electrical resistivity measurement is carried out.

3) Bridgman anvil cell

Figure 3.47(a) shows the schematic view of a Bridgman-anvil-type pressure cell in order to measure the electrical resistivity. The Bridgman anvil cell consists of two opposite anvil faces, with a sample compressed in between. The cylinder body and clamping nuts are made of MP35N, and anvils are made of a non-magnetic WC. A Cu-Be

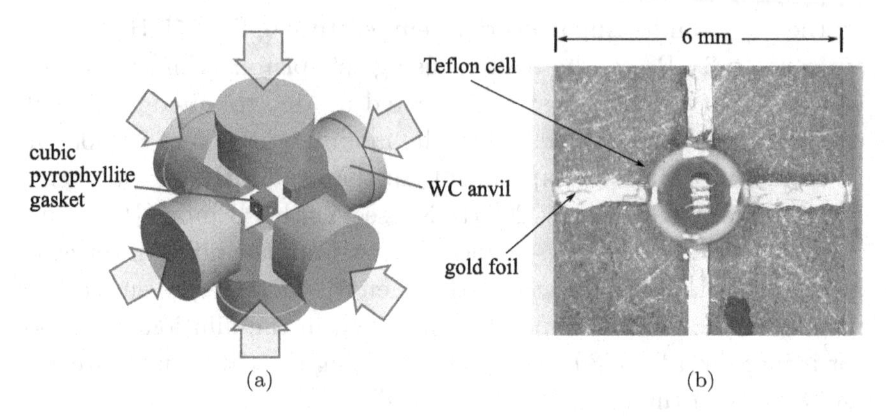

Figure 3.46: (a) Schematic view of the cubic anvil cell and (b) a picture of the sample setting in a Teflon cell of the cubic pyrophyllite gasket.

pressure gasket (inner gasket) is supported by the MP35N outer ring (gasket support) in order to tighten the inner gasket. The size of the anvil top and outer diameter of the inner gasket is 3 mm, for example. The inner diameter and the thickness of the gasket are 1.8 mm and 0.5 mm, respectively. A photograph of a sample setting and schematic diagram are shown in Fig. 3.47(b). Electrical leads are introduced into the sample space through a hole of the WC anvil. Daphne oil (7373) is used as a pressure-transmitting medium. Pressure inside the cell is calibrated by the superconducting transition temperature of Pb [86] and high pressures up to 7 GPa can be obtained.

4) Diamond anvil cell

The main concept of the diamond anvil cell is similar to that of the Bridgman-anvil-type pressure cell. A pair of diamonds is used instead of the non-magnetic WC as anvils. Figures 3.48(a) and (b) are schematic views of the diamond anvil cell. Most of the parts, including bolts and clamping nuts, are made of Cu-Be alloy. A gasket made of rhenium (Re) or stainless steel (SUS310S) is covered with an insulating layer of cubic boron nitride (c-BN) epoxy. The inner diameter of the gasket hole and the thickness of the

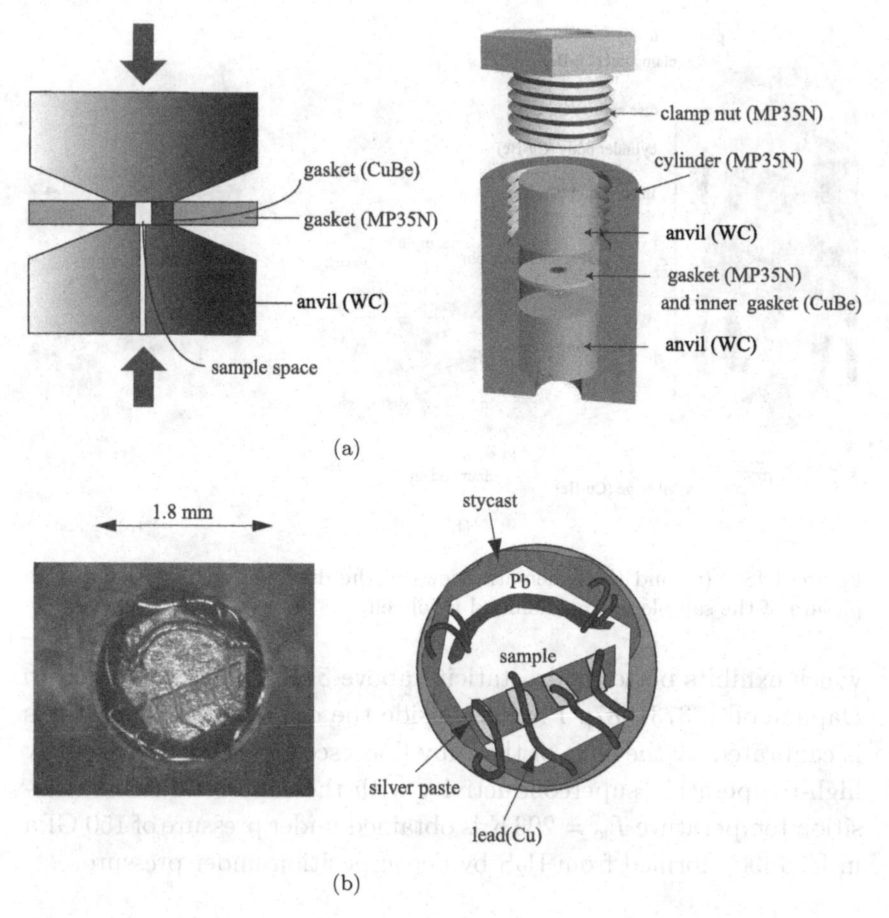

Figure 3.47: (a) Schematic view of a Bridgman anvil cell and (b) top view of the sample setting on the WC anvil.

gasket are about 0.2–0.3 mm 'depending on the culet size of a diamond anvil', and about 0.1 mm, respectively. Due to the small sample space, the typical sample size for the pressure cell is about 0.1 mm × 0.05 mm × 0.03 mm, as shown in Fig. 3.48(c). The electrical leads (Au or Pt foil with the thickness of 10 microns) are introduced through the surface of the gasket with the insulating layer. Several disc springs are used in order to reduce pressure change due to thermal shrinkage at low temperatures. In the experiment, glycerol or sodium chloride (NaCl) is used as a pressure-transmitting medium

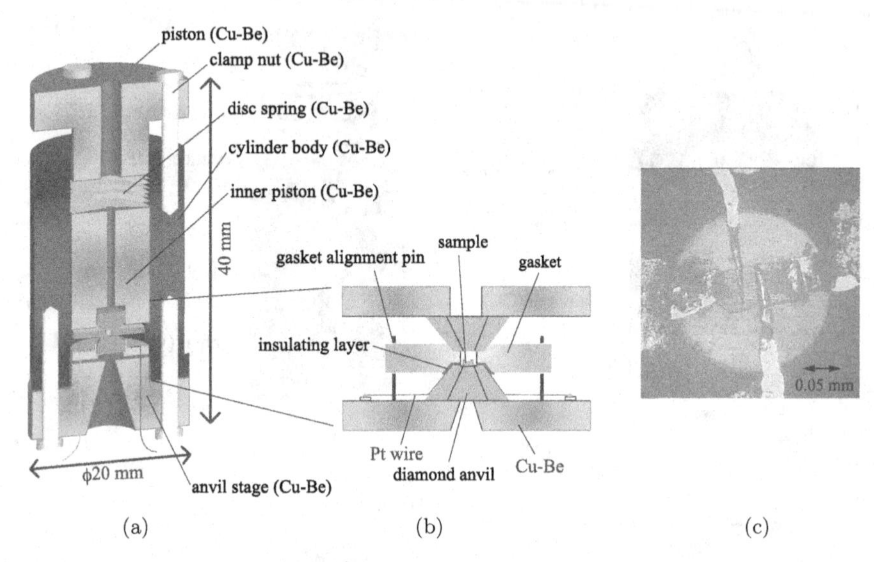

Figure 3.48: (a) and (b) Schematic views of the diamond anvil cell, and (c) a picture of the sample on the diamond anvil cell.

which exhibits better hydrostaticity above 5 GPa when compared to Daphne oil (7373) [87]. Pressure inside the cell at low temperatures is calibrated by the shift of the ruby fluorescence [88, 89]. Extremely high-temperature superconductivity with the superconducting transition temperature $T_{sc} = 203$ K is obtained under pressure of 150 GPa in H_3S [90], formed from H_2S by decomposition under pressure.

Chapter 4

Itinerant 3*d* Electrons

The 3*d* electrons in transition metal compounds are itinerant and become conduction electrons. The Fermi surfaces are thus mainly composed of the 3*d* electrons. Correlations between these conduction electrons are reflected in a relatively large electronic specific heat coefficient and cyclotron effective mass, together with the magnetic moment and magnetic ordering. This is derived from a large partial density of states in the 3*d* electrons. The ordered moment is thus produced from a difference between the spin-up(majority) and spin-down(minority) densities of states containing the 3*d* electrons.

4.1 Itinerant magnetism

Observed magnetic moments in metals containing 3*d* electrons are usually considerably smaller than the theoretical values calculated from the free ions in Table 2.1 and are frequently non-integral. It is thus natural to adopt a band model on which the 3*d* electrons are visualized to be in two energy bands, as schematically shown in Fig. 4.1: one band for electrons with spin up (↑) and the other band for 3*d* electrons with spin down (↓). The bands are separated in energy by the exchange interaction. The magnetic moment is deduced from two different bands.

The 3*d* electron system is often described by the following Hubbard model

$$\mathcal{H} = \sum_{ij\sigma} t_{ij} c_{i\sigma}^{\dagger} c_{j\sigma} + U \sum_{j} n_{j\uparrow} n_{j\downarrow} \tag{4.1}$$

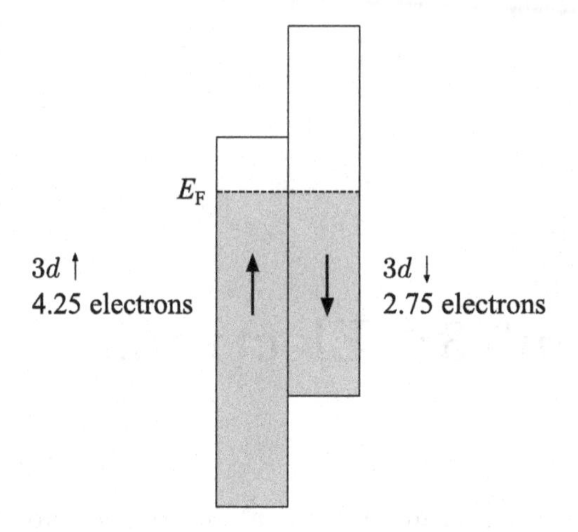

Figure 4.1: Schematic view of $3d$ electrons with a non-integral magnetic moment based on the band model. The ↑ spin(majority) and ↓ spin(minority) bands contain 5 states per atom, respectively. The case of 7 electrons, 4.25 in one band and 2.75 in the other, is shown, which results in the effective magnetic number of 1.5, or the magnetic moment of 1.5 μ_B/atom.

where t_{ij} is a transfer integral, U is a Coulomb potential, $c^{\dagger}_{i\sigma}$ and $c_{i\sigma}$ are annihilation and creation operators of the $3d$ electrons, respectively, and n_j is a number operator denoted as $n_j = c^{\dagger}_j c_j$. The first term can be attributed to the kinetic energy when the electron moves into the neighbor site, which is proportional to a band width W. The second term is the Coulomb repulsive energy when two electrons set on one site. The relative difference between two terms provides the different character for the $3d$ electrons. For example, the $3d$ electrons are localized on the lattice and show a ferro- or antiferromagnetism when U is extremely large. The limit $U/t \gg 1$ provides the Heisenberg Hamiltonian $\mathcal{H} = \sum_{ij} J_{ij} \boldsymbol{S}_i \cdot \boldsymbol{S}_j$, which is a useful model for magnetism of an insulator, where $\boldsymbol{S}$ is the spin of the localized electron and J_{ij} is an exchange interaction between i and j sites.

From the energy band theory and dHvA experiment, however, the $3d$ electrons in Fe and Ni can be described as a one-electron band picture in the ground state, that is, the $3d$ electrons are itinerant. Here, the one-electron band picture means the theory which treats the correlated electrons in the mean field approximation. This contradiction was explained by Stoner as given next.

First, we calculate the Pauli paramagnetic susceptibility of conduction electrons with ↑ and ↓ spins. The spin moment of an electron is $\boldsymbol{\mu} = -g\mu_B \boldsymbol{s}$ or $\mu_z = -\mu_B$ for the ↑ spin and $\mu_z = \mu_B$ for the ↓ spin because of $g = 2$ and $s_z = \pm 1/2$. The corresponding Zeeman energy is $\mathcal{H}_Z = -\boldsymbol{\mu}\cdot\boldsymbol{H}$ or $+\mu_B H$ for the ↑ spin and $-\mu_B H$ for the ↓ spin from Eqs. (2.3b) and (2.5). The kinetic energy is $\varepsilon_\uparrow = \varepsilon(k) + \mu_B H$ for the ↑ spin and $\varepsilon_\downarrow = \varepsilon(k) - \mu_B H$ for the ↓ spin. The corresponding density of states is $D_\uparrow = D(\varepsilon - \mu_B H)$ for the ↑ spin and $D_\downarrow = D(\varepsilon + \mu_B H)$ for the ↓ spin. The cross-sectional area of the Fermi surface, which is based on $\hbar^2 k_\uparrow^2/2m^* + \mu_B H = \hbar^2 k_F^2/2m^*$ or $\pi k_\uparrow^2 = \pi k_F^2 - (2\pi m^*/\hbar^2)\mu_B H$ for the ↑ spin, is $S_\uparrow = S_F - (2\pi m^*/\hbar^2)\mu_B H$ for the ↑ spin and $S_\downarrow = S_F + (2\pi m^*/\hbar^2)\mu_B H$ for the ↓ spin, where S_F is a cross-sectional area of the Fermi surface at $H = 0$. The energy band, Fermi surface, and cross-sectional area of the Fermi surface are shown later in Fig. 6.35 for the ↑ and ↓ spins of the conduction electrons. The direction of the magnetic moment for the magnetic field along the z-axis is, however, opposite; $\mu_z = -\mu_B$ for the ↑ spin. On the other hand, the ↓ spin is directed along the field direction. Moreover, the ↓ spins, where the corresponding arrow is ↑ in Fig. 4.2(b), occupy the majority of spin states and contribute to the Pauli paramagnetism. In magnetism, the arrows (↑, ↓) do not mean the spin states, but the directions of the magnetic moments or correspond to the majority or minority of spin states. The majority spin states correspond to the ↑ spins. The number of electrons with the ↑ spins, n_+ and the number of electrons with the ↓ spins, n_- in Fig. 4.2(b) are calculated as

$$n_+ = \int_{-\mu_B H}^{\varepsilon_F} D(\varepsilon + \mu_B H) n(\varepsilon)\, d\varepsilon \simeq \int_0^{\varepsilon_F} D(\varepsilon) n(\varepsilon)\, d\varepsilon + \mu_B H D(\varepsilon_F)$$

and

$$n_- = \int_{\mu_B H}^{\varepsilon_F} D(\varepsilon - \mu_B H) n(\varepsilon)\, d\varepsilon \simeq \int_0^{\varepsilon_F} D(\varepsilon) n(\varepsilon)\, d\varepsilon - \mu_B H D(\varepsilon_F)$$

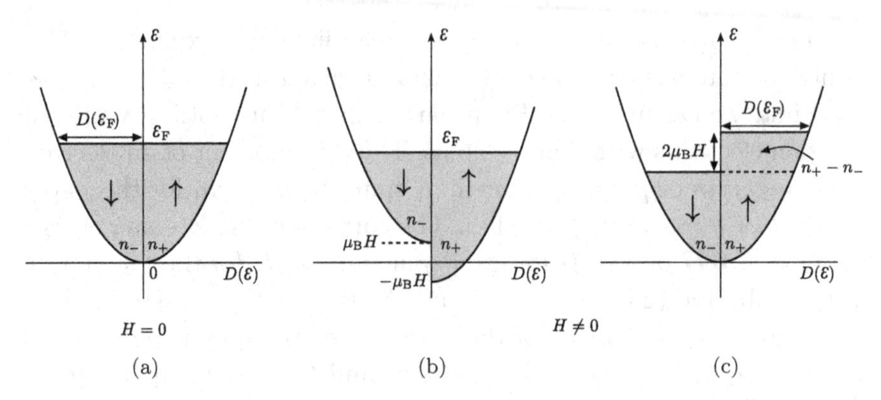

Figure 4.2: Pauli magnetism. The density of states in the shaded region is occupied by electrons with ↑ and ↓ spins. Note that the arrows (↑, ↓) do not refer to the spin states, but correspond to the directions of the magnetic moments.

where $D(\varepsilon)$ is the density of states and $n(\varepsilon)$ is the Fermi–Dirac distribution function. The magnetic moment M is obtained

$$\begin{aligned} M &= (n_+ - n_-)\mu_B \\ &\simeq 2\mu_B^2 D(\varepsilon_F) H \end{aligned} \tag{4.2a}$$

or the Pauli paramagnetic susceptibility is

$$\chi_p = \frac{M}{H} = 2\mu_B^2 D(\varepsilon_F). \tag{4.2b}$$

Itinerant $3d$ electron systems such as Fe or Ni demonstrate Curie–Weiss behavior at high temperatures, just like a localized moment, whereas it is described on the basis of the itinerant band picture, as mentioned above. Stoner added the molecular field into the Pauli paramagnetization and introduced a spontaneous magnetization

$$M = \chi_p(H + \alpha M). \tag{4.3}$$

This is transformed into

$$\chi = \frac{M}{H} = \frac{\chi_p}{1 - \alpha\chi_p}. \tag{4.4}$$

where M is the magnetization, χ_p is the Pauli paramagnetic susceptibility and H is a magnetic field. The magnetic susceptibility is thus

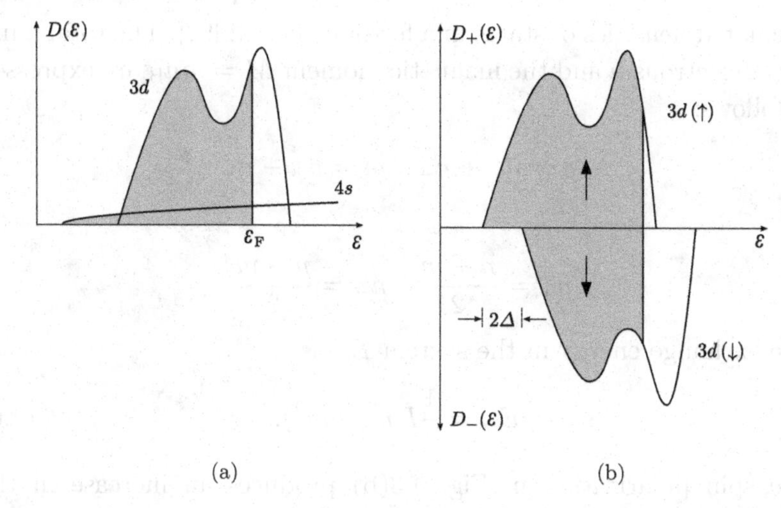

Figure 4.3: (a) Schematic representation of the $3d$ and $4s$ densities of states in the paramagnetic state and (b) the densities of states for $3d$ electrons with ↑ and ↓ spin states in the ferromagnetic state.

enhanced by the molecular field. Ferromagnetism appears under the condition that the denominator $1 - \alpha\chi_{\mathrm{p}} < 0$, namely

$$2\alpha\mu_{\mathrm{B}}^2 D(\varepsilon_{\mathrm{F}}) > 1. \tag{4.5}$$

This is called the Stoner's condition [91], and a large α or $D(E_{\mathrm{F}})$ value is a condition for ferromagnetism.

Here, we explain again the present Stoner model on the basis of the Hubbard model and the corresponding ferromagnetic density of states, where the density of states for electrons with the ↑ spin is shifted by $-\Delta$ and that for electrons with the ↓ spin is shifted by Δ due to the exchange interaction, as shown in Fig. 4.3(b). The effective value of U is smaller than the band width W. This is because the itinerant $3d$ electrons move in the crystal, avoiding one other. This means that U corresponds to the exchange interaction constant between the itinerant $3d$ electrons. Hereafter, we use the notation I instead of U. Figures 4.3(a) and (b) show the schematic densities of states for the $3d$ and $4s$ electrons in the paramagnetic state and the densities of states of $3d$ electrons with ↑ and ↓ spins in the ferromagnetic state, respectively, which, for example, correspond to the calculated densities of states in a ferromagnet Ni [92]. The total number of electrons n and the magnetic moment $M = m\mu_{\rm B}$ are expressed as follows:

$$n = n_+ + n_- \quad m = n_+ - n_-$$

or

$$n_+ = \frac{n+m}{2} \quad n_- = \frac{n-m}{2}.$$

The exchange energy in the system $E_{\rm ex}$ is

$$E_{\rm ex} = \frac{1}{4} I(n^2 - m^2). \tag{4.6}$$

The spin polarization in Fig. 4.3(b) produces an increase in the kinetic energy because the electrons are transferred from the ↓ spin band to the ↑ spin band. The kinetic energy with $m \neq 0$ $(2\Delta > 0)$ is expressed as

$$E_{\rm kin}(m) = \int_0^{\varepsilon_{\rm F}+\Delta} \varepsilon D(\varepsilon) d\varepsilon + \int_0^{\varepsilon_{\rm F}-\Delta} \varepsilon D(\varepsilon) d\varepsilon. \tag{4.7}$$

An increase of the kinetic energy due to the polarization is obtained as

$$\begin{aligned} E_{\rm kin}(m) - E_{\rm kin}(m=0) &= \int_0^{\varepsilon_{\rm F}+\Delta} \varepsilon D(\varepsilon)\, d\varepsilon + \int_0^{\varepsilon_{\rm F}-\Delta} \varepsilon D(\varepsilon)\, d\varepsilon \\ &\quad - 2\int_0^{\varepsilon_{\rm F}} \varepsilon D(\varepsilon)\, d\varepsilon \\ &= D(\varepsilon)\Delta^2 \left[1 - \frac{3}{4}\left(\frac{D'^2}{D^2} - \frac{D''}{3D}\right)\Delta^2 + \cdots\right]. \end{aligned}$$

Note that $D(\varepsilon)$ is expanded in the Taylor series around $\varepsilon_{\rm F}$,

$$D(\varepsilon) = D(\varepsilon_{\rm F}) + D'(\varepsilon - \varepsilon_{\rm F}) + \frac{1}{2} D''(\varepsilon - \varepsilon_{\rm F})^2 + \cdots$$

where

$$D' = \left.\frac{\partial D(\varepsilon)}{\partial \varepsilon}\right|_{\varepsilon=\varepsilon_{\rm F}} \quad \text{and} \quad D'' = \left.\frac{\partial^2 D(\varepsilon)}{\partial \varepsilon^2}\right|_{\varepsilon=\varepsilon_{\rm F}}.$$

Finally, the kinetic energy is expressed using the relation $m = 2\Delta D(\varepsilon_F)$ so that

$$E_{kin}(m) - E_{kin}(m=0) = \frac{1}{4D(\varepsilon_F)}m^2 + \frac{1}{64D(\varepsilon_F)^3} \times \left(\frac{D'^2}{D(\varepsilon_F)^2} - \frac{D''}{3D(\varepsilon_F)}\right)m^4. \quad (4.8)$$

The total energy is

$$E_{tot} = E_{ex} + E_{kin} \quad (4.9a)$$

$$\delta E_{tot} = -\frac{I}{2}m + \frac{1}{2D(\varepsilon_F)}m = \frac{1 - ID(\varepsilon_F)}{2D(\varepsilon_F)}m. \quad (4.9b)$$

When $\delta E_{tot} < 0$, the paramagnetic state ($m = 0$) becomes unstable and the ferromagnetic state is realized, namely under the condition

$$ID(\varepsilon_F) - 1 > 0. \quad (4.10)$$

When $ID(\varepsilon_F) - 1 < 0$, the paramagnetic susceptibility is obtained by differentiating the free energy F_m by the magnetic moment m, defined as follows

$$F_m = E_{tot} - m\mu_B H \quad (4.11)$$

$$\frac{\partial F_m}{\partial m} = -\frac{I}{2}m + \frac{1}{2D(\varepsilon_F)}m - \mu_B H = 0$$

$$\frac{m\mu_B}{H} = \chi = \frac{2\mu_B^2 D(\varepsilon_F)}{1 - ID(\varepsilon_F)} = \frac{\chi_P}{1 - ID(\varepsilon_F)}. \quad (4.12)$$

The paramagnetic susceptibility is enhanced by $[1 - ID(\varepsilon_F)]^{-1}$, where $[1 - ID(\varepsilon_F)]^{-1}$ is called the "Stoner enhancement factor".

The enhancement of the magnetic susceptibility is clearly related to the density of states $D(\varepsilon_F)$ and the exchange interaction constant I. Figure 4.4(a) shows the theoretically calculated partial density of states for 3*d* electrons of YT_2 (T: Fe, Co, and Ni) in the paramagnetic state [93]. The $ID(\varepsilon_F)$ factor is estimated to be 2.6 for

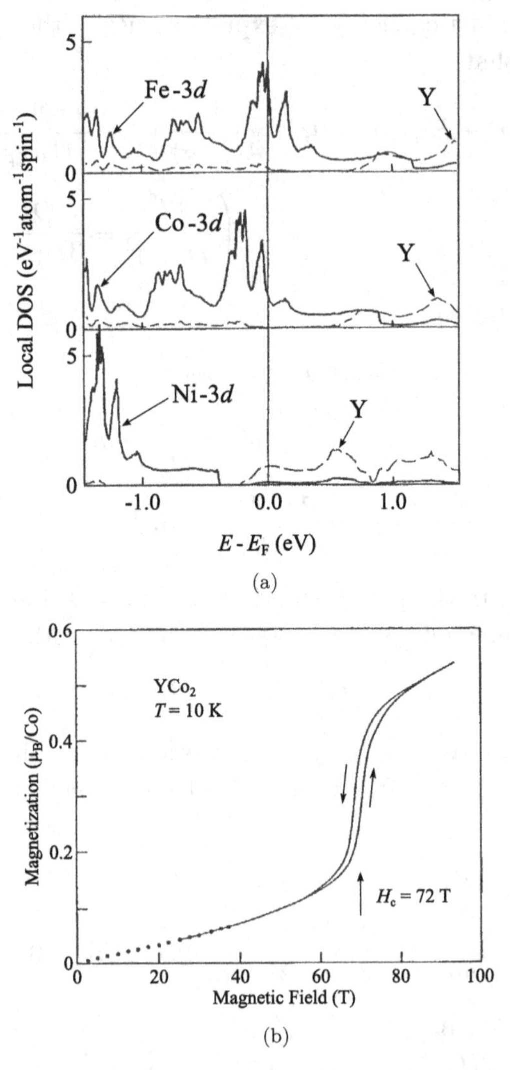

Figure 4.4: (a) Calculated partial density of states for T-3d electrons in YT_2 (T: Fe, Co, and Ni) in the paramagnetic states and (b) the high-field magnetization curve in YCo_2, courtesy of H. Yamada and T. Sakakibara, cited from Refs. [93, 95].

a ferromagnet YFe_2, 0.9 for a near ferromagnet YCo_2, and 0.2 for a Pauli paramagnet YNi_2. Here, we shall remark on the magnetic susceptibility of YCo_2 with the cubic Laves-phase crystal structure [94]. The magnetic susceptibility of YCo_2 possesses a broad maximum around 250 K. The susceptibility at zero temperature $\chi(0)$ is also enhanced. Therefore, YCo_2 is called an exchange-enhanced paramagnet. With an increasingly strong applied magnetic field, the magnetization indicates a non-linear increase in magnetization, so-called the metamagnetic transition at a critical field $H_c = 72\,\mathrm{T}(=720\,\mathrm{kOe})$, as shown in Fig. 4.4(b) [95]. YCo_2 is therefore close to a ferromagnet. Next, we will explain the mechanism of the metamagnetic transition in an itinerant 3*d* electron system.

4.2 Metamagnetic transition in an itinerant 3*d*-electron system

The metamagnetic transition in an itinerant 3*d*-electron system is observed when the Stoner factor "$ID(\varepsilon_F) - 1$" is close to zero and the Fermi energy ε_F is also situated between two peaks in the density of states, or a sharp peak of the density of states exists just below the Fermi energy ε_F as in YCo_2, shown in Fig. 4.4(a). Here, we explain the metamagnetism [96–98].

We again represent the free energy F_m in Eq. (4.11)

$$F_m = \frac{1}{4D}(1 - ID)m^2 + \frac{1}{64D^3}\left(\frac{D'^2}{D^2} - \frac{D''}{3D}\right)m^4 + \cdots - m\mu_B H$$

or

$$F_M = \frac{1}{2}aM^2 + \frac{1}{4}bM^4 + \frac{1}{6}cM^6 - MH. \tag{4.13}$$

The ferromagnetic condition in the Stoner model under $H = 0$ is schematically shown in Fig. 4.5[curve (A)]. Note the condition of $a < 0$ and $b > 0$ or $1 - ID < 0$ is necessary. In the nearly ferromagnet, the curve (B) is expected phenomenologically for the metamagnetic transition. The characteristic conditions are as follows:

(1) The F_M vs M curve possesses two minima and one maximum, which requires $a > 0$, $b < 0$, and $c > 0$.
(2) $M = 0$ becomes a minimum when $H = 0$.
(3) When the magnetic field is applied and curve (C) is obtained, the metamagnetic transition is realized at M_c or the critical field H_c.

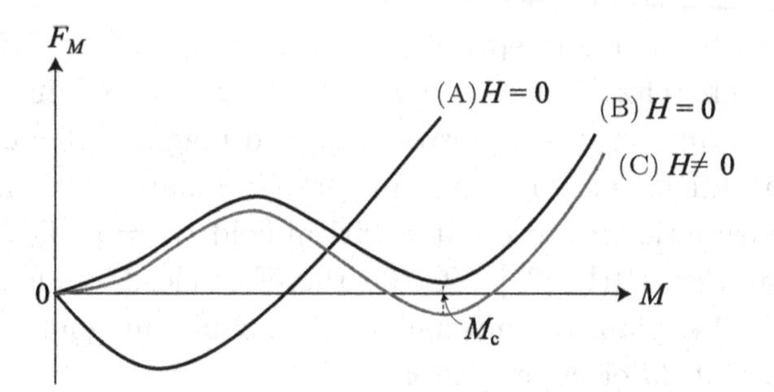

Figure 4.5: M-dependence of the free energy F_M. Curve (A) presents the ferromagnetic state in the Stoner model. Curve (B) presents the nearly ferromagnetic state as in YCo_2 under $H = 0$. Under magnetic fields, the ferromagnetic state is realized at M_c, or at a critical field H_c [curve (C)].

First, we consider the condition (2) by calculating $\partial F_M/\partial M = 0$ at $H = 0$.

$$\frac{\partial F_M}{\partial M} = aM + bM^3 + cM^5 = M(a + bM^2 + cM^4) = 0.$$

F_M thus becomes minimum at

$$M_c = \left(\frac{-b + \sqrt{b^2 - 4ac}}{2c}\right)^{1/2}. \tag{4.14}$$

Here, the following conditions are necessary:

$$F_M(H = 0) > 0 \quad \text{at} \quad M = M_c, \quad \text{namely}$$

$$F_M(H = 0, M = M_c) = \frac{-b + \sqrt{b^2 - 4ac}}{4c}$$

$$\times \left(\frac{8ac - b^2 + b\sqrt{b^2 - 4ac}}{12c}\right) > 0.$$

This means $\sqrt{b^2 - 4ac} > 0$ and $8ac - b^2 + b\sqrt{b^2 - 4ac} > 0$, namely

$$\frac{ac}{b^2} > \frac{3}{16}. \tag{4.15}$$

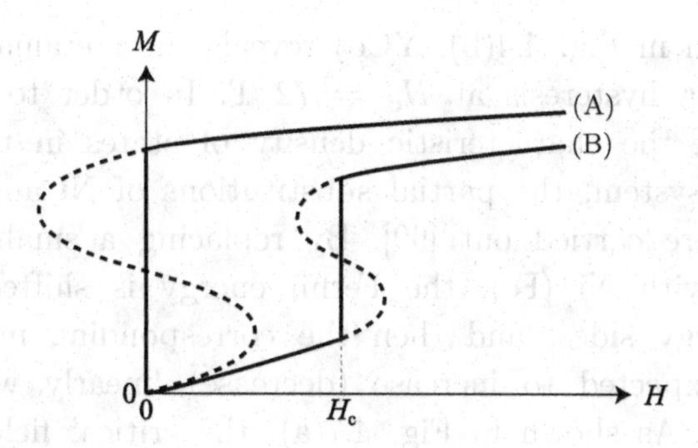

Figure 4.6: Magnetization curves in (A) the ferromagnetic state, (B) the metamagnetic transition with a hysteresis.

Next, we consider the condition (3) by calculating $\partial F_M/\partial M$, as follows:

$$\frac{\partial F_M}{\partial M} = aM + bM^3 + cM^5 - H = 0. \tag{4.16}$$

The magnetization curves in Fig. 4.6 represents various electronic states, depending on the a, b, and c values in Eq. (4.16). The magnetization curve (A) indicates the ferromagnetic state. The curve (B) presents the metamagnetic transition with a hysteresis, where the ground state is paramagnetic and the electronic state becomes ferromagnetic for $H > H_c$. We can obtain curve (B) by calculating $\partial H/\partial M$ in Eq. (4.16), as

$$\frac{\partial H}{\partial M} = a + 3bM^2 + 5cM^4 = 0, \tag{4.17}$$

then

$$M^2 = \frac{-3b \pm \sqrt{9b^2 - 20ac}}{10c}$$

or

$$9b^2 - 2ac > 0 \quad \text{or} \quad \frac{ac}{b^2} < \frac{9}{20}. \tag{4.18}$$

Finally, we have the condition

$$\frac{3}{16} < \frac{ac}{b^2} < \frac{9}{20} \tag{4.19}$$

for the metamagnetic transition.

As shown in Fig. 4.4(b), YCo_2 reveals the metamagnetic transition with a hysteresis at H_c = 72 T. In order to test a possible role of the characteristic density of states in the itinerant 3d-electron system, the partial substitutions of Ni and Fe for Co in YCo_2 were carried out [99]. By replacing a small amount of Co atoms with Ni (Fe), the Fermi energy is shifted to higher (lower) energy sides, and then the corresponding metamagnetic fields are expected to increase (decrease) linearly with increasing Ni (Fe). As shown in Fig. 4.7(a), the critical field H_c in the metamagnetic transition is found to increase linearly with increasing Ni concentration x in $Y(Co_{1-x}Ni_x)_2$. The dependence of concentration x on H_c in $Y(Co_{1-x}Ni_x)_2$ and $Y(Co_{1-x}Fe_x)_2$ is shown in Fig. 4.7(b).

Later, we present the dHvA experimental results and the corresponding theoretical ones for a new nearly ferromagnetic compound

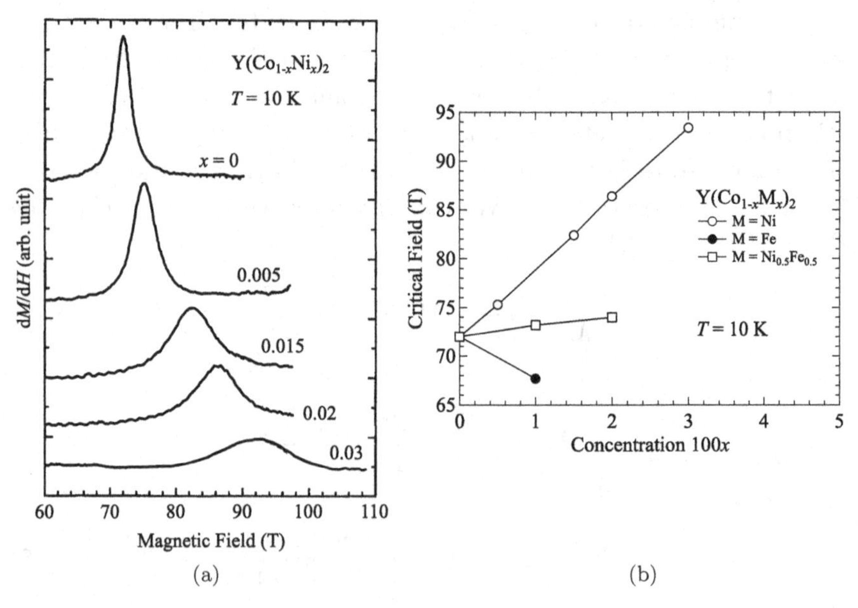

Figure 4.7: (a) Differential magnetization dM/dH for $Y(Co_{1-x}Ni_x)_2$ ($0 \le x \le 0.03$) around the metamagnetic transition and (b) the concentration dependence of the metamagnetic critical field H_c for the substituted compounds $Y(Co_{1-x}M_x)_2$ (M = Ni, Fe, and $Ni_{0.5}Fe_{0.5}$, $0 \le x \le 0.03$), courtesy of T. Sakakibara, cited from Ref. [99].

$SrCo_2P_2$. This compound also displays a metamagnetic transition at $H_c \simeq 60$ T, with hysteresis as in YCo_2 [100].

4.3 SCR theory

Two key laws of the Stoner model is that it cannot explain the Curie-Weiss law at high temperatures, and gives a large Curie temperature T_C. These were eliminated by Moriya on the basis of SCR (Self Consistent Renormalization) theory, which treats the spin fluctuations for the band electrons [101–103].

We show simply in Fig. 4.8 the spatial distribution of the spin density $\rho_s(r)$ [103]. $\rho_s(r)$ can be expressed as a deviation between the wave functions of up-spin and down-spin, namely, $\rho_s(r) = |\psi_\uparrow(r)|^2 - |\psi_\downarrow(r)|^2$. In the Stoner model, there is no difference between the density of states of up- and down-spins above T_C, namely in the paramagnetic region, and thus $\rho_s(r) = 0$ (see Fig. 4.8(a)). On the other hand, in the localized spin model, the magnitude of each moment is constant, but the spin takes any direction by the thermal motion above T_C (see Fig. 4.8(b)). Considering the spin fluctuation effect, the spin at the finite temperature varies not only in direction but also in amplitude in the SCR theory, as shown in Fig. 4.8(c).

Note that the $3d$ electrons are localized at the atomic sites in the localized spin model, while the $3d$ electrons are itinerant in the spin fluctuation model. Moreover, the magnetic moments at the atomic sites exist below and above T_C, although the directions of the magnetic moments change randomly in space and time. The temperature dependence of the magnetic susceptibility in the spin fluctuation model reveals the Curie-Weiss law above T_C. In the Stoner model, the polarization of bands with up- and down-spins becomes zero above T_C, which corresponds to the usual Pauli paramagnetism, revealing no Curie-Weiss law. Therefore, the magnetic moments disappear completely at the atomic sites above T_C.

In the spin fluctuation model, the following quantity of the spin density S_q is important [102]:

$$S_L^2 = N^{-1} \sum_q \langle |S_q|^2 \rangle. \tag{4.20}$$

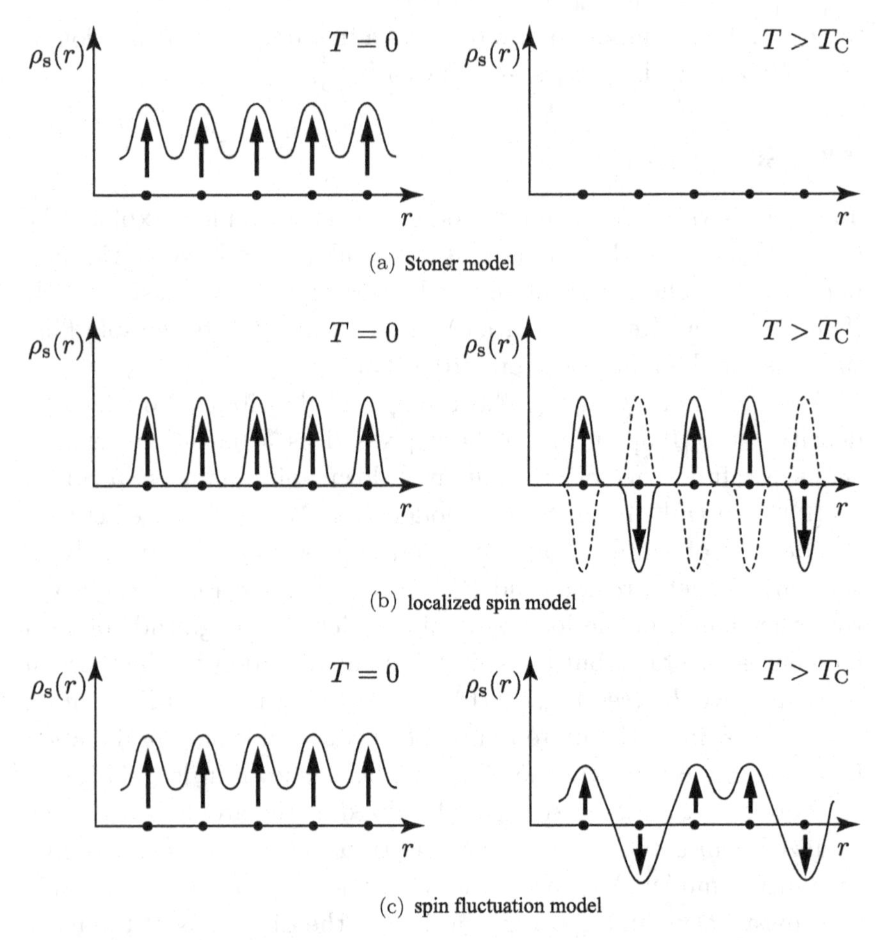

Figure 4.8: Spatial distribution of the spin density $\rho_s(r)$ for (a) Stoner model, (b) localized spin model, and (c) spin fluctuation model. Left side is at $T = 0$, right side is $T > T_C$. In (a) $\rho_s(r)$ vanishes, in (b) $\rho_s(r)$ is unchanged and in (c) the local density of spin is slightly diminished at $T > T_C$, courtesy of M. Shiga, cited from Ref. [103].

Here, the magnetic excitation, namely the magnon depends on the wave number q. The corresponding spin density $\boldsymbol{S}_q$ also depends on q. Note that S_L^2 is constant in the whole temperature region for the localized spin model. In the spin fluctuation model, the magnitude of S_L^2 depends on the compound, but approaches the magnitude of the localized spin model with increasing temperature, as

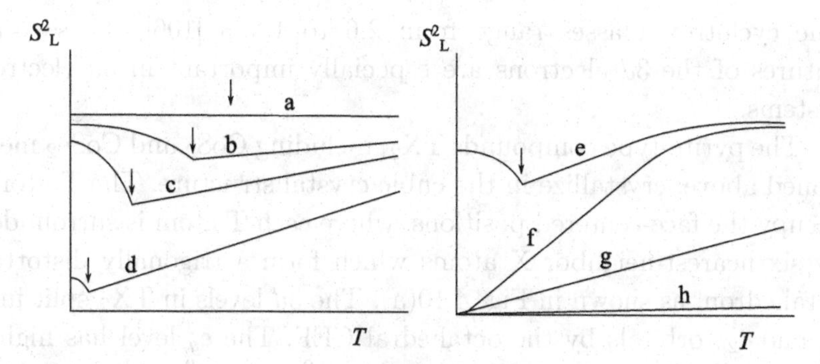

Figure 4.9: Temperature dependence of S^2_L for several cases, where case a corresponds to the localized spin system and case d a weak ferromagnet, cited from Ref. [102]. Arrows indicate Curie temperatures.

shown in Fig. 4.9. Line a is the case of local moment where S^2_L is constant as a function of temperature and line d corresponds to weakly ferromagnetic metals such as $ZrZn_2$ and Sc_3In. The Curie temperature and ordered moment are 25 K and 0.12 μ_B/Zr in $ZrZn_2$ and 6 K and 0.04 μ_B/Sc in Sc_3In [104, 105]. Lines b and c represent an intermediate case. Lines e and f correspond to the cases of CoS_2 and $CoSe_2$, respectively. Lines g and h correspond to a nearly ferromagnetic metal, such as Pd, and a Pauli paramagnet, respectively.

The local moments of CoS_2 and $CoSe_2$ are enhanced with increasing temperature. The magnetic susceptibilities χ in these compounds follow the Curie-Weiss law at high temperatures with an effective magnetic moment of $\mu_{eff} = 1.76\ \mu_B$/Co, which corresponds to 1.73 μ_B/Co of Co^{2+} ($S = 1/2$, $g = 2$, $\mu_{eff} = 1.73\ \mu_B$/Co, and $\mu_s = 1\ \mu_B$/Co). With decreasing temperature, CoS_2 orders ferromagnetically at $T_C = 122$ K, with an ordered moment of 0.93 μ_B/Co, while $CoSe_2$ becomes an exchange-enhanced paramagnet at low temperatures. Co-3*d* electrons thus produce the ferromagnetic ordering in CoS_2 and a relatively large magnetic susceptibility in $CoSe_2$. Note that the large ordered moment of CoS_2 is due to a half-metallic spin state of CoS_2. Namely, only the Co-3*d* electrons with the spin-up state contribute to the ordered moment and the Fermi surface. At the same time, Co-3*d* electrons become conduction electrons with large cyclotron masses. The γ value in CoS_2 is 21 mJ/(K^2·mol).

The cyclotron masses range from 2.6 to $19m_0$ [106]. These dual natures of the $3d$ electrons are especially important in $3d$ electron systems.

The pyrite-type compounds TX_2, including CoS_2 and $CoSe_2$ mentioned above, crystallize in the cubic crystal structure. The T atoms occupy the face-centered positions, where each T atom is surrounded by six nearest-neighbor X atoms which form a trigonally distorted octahedron, as shown in Fig. 4.10(a). The $3d$ levels in TX_2 split into e_g and t_{2g} orbitals by the octahedral CEF. The e_g level has higher energy than the t_{2g} level. The T-e_g ($d\gamma$: x^2-y^2, $3z^2-r^2$) orbitals in TS_2 expand along the S atoms in the octahedron, hybridize with S-$3p$ orbitals, and form the antibonding and bonding orbitals, separated by a large gap. On the other hand, the T-t_{2g} ($d\varepsilon$: xy, yz, zx) orbitals scarcely possess a hybridized partner, forming a localized narrow-band. The electronic properties of the pyrite-type compounds are thus understood by considering progressive filling of the antibonding $3d$-e_g band. Namely, the filling of the e_g band due to $3d$ electrons is empty in a band gap insulator (nonmagnetic Fe) FeS_2 ($t_{2g}^6e_g^0$), a quarter full in a metallic ferromagnet CoS_2 ($t_{2g}^6e_g^1$), half full in a Mott-Hubbard insulator NiS_2 ($t_{2g}^6e_g^2$) (complicated antiferromagnet) based on the electron correlations, three quarters full in a superconductor CuS_2 ($t_{2g}^6e_g^3$), and completely full in a diamagnetic wide-band-gap insulator ZnS_2 ($t_{2g}^6e_g^4$). Figure 4.10(b) shows the theoretical total and partial ($3d$ electrons) densities of states in CoS_2, NiS_2, and CuS_2 in the paramagnetic state. The corresponding electrical resistivities are shown in Fig. 4.10(c), revealing a ferromagnetic ordering at $T_C = 122$ K in CoS_2, the Mott-Hubbard insulating property in NiS_2, and a superconducting transition at $T_{sc} = 1.5$ K in CuS_2.

4.4 Nearly ferromagnet $SrCo_2P_2$ and ferromagnet $LaCo_2P_2$

In this section, we present the electronic properties of an exchange-enhanced paramagnetic material or a nearly ferromagnetic $ScCo_2P_2$ [107] and a ferromagnetic $LaCo_2P_2$ [108] with the

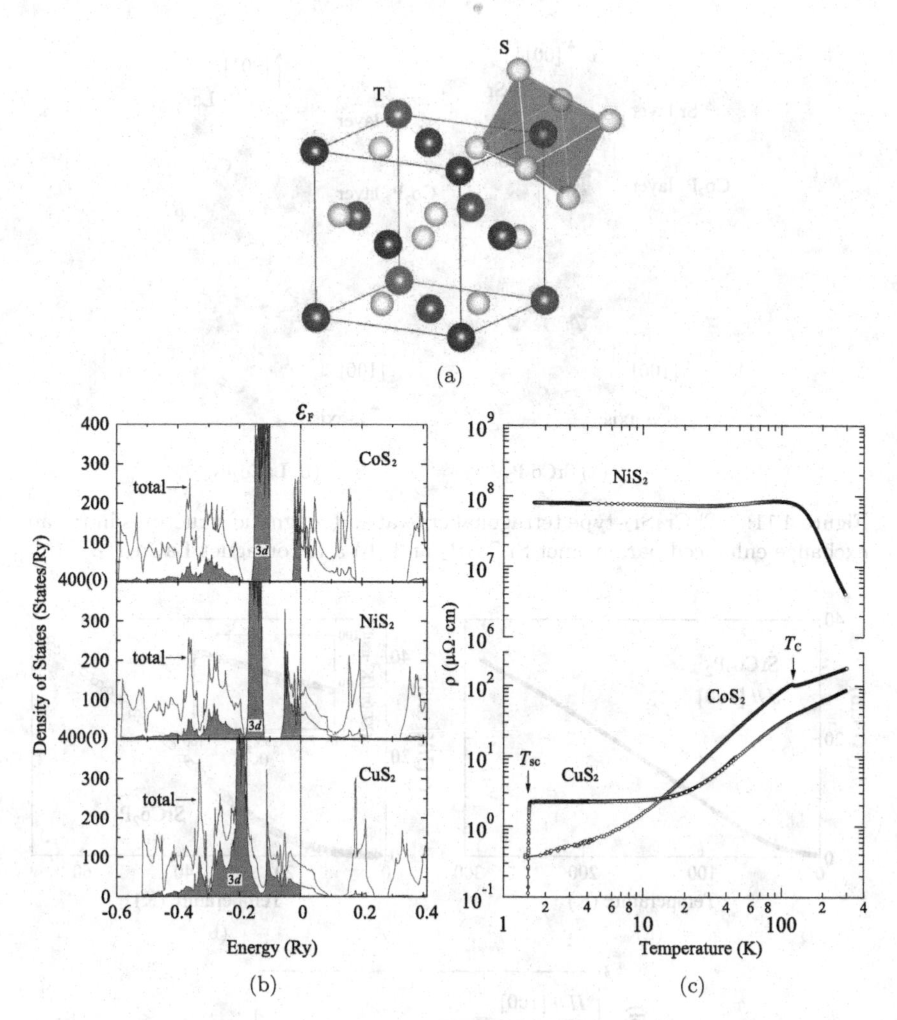

Figure 4.10: (a) Pyrite-type cubic crystal structure in TX_2 (T: transition metal, X: S, Se, Te), (b) theoretical total and partial (3*d* electrons) densities of states in CoS_2, NiS_2, and CuS_2 in the paramagnetic state, courtesy of H. Harima, and (c) the temperature dependence of electrical resistivities in a ferromagnet CoS_2, a Mott-Hubbard insulator NiS_2, and a superconductor CuS_2.

$ThCr_2Si_2$-type tetragonal structure, as shown in Fig. 4.11. 3*d* electrons with a characteristic density of states contribute to the Fermi surface and also a magnetic moment.

Figure 4.12(a) shows the temperature dependence of the electrical resistivity ρ for the current J along the [100] direction for $ScCo_2P_2$.

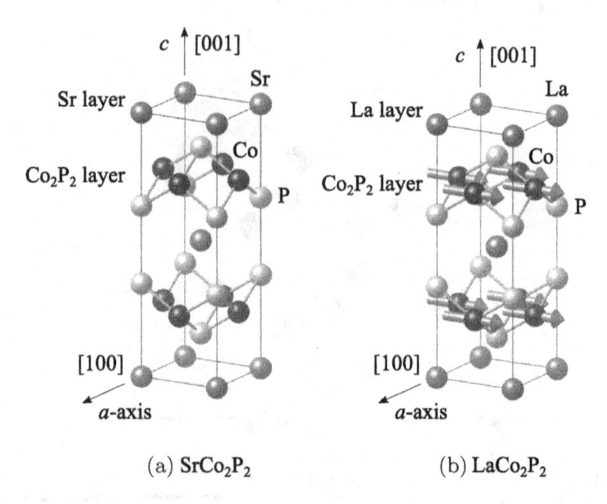

Figure 4.11: $ThCr_2Si_2$-type tetragonal crystal and magnetic structures in (a) an exchange-enhanced paramagnet $SrCo_2P_2$ and (b) a ferromagnet $LaCo_2P_2$.

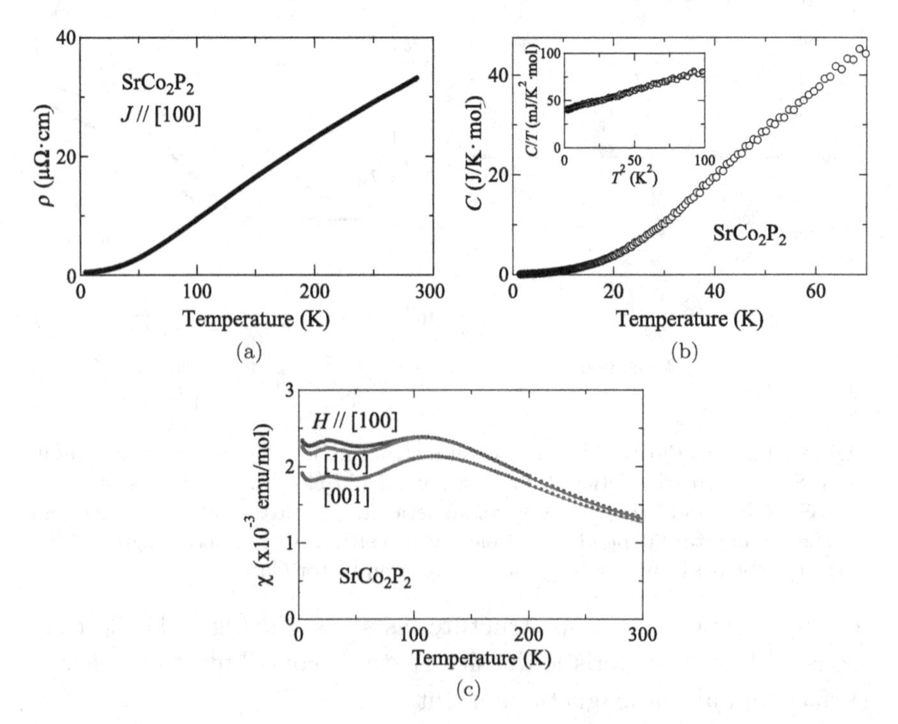

Figure 4.12: Temperature dependence of (a) the electrical resistivity, (b) the specific heat, and (c) the magnetic susceptibility in $SrCo_2P_2$, cited from Ref. [107]. The inset of (b) shows a T^2-dependence of C/T.

The resistivity is typical in a $3d$ electron system, revealing a negative curvature of the resistivity against temperature, $d^2\rho/dT^2 < 0$ at high temperatures. Below 70 K, the resistivity follows the Fermi liquid relation of $\rho = \rho_0 + AT^2$ ($A = 0.97 \times 10^{-3}$ $\mu\Omega$·cm/K^2, and the residual resistivity $\rho_0 = 0.42$ $\mu\Omega$·cm). The residual resistivity ratio RRR (ρ_{RT}/ρ_0, ρ_{RT}: resistivity at room temperature) is 75, indicating a high-quality sample.

The specific heat C is also shown in Fig. 4.12(b). The inset of Fig. 4.12(b) shows the T^2-dependence of C/T, revealing a relatively large electronic specific heat coefficient $\gamma = 39.6$ mJ/(K^2·mol). The magnetic susceptibility follows the Curie-Weiss law at temperatures higher than about 200 K, revealing the effective magnetic moment $\mu_{eff} = 1.72$ μ_B/Co. The susceptibility increases with decreasing temperature, and becomes almost constant below about 100 K. Note that the susceptibility possesses two broad peaks at 110–120 K and 25 K, as shown in Fig. 4.12(c), revealing that $SrCo_2P_2$ is an exchange-enhanced paramagnet as in YCo_2.

Figures 4.13(a) and (b) show the typical dHvA oscillations in the field ranging from 100 to 145 kOe for the magnetic field along the [001] direction, and the corresponding fast Fourier transformation (FFT) spectrum, respectively. The dHvA frequency F $(= c\hbar S_F/2\pi e)$ in Fig. 4.13(b) is proportional to the extremal cross-sectional area of the Fermi surface S_F, which is shown as a unit of magnetic field. Several dHvA branches are observed, ranging from 1.187 to 6.941 $\times$ 10^7 Oe, which are denoted by α (α_1, α_2, α_3), β, and γ.

Figure 4.14(a) shows the angular dependence of the dHvA frequencies. The solid lines are guides connecting the data. The dHvA branches α (α_1, α_2, and α_3) correspond to cylindrical parts of a Fermi surface, and the branch β is due to an ellipsoidal Fermi surface, which is flat along the [001] direction.

The cyclotron masses m_c^* for the dHvA branches were determined from the temperature dependence of the dHvA amplitudes. The cyclotron masses are relatively large, ranging from 0.87 to $7.2m_0$. We summarize in Table 4.1 the dHvA frequencies F and the corresponding cyclotron masses m_c^*, together with the theoretical dHvA frequencies F_b and the band masses m_b.

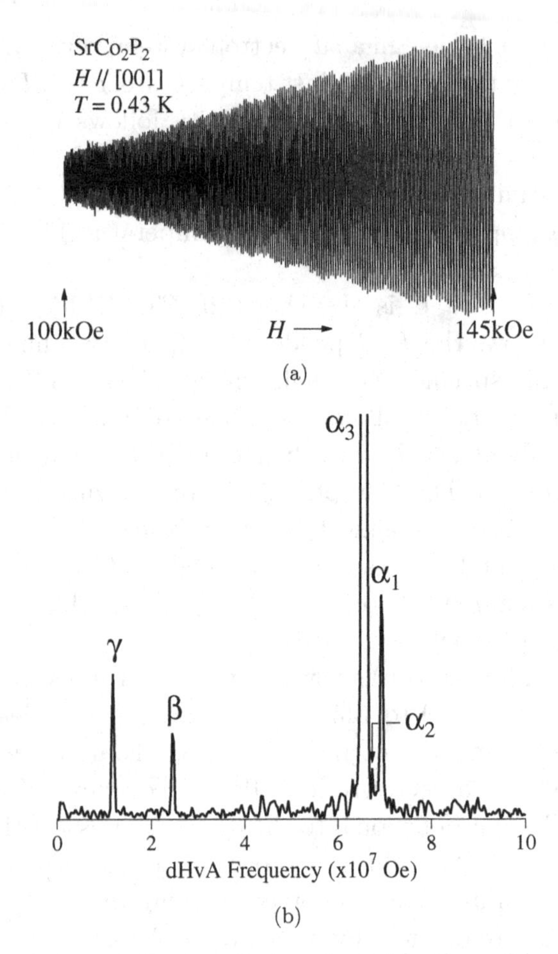

Figure 4.13: (a) Typical dHvA oscillations for $H \parallel [001]$ and (b) the corresponding FFT spectrum in $SrCo_2P_2$ with the $ThCr_2Si_2$-type tetragonal structure, cited from Ref. [107].

The topology of the Fermi surface and the energy band structure for $SrCo_2P_2$ were calculated using the full potential LAPW method with local density approximation (LDA) for the exchange correlation potential. In the present band calculations, the scalar relativistic effect was taken into account for all the electrons, and the spin-orbit interaction was included self-consistently for all the valence electrons as in a second variational procedure. $4s^2 4p^6 5s^2$ electrons for Sr, $3p^6 3d^7 4s^2$ electrons for Co, and $3s^2 3p^3$ electrons for P are treated as

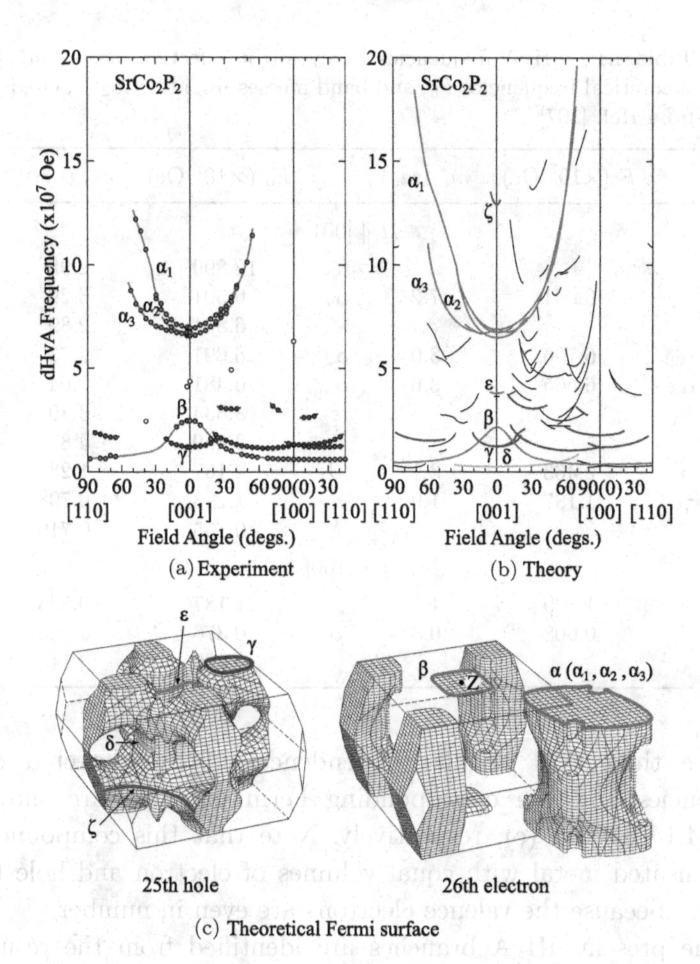

Figure 4.14: Angular dependence of (a) detected dHvA frequencies and (b) theoretical ones in $SrCo_2P_2$, together with (c) the corresponding theoretical Fermi surfaces, cited from Ref. [107].

valence electrons in the calculations. Sr-$4s^2 4p^6$ and Co-$3p^6$ electrons are treated as semi-core states to be calculated using another energy window. The space group and lattice parameters are adopted from Ref. [109]: $a = 3.794$ Å, $c = 11.610$ Å, and the z-parameter of 0.3525 for P. Muffin-tin (MT) radii were set as $0.4080a$ for Sr and $0.2803a$ for Co and P. The LAPW basis functions were truncated at $|\boldsymbol{k} + \boldsymbol{G}_i| \leq 4.85 \times 2\pi/a$, corresponding to 731 LAPW functions at the Γ point.

Table 4.1: dHvA frequencies F, cyclotron effective masses m_c^*, theoretical frequencies F_b, and band masses m_b in $SrCo_2P_2$, cited from Ref. [107].

	F ($\times 10^7$ Oe)	m_c^* (m_0)		F_b ($\times 10^7$ Oe)	m_b (m_0)
		$H \parallel [001]$			
			ζ	12.899	2.98
α_1	6.941	7.2	α_1	6.901	3.29
				6.801	2.89
α_2	6.736	3.0	α_2	6.691	2.73
α_3	6.565	3.6	α_3	6.481	2.64
			ε	3.834	3.40
				3.659	4.87
β	2.463	3.1	β	2.167	1.23
γ	1.187	1.9	γ	1.223	0.798
			δ	0.255	0.719
		$H \parallel [100]$			
	1.309	1.1		1.187	0.533
β	0.608	0.87	β	0.376	0.406
				0.289	2.11

The theoretical angular dependences of the detected dHvA frequencies and the corresponding Fermi surfaces are shown in Figs. 4.14(b) and (c), respectively. Note that this compound is a compensated metal with equal volumes of electron and hole Fermi surfaces because the valence electrons are even in number.

The present dHvA branches are identified from the results of energy band calculations, as follows:

1) Branches α (α_1, α_2, and α_3) are due to the band 26th electron Fermi surface, which is corrugated but cylindrical along the [001] direction, with many concave and convex shapes.
2) Branch β is due to the band 26th flat electron Fermi surface at the Z point.
3) Branch γ is due to the band 25th hole Fermi surface, which is a multiply-connected Fermi surface with cylindrical parts.

Theoretically, branches ζ, ε, and δ exist for $H \parallel [001]$, and many other branches also exist in the other field directions, as shown in

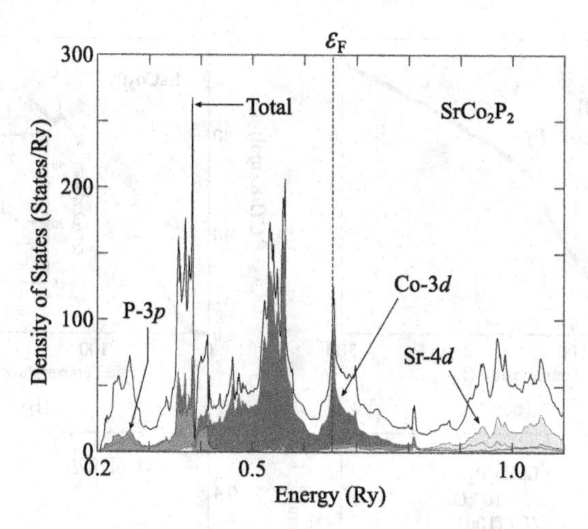

Figure 4.15: Total and partial densities of states in $SrCo_2P_2$, cited from Ref. [107].

Fig. 4.14(b). These have not been observed experimentally, which might be due to a damping effect of the dHvA amplitude based on a so-called curvature factor of the Fermi surface. However, the detected dHvA branches are consistent with the present results of band calculations.

Figure 4.15 shows the total and partial densities of states. The Co-3*d* electrons mainly contribute to the Fermi surfaces and the density of states, although there are also small contributions from Sr-4*d* and P-3*p* electrons. Note that the contribution of P-3*p* electrons to the Fermi surface is small in $SrCo_2P_2$. Furthermore, the Sr-4*d* electrons do not exist in the initial state but form a wide band in the final state of calculations. The occupation number of Sr-4*d* electrons below the Fermi energy is 0.417/Sr. It must be pointed out that the Fermi energy ε_F is situated at a peak of the density of states. From the total density of states in Fig. 4.15, the electronic specific heat coefficient γ_b is calculated as γ_b = 83.90 states/(Ry·cell) or 14.54 mJ/(K^2·mol), which is compared with the experimental value γ = 39.6 mJ/(K^2·mol). Correspondingly, the cyclotron effective mass m_c^* in Table 4.1 is roughly twice as large as the corresponding band masses m_b.

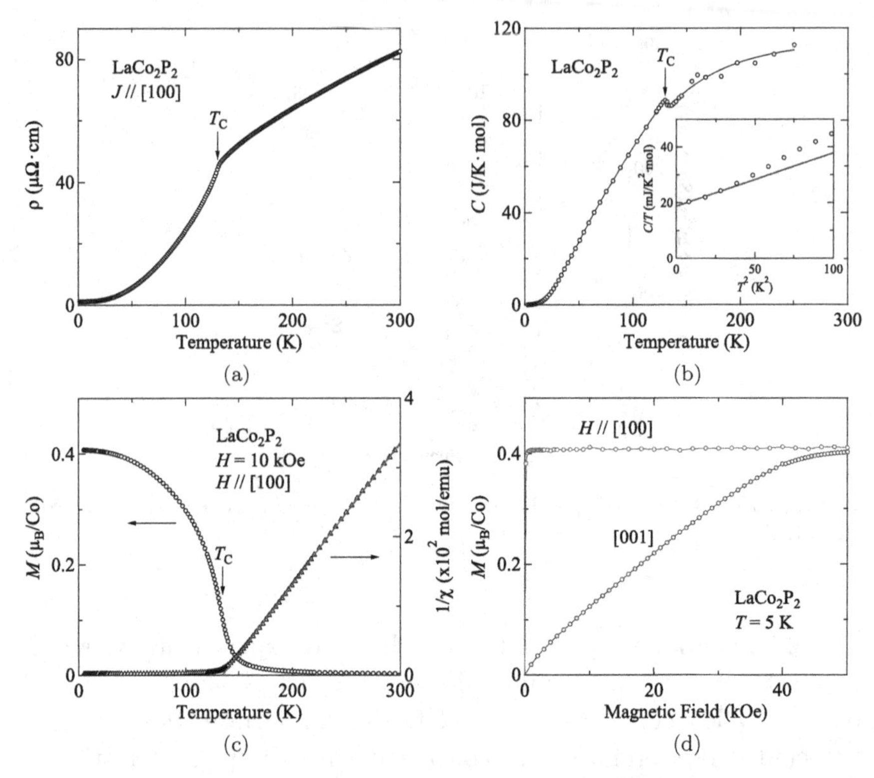

Figure 4.16: Temperature dependences of (a) electrical resistivity ρ, (b) specific heat C, (c) magnetization M and the inverse magnetic susceptibility $1/\chi$, and (d) magnetization curves at 5 K for $H \parallel [100]$ and [001] in a ferromagnet $LaCo_2P_2$.

Next, we explain the ferromagnetic properties of $LaCo_2P_2$. Figure 4.16(a) shows the temperature dependence of the electrical resistivity, indicating a steep decrease of the resistivity below T_C = 130 K. ρ_0 = 1.2 $\mu\Omega$·cm and RRR = 69, revealing a good single crystal sample. The present ferromagnetic ordering is observed at T_C = 130 K in the specific heat, as shown in Fig. 4.16(b). The γ value is estimated as γ = 19 mJ/(K^2·mol) from a C/T vs T^2 plot, as shown in the inset of Fig. 4.16(b).

Figure 4.16(c) shows the temperature dependence of the magnetic susceptibility χ and magnetization M for $H \parallel [100]$. The effective magnetic moment $\mu_{\rm eff}$ and the paramagnetic Curie temperature are obtained respectively as 1.0 μ_B/Co and 135 K. The

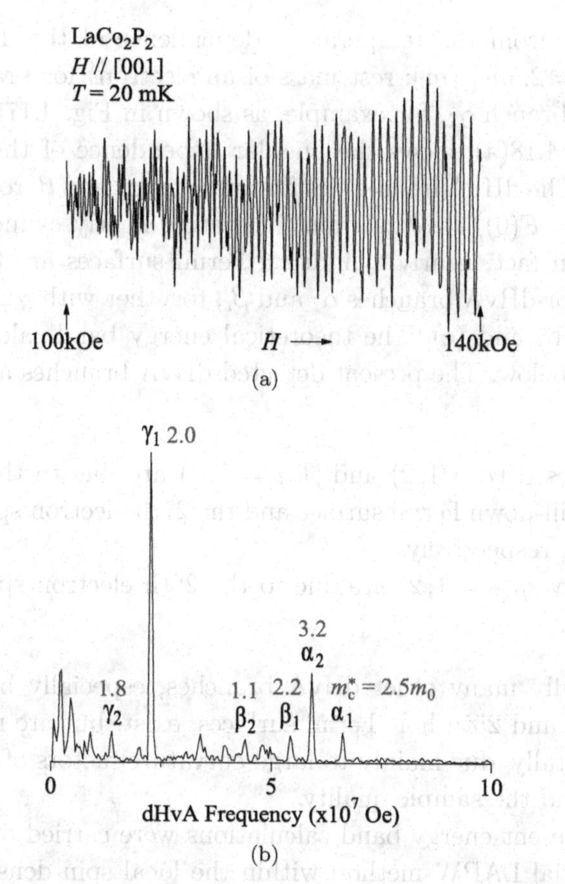

Figure 4.17: (a) Typical dHvA oscillations and (b) the corresponding FFT spectrum in $LaCo_2P_2$, cited from Ref. [108].

magnetization at $H = 10\,\text{kOe}$ for $H \parallel [100]$ increases steeply below 150 K approximately and saturates at lower temperatures. Here, the magnetizations for $H \parallel [100]$ and [001] correspond to easy-axis and hard-axis magnetizations, respectively, as shown in Fig. 4.16(d). The saturated moment is 0.4 μ_B/Co. The magnetic structure is shown in Fig. 4.11(b).

Figure 4.17 shows the typical dHvA oscillations and the corresponding FFT spectrum. The dHvA frequency is in the range of 6.627×10^7 Oe denoted by α_1 and 1.390×10^7 Oe, denoted by γ_2. For these dHvA branches, we determined the cyclotron effective masses m_c^* from the temperature dependence of the dHvA amplitude: $m_c^* = 2.5m_0$ (m_0: rest mass of an electron) for branch α_1 and $1.8m_0$ for branch γ_2, for example, as shown in Fig. 4.17(b).

Figure 4.18(a) shows the angular dependence of the dHvA frequencies. The dHvA frequencies for branches α_i and β_i roughly follow the $F(\theta) = F(0)/\cos\theta$ relation, indicating nearly cylindrical Fermi surfaces. In fact, nearly cylindrical Fermi surfaces are theoretically obtained for dHvA branches α_i and β_i, together with γ_i, as shown in Figs. 4.18(b) and (c). The theoretical energy band calculations are described below. The present detected dHvA branches are identified as follows:

1) branches $\alpha_i(i = 1, 2)$ and $\beta_i(i = 1, 2)$ are due to the 26th electron spin-down Fermi surface and the 27th electron spin-up Fermi surface, respectively.
2) branches $\gamma_i(i = 1, 2)$ are due to the 28th electron spin-up Fermi surface.

Theoretically many other dHvA branches, especially based on the band 26th and 25th hole Fermi surfaces, exist, but are not detected experimentally, due mainly to large curvature factors of these Fermi surfaces and the sample quality.

The present energy band calculations were carried out using the full potential LAPW method within the local spin density approximation (LSDA), where the ferromagnetic state is assumed, and calculations were performed self-consistently. The spin-orbit interaction is neglected for simplicity. The total and partial densities of states are shown in Figs. 4.19(a) and (b), respectively. The Fermi surfaces are mainly due to Co-3d electrons. The Fermi surfaces consist of the 26th hole spin-up, 25th hole spin-down, 27th electron spin-up, 26th electron spin-down, and 28th electron spin-up Fermi surfaces, as shown in Fig. 4.18(c). From the self-consistent calculations, Co-3d electrons produce a magnetic moment of 0.4 μ_B/Co.

In summary, we again show in Fig. 4.20 the partial densities of states of $SrCo_2P_2$ and $LaCo_2P_2$ to understand the relation between a nearly ferromagnetic $SrCo_2P_2$ and a ferromagnetic $LaCo_2P_2$. The Fermi energy ε_F is situated at a peak of the partial density of states

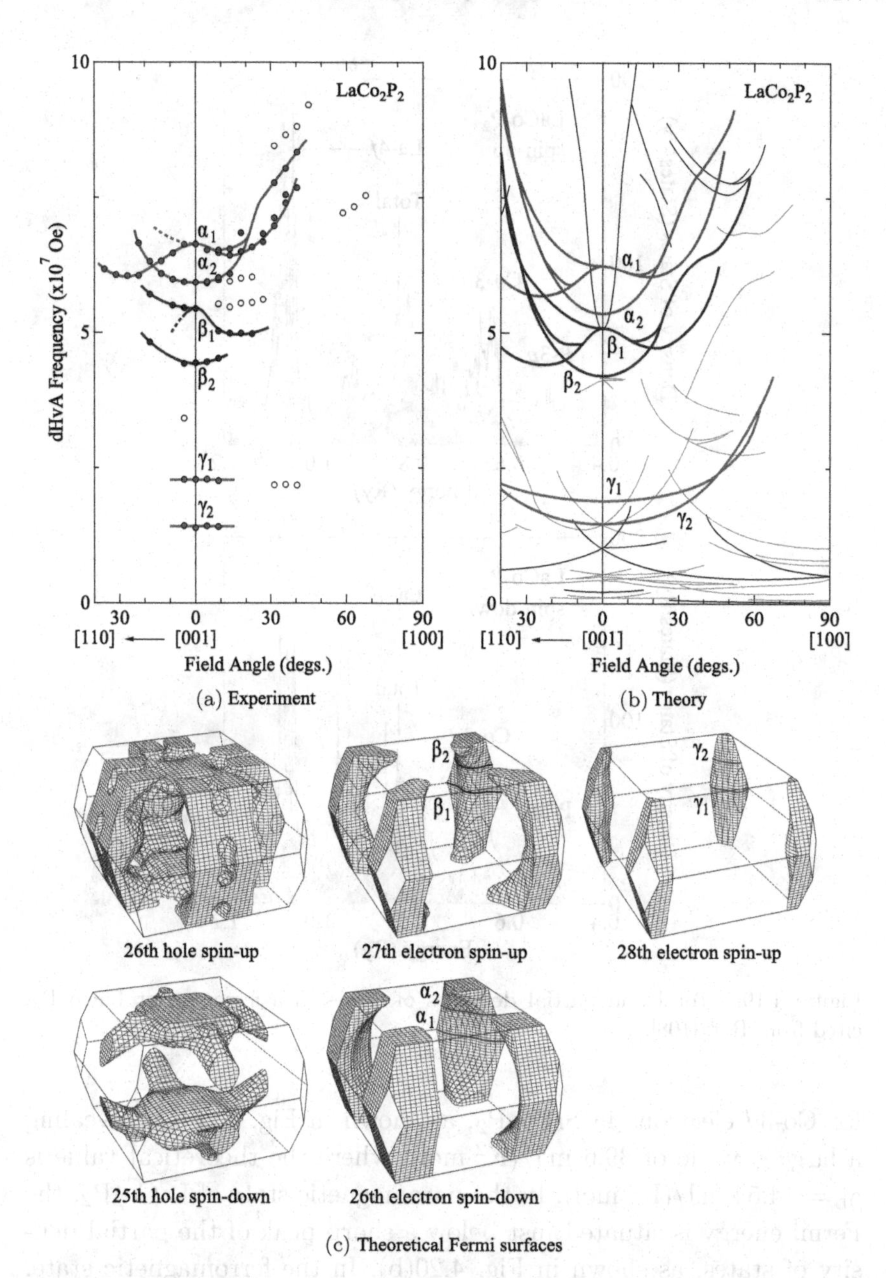

Figure 4.18: (a) Angular dependence of the dHvA frequencies and (b) theoretical ones, and (c) the corresponding theoretical Fermi surfaces with up- and down-spin states in a ferromagnet $LaCo_2P_2$, cited from Ref. [108].

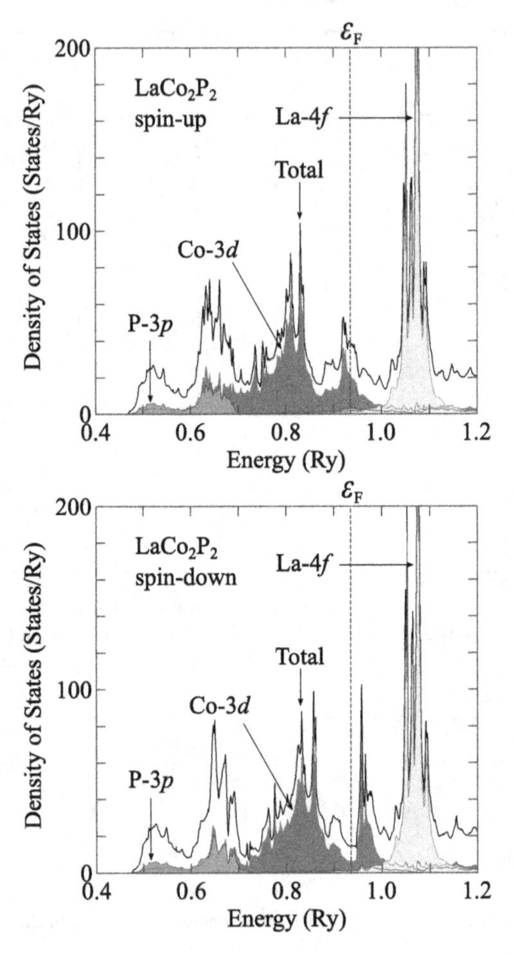

Figure 4.19: Total and partial densities of states in a ferromagnet $LaCo_2P_2$, cited from Ref. [108].

for Co-3d electrons in $SrCo_2P_2$, as shown in Fig. 4.20(a), revealing a large γ value of 39.6 mJ/(K^2·mol), where the theoretical value is $\gamma_b = 14.54$ mJ/(K^2·mol). In the paramagnetic state of $LaCo_2P_2$, the Fermi energy is situated just below a sharp peak of the partial density of states, as shown in Fig. 4.20(b). In the ferromagnetic state, as shown in Fig. 4.20(c), the Fermi energy is shifted to a higher energy from a sharp peak of the spin-up density of states, while it is shifted to a lower energy from the sharp peak of the spin-down

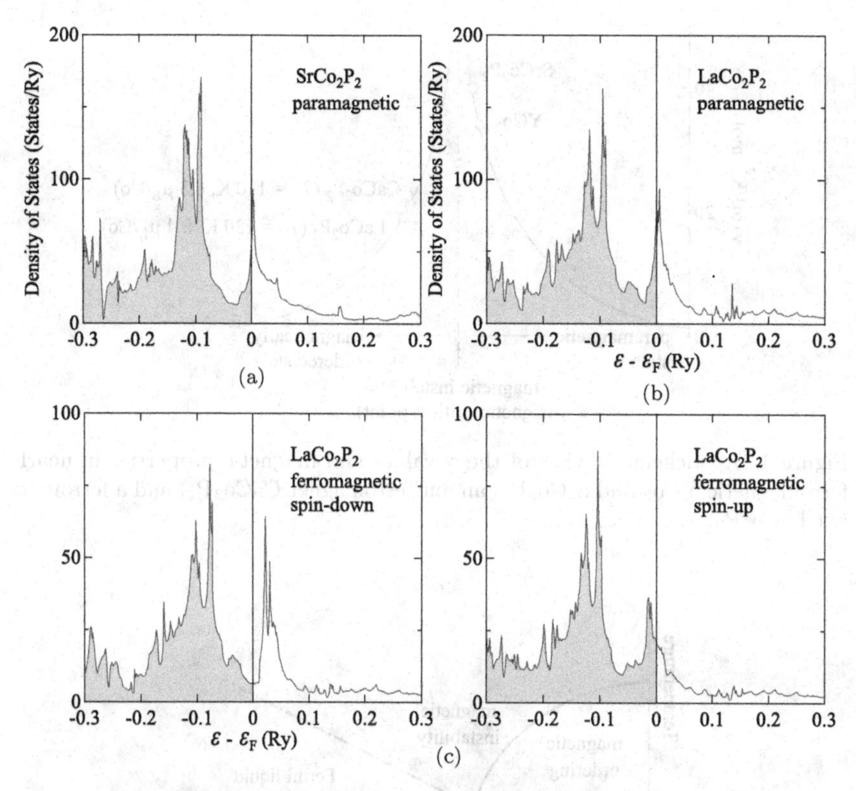

Figure 4.20: Partial densities of states for Co-3*d* electrons in (a) a nearly ferromagnetic $SrCo_2P_2$, (b) in the paramagnetic state of $LaCo_2P_2$, and (c) in the ferromagnetic state of $LaCo_2P_2$ (spin-up and spin-down), cited from Refs. [107, 108].

density of states. The present difference between the two densities of states produces a saturated magnetic moment of 0.4 μ_B/Co in the ferromagnetic ordering of $LaCo_2P_2$. The γ value in a ferromagnet $LaCo_2P_2$ is 19 mJ/(K^2·mol), smaller than the 39.6 mJ/(K^2·mol) observed in $SrCo_2P_2$.

The γ value is large in the nearly ferromagnetic $SrCo_2P_2$ and becomes small when the electronic states are magnetically ordered, as in $LaCo_2P_2$ (T_C = 130 K, μ_s = 0.4 μ_B/Co). Magnetic instability is now defined as the electronic state revealing the ordered moment $\mu_s \to 0$, as schematically shown in Fig. 4.21. The corresponding electronic state is called the magnetic critical point or the quantum

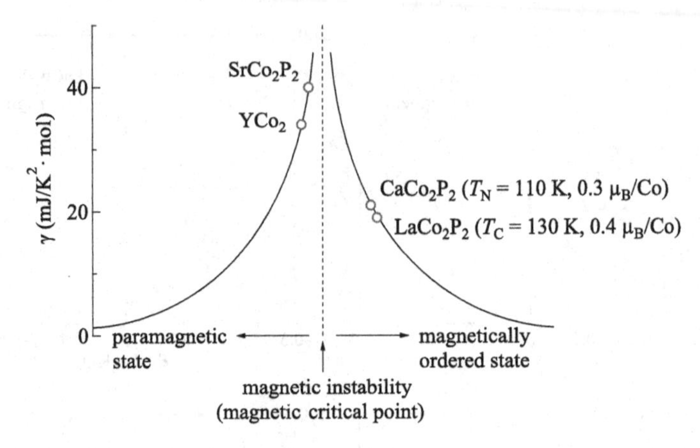

Figure 4.21: Schematic view of the γ values and magnetic properties in nearly ferromagnetic YCo_2 and $SrCo_2P_2$, an antiferromagnet $CaCo_2P_2$, and a ferromagnet $LaCo_2P_2$.

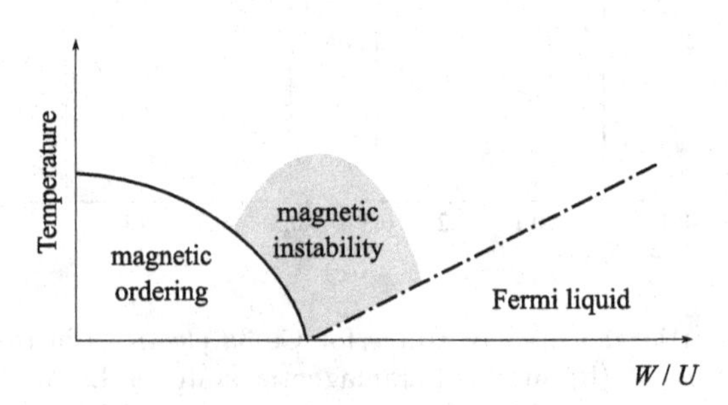

Figure 4.22: Electronic phase diagram close to the magnetic instability, courtesy of T. Moriya and K. Ueda, cited from Ref. [110].

Table 4.2: Quantum critical phenomena based on the SCR theory, courtesy of T. Moriya and K. Ueda, cited from Ref. [110].

	Ferromagnetism		Antiferromagnetism		
	2dim	3dim	2dim	3dim	Fermi liquid
$\chi(Q)^{-1}$	$-T\log T$	$T^{4/3}$	T	$T^{3/2}$	$a+bT^2$
C/T	$T^{-1/3}$	$-\log T$	$-\log T$	$a-bT^{1/2}$	a
ρ	$T^{4/3}$	$T^{5/3}$	T	$T^{3/2}$	T^2
$1/T_1T$	$\chi(Q)^{3/2}$	$\chi(Q)$	$\chi(Q)$	$\chi(Q)^{1/2}$	a

critical point. The electronic phase diagram close to the magnetic critical point is shown in Fig. 4.22, revealing the T vs W/U phase diagram [110]. Here, W is the band-width and U is the Coulomb potential. W/U corresponds to $|J_{\rm cf}|D(\varepsilon_{\rm F})$ in the f electron systems, which will be shown in Chap. 5. At the quantum critical point, the electronic state is very different from the Fermi liquid state. Characteristic electronic properties are calculated on the basis of the SCR theory, as summarized in Table 4.2.

Chapter 5

Heavy Fermions

Localized-f electrons at high temperatures in some Ce and Yb compounds or actinide compounds change into f-itinerant electrons at low temperatures due to the Kondo effect, forming heavy fermions. Correspondingly, the energy scale is changed from the Fermi energy of $\varepsilon_{\mathrm{F}} = 10^4$ K to the Kondo temperature of $T_{\mathrm{K}} = 0\text{–}100\,\mathrm{K}$. Since the energy scale is small in magnitude, the electronic state can be appreciably changed by temperature, magnetic field, and pressure.

5.1 Kondo effect and singlet ground state

Kondo showed that the third order scattering due to the s–d exchange interaction diverges logarithmically with decreasing temperature and determined the origin of a long standing resistivity minimum problem which was found in Cu or Au with small amounts of Fe impurity [111]. This became the start of the Kondo problem, and it took ten years for theorists to solve this divergence problem at the Fermi energy [112].

The electrical resistivity ρ is described by

$$\rho = \frac{1}{nq\mu} = \frac{m^*}{nq^2}\frac{1}{\langle\tau\rangle} \tag{5.1}$$

$$\frac{1}{\tau(\varepsilon)} = \sum_{\boldsymbol{p}'} W(\boldsymbol{p} \to \boldsymbol{p}') \tag{5.2a}$$

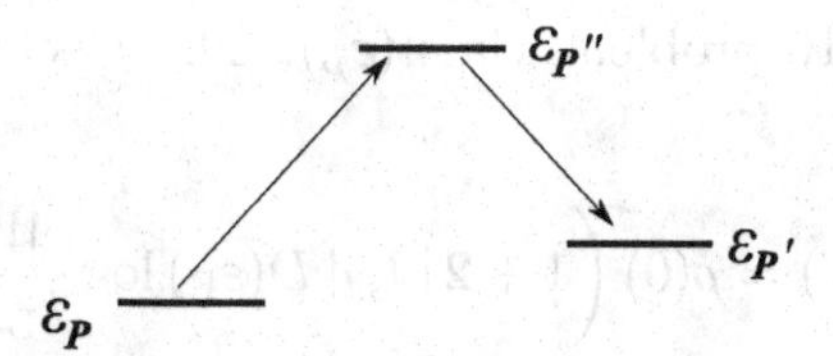

Figure 5.1: Virtual scattering process in the Kondo effect.

where $\langle\tau\rangle$ is the average of the scattering lifetime $\tau(\varepsilon)$ and $\tau(\varepsilon)^{-1}$ is proportional to the scattering probability $W(\boldsymbol{p} \to \boldsymbol{p}')$. An electron is scattered by a magnetic impurity from an initial electronic state $\varepsilon_{\boldsymbol{p}}$ $(\boldsymbol{p} = \hbar\boldsymbol{k}, \boldsymbol{\sigma})$ to a final electronic state $\varepsilon_{\boldsymbol{p}'}$ $(\boldsymbol{p}' = \hbar\boldsymbol{k}', \boldsymbol{\sigma}')$, revealing the spin-flip scattering of the conduction electron. At the same time, the spin state of the magnetic impurity M is changed to M'. The third order scattering involves the scattering process $\varepsilon_{\boldsymbol{p}} \to \varepsilon_{\boldsymbol{p}''} \to \varepsilon_{\boldsymbol{p}'}$, as shown in Fig. 5.1, where the electronic state $\varepsilon_{\boldsymbol{p}''}$ corresponds to an intermediate electronic state or the virtual state. An exact calculation of $W(\boldsymbol{p} \to \boldsymbol{p}')$ contains a term of

$$W_{\boldsymbol{p}'}(\boldsymbol{p} \to \boldsymbol{p}') \sim \sum_{\boldsymbol{p}''} \frac{n(\varepsilon_{\boldsymbol{p}''})}{\varepsilon_{\boldsymbol{p}} - \varepsilon_{\boldsymbol{p}''}}. \tag{5.2b}$$

This term is important for a good understanding of the Kondo effect. The magnetic exchange interaction between the spin $\boldsymbol{s}$ of the conduction electron (s) and the spin $\boldsymbol{S}$ of the magnetic impurity (d) is

$$H_{sd} = -2J_{sd}\boldsymbol{s} \cdot \boldsymbol{S} \tag{5.3}$$

where the Kondo effect occurs when the exchange integral J_{sd} becomes negative; $J_{sd} < 0$. The electrical resistivity based on the second order scattering is proportional to $J_{sd}^2 S(S+1)$ which is temperature-independent. Here, $S = 1/2$ in the present case. With antiferromagnetic coupling, one conduction electron is scattered by one magnetic impurity, but the exact calculation of the third order scattering involves the Fermi-Dirac distribution function $n(\varepsilon_{\boldsymbol{p}}) = 1/[(\varepsilon_{\boldsymbol{p}} - \mu)/k_{\mathrm{B}}T + 1]$ in Eq. (5.2b). This does not become the simple scattering problem of the conduction electron but corresponds to the many-body problem via $n(\varepsilon_{\boldsymbol{p}})$. The electrical resistivity is expressed as

$$\begin{aligned}\rho(T) &= \rho(0)\left(1 + 2\,|J_{sd}|\,D(\varepsilon_{\mathrm{F}})\log\frac{W}{k_{\mathrm{B}}T}\right)\\ &= \rho(0)\left(1 - 2\,|J_{sd}|\,D(\varepsilon_{\mathrm{F}})\log\frac{k_{\mathrm{B}}T}{W}\right)\end{aligned} \tag{5.4a}$$

where $\rho(0)$ is due to the second order scattering and W is the band width. The electrical resistivity thus increases logarithmically with decreasing temperature, following $-\log T$ dependence. This means $\rho(T) \to \infty$ for $T \to 0$. Here, the fourth, fifth, ... order scatterings lead to additional terms of $(\log T)^2$, $(\log T)^3$, ..., respectively [113], resulting in

$$\begin{aligned}\rho(T) &= \rho(0)\left[1 + 2\,|J_{sd}|\,D(\varepsilon)\log\frac{W}{k_{\mathrm{B}}T}\right.\\ &\qquad \left. +3\left(|J_{sd}|\,D(\varepsilon)\log\frac{W}{k_{\mathrm{B}}T}\right)^2 + \cdots\right]\\ &= \frac{\rho(0)}{\left(1 - |J_{sd}|\,D(\varepsilon)\log\frac{W}{k_{\mathrm{B}}T}\right)^2}.\end{aligned} \tag{5.4b}$$

The electrical resistivity $\rho(T)$ becomes infinite at a characteristic temperature

$$T = T_{\mathrm{K}} = \frac{W}{k_{\mathrm{B}}}e^{-\frac{1}{|J_{sd}|D(\varepsilon)}}. \tag{5.5}$$

Note that T_{K} in Eq. (5.5) is not a transition temperature but is a cross-over temperature, meaning that the electronic state gradually changes in nature as a function of temperature.

For the simplest case with no orbital degeneracy, the localized spin $\uparrow$(l) of the magnetic impurity is compensated by the spin $\downarrow$(c) of the conduction electron at $T = 0$, forming the Kondo cloud. Consequently, the singlet state $\uparrow$ (l) $\downarrow$ (c) or $\downarrow$ (l) $\uparrow$ (c) is formed with a binding energy $k_{\mathrm{B}}T_{\mathrm{K}}$, relative to the magnetic state [114]. Here, T_{K} is called the Kondo temperature. The logarithmic divergence which

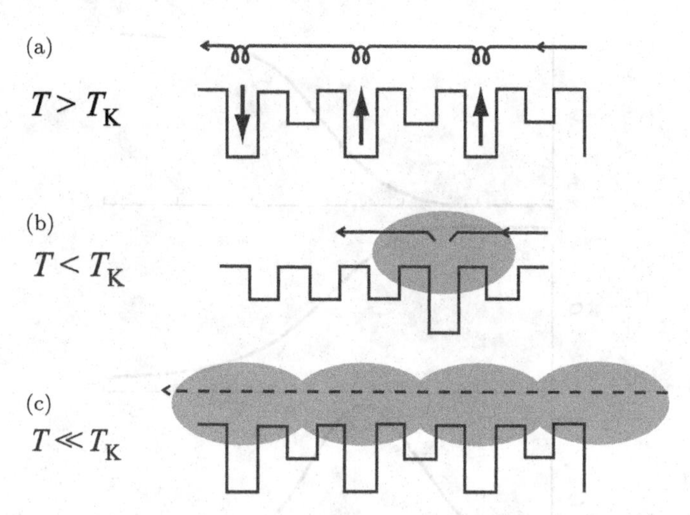

Figure 5.2: (a) Kondo scattering in the impurity system at $T > T_{\mathrm{K}}$, (b) the corresponding Kondo cloud at $T < T_{\mathrm{K}}$, and (c) the f-derived band in the periodic Kondo lattice.

appears in the perturbation is thus changed into this singlet bound state at low temperatures. The Kondo scattering, the Kondo cloud, and the f-derived band are schematically shown in Figs. 5.2(a), (b), and (c), respectively.

The temperature dependence of magnetic moment M, electrical resistivity ρ, specific heat C, and thermoelectric power S are schematically shown in Fig. 5.3. The magnetic moment is reduced below T_{K} and becomes zero at $T = 0$ due to the formation of the Kondo singlet bound state, whereas the corresponding magnetic susceptibility follows the Curie-Weiss law at temperatures higher than T_{K}. The electrical resistivity increases logarithmically with decreasing temperature and reaches a constant residual resistivity, known as the "unitarity limit". T_{K} corresponds to the temperature that demonstrates half of the unitarity limit. The specific heat possesses a peak at $T \simeq T_{\mathrm{K}}/3$, revealing

$$\int_0^{T_{\mathrm{K}}} C dT \simeq k_{\mathrm{B}} T_{\mathrm{K}}. \tag{5.6}$$

The thermoelectric power also possesses a peak around T_{K}.

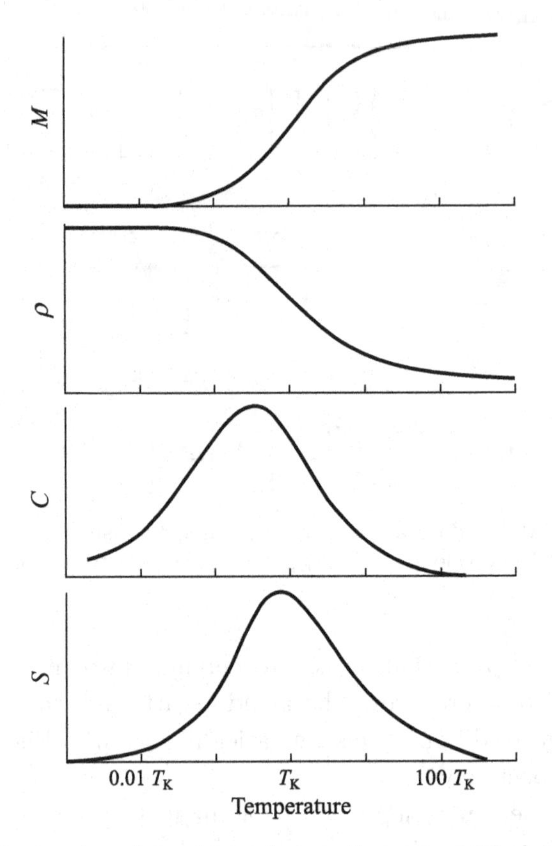

Figure 5.3: Temperature dependence of magnetic moment M, electrical resistivity ρ, specific heat C, and thermoelectric power S in the impurity Kondo effect.

In the impurity Kondo state, the low-energy excitation at low temperatures is described on the basis of the local Fermi liquid. The linear T dependence of the specific heat C is

$$C = \gamma T. \tag{5.7}$$

The T^2 dependence of the electrical resistivity ρ and the magnetic susceptibility χ is

$$\rho = \rho_0 - AT^2 \tag{5.8}$$

and

$$\chi = \chi_0 - BT^2 \tag{5.9}$$

in which the ratios of $\sqrt{A}/\gamma$ and χ_0/γ are constant. In particular, $\chi_0/\gamma(=R_W)$ is called the Wilson ratio, which is $R_W = 1$ for the usual electron, and $R_W \simeq 2$ for strongly correlated electrons.

The magnetoresistance becomes negative, revealing

$$\begin{aligned}\Delta\rho/\rho &= \{\rho(H) - \rho(H=0)\}/\rho(H=0) \\ &= -\sin^2(\pi M/2M_0)\end{aligned} \tag{5.10}$$

where M_0 is the magnetization in infinite magnetic field at 0 K [115]. Equation (5.10) is simply expressed as $\Delta\rho/\rho \sim -M^2 \sim -H^2$ if $M/M_0 \ll 1$ and $M = \chi H$.

The Kondo temperature is found to depend on the number of $3d$ electrons in the transition metal. The highest Kondo temperature $T_K \simeq 1000$ K in the Ti($3d^2$) impurity decreases down to $T_K \simeq 0.01$ K in the Mn($3d^5$) impurity, but increases up to $T_K \simeq 1000$ K in the Ni($3d^8$) impurity [116].

These experimental results suggest that the Kondo effect might occur in Ce and Yb impurities among the rare earths. Here, the electronic configuration is $4f^1$ in Ce^{3+} and $4f^{13}$ in Yb^{3+}, where the CEF ground state is the Kramers doublet, as shown in Fig. 2.14. In the case of Ce^{3+}, the CEF splitting energies of three doublets are defined as Δ_1 and Δ_2. The Kondo temperature T_K^h at high temperatures ($T > \Delta_1$, Δ_2) is obtained as

$$T_K^h = \frac{W}{k_B} e^{-\frac{1}{3|J_{cf}|D(\varepsilon_F)}} \tag{5.11}$$

where J_{sd} in Eq. (5.5) is changed to J_{cf} in the f-electron systems. On the other hand, at low temperatures ($T < \Delta_1$, Δ_2), the Kondo temperature is

$$T_K = \frac{W^2}{\Delta_1\Delta_2} \cdot \frac{W}{k_B} e^{-\frac{1}{|J_{cf}|D(\varepsilon_F)}}. \tag{5.12}$$

To simplify our thinking, T_K is defined as

$$T_K^0 = \frac{W}{k_B} e^{-\frac{1}{|J_{cf}|D(\varepsilon_F)}}. \tag{5.13}$$

This is the same as the Kondo temperature in Eq. (5.5). If we consider only the doublet ground state, the Kondo temperature is T_K^0 in Eq. (5.13), but the exact Kondo temperature is enhanced to T_K in Eq. (5.12) due to low-lying CEF excited doublets [117]. For example, if we assume $T_K = 5$ K under $\Delta_1 = 100$ K and $\Delta_2 = 180$ K in $Ce_xLa_{1-x}Al_2$, $T_K^h = 45$ K is obtained from a relation of $(k_BT_K^h)^3 = \Delta_1\Delta_2(k_BT_K)$, whereas $T_K^0 \simeq 10^{-3}$ K. The electrical resistivity in $Ce_xLa_{1-x}Al_2$ possesses a two-peak structure at 5 K and 60 K, which roughly corresponds to T_K and T_K^h, respectively [118].

In the 1970s, we suddenly discovered various anomalous rare earth compounds, typically Ce compounds, in which Kondo-like behavior was observed even in the dense system. In these Ce compounds, different from the $3d$ case, the ground state cannot be a scattering state but is a coherent Kondo lattice state. The Kondo lattice state possesses an extremely large γ value of 1000 mJ/(K^2·mol) or an extremely large cyclotron mass of $100m_0$.

5.2 Impurity Kondo effect and Kondo lattice

In this section, we present a change of electronic states in $Ce_xLa_{1-x}Cu_6$ as a function of the constitution x [119–121]. $Ce_xLa_{1-x}Cu_6$ crystallizes in the monoclinic structure at low temperature, as shown in Fig. 5.4. Note that $CeCu_6$ crystallizes in

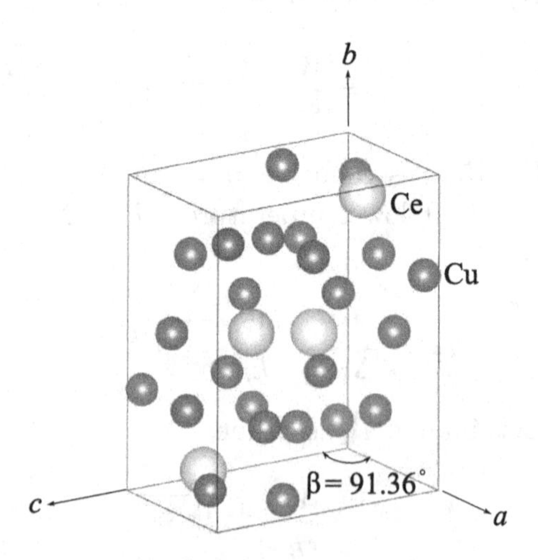

Figure 5.4: Monoclinic structure of $CeCu_6$.

the orthorhombic structure (No. 62, a = 8.105 Å, b = 5.105 Å, c = 10.159 Å, Z = 4) at room temperature, but is changed into the monoclinic one (No. 14, a = 5.080 Å, b = 10.121 Å, c = 8.067 Å, $\beta = 91.36°$) below about 200 K [122]. Similarly this happens below about 500 K in $LaCu_6$ (a = 5.143 Å, b = 10.204 Å, c = 8.144 Å, $\beta = 91.40°$). In the following, we use the notation of the monoclinic structure.

$Ce_xLa_{1-x}Cu_6$ ($x = 0.094$) corresponds to a Kondo system with impurity, while $CeCu_6$ corresponds to the coherent Kondo lattice or a typical heavy fermion. Figures 5.5(a) and (b) show the temperature dependences of electrical resistivities in $Ce_xLa_{1-x}Cu_6$. We define here the magnetic resistivity $\rho_{\mathrm{m}} = \rho_{\mathrm{Ce}_x\mathrm{La}_{1-x}\mathrm{Cu}_6} - \rho_{\mathrm{LaCu}_6}$. The electrical resistivity for $x = 0.094$ corresponds to the impurity Kondo system. As shown in Fig. 5.5(a), the electrical resistivity for $x = 0.094$ shows a resistivity minimum at around 30 K, a $-\log T$ dependence at lower temperatures, and finally, a constant resistivity, the unitarity limit. The Kondo temperature for the $x = 0.094$ sample is estimated to be 3.7 K from the resistivity below 20 K on the basis of the Suhl-Nagaoka theory [123].

Figures 5.6(a) and (b) show the cerium concentration dependence of the residual resistivities ρ_{m_0} in $Ce_xLa_{1-x}Cu_6$. The residual resistivity in $CeCu_6$ is larger than that of $LaCu_6$, but is of the same order as $LaCu_6$. Later, we will discuss the residual resistivity of $CeCu_6$, which is based on a $4f$-derived heavy band. Here, we emphasize that the residual resistivity of $Ce_xLa_{1-x}Cu_6$ can be simply explained by an alloy model, with each Ce atom contributing a resistivity equal to the unitarity limit, 320 $\mu\Omega$·cm. As shown in Fig. 5.6(b), the residual resistivity per mole of cerium decreases almost linearly with concentration x and tends to zero in $CeCu_6$: ρ_0 = 5.7 $\mu\Omega$·cm in the present sample of $CeCu_6$. The residual resistivity thus follows the so-called Nordheim law, $320x(1-x)$, as expected for an alloy model.

Next, we show in Fig. 5.7 the low-temperature resistivity of $CeCu_6$. The Fermi liquid relation $\rho = \rho_0 + AT^2$ is applicable below 0.1 K, as shown in Fig. 5.7. The A(and ρ_0) values are 41.6(5.71), 76.3(10.6), and 142.7 $\mu\Omega$·cm/K^2 (18.1 $\mu\Omega$·cm) for $J \| a$, b, and c-axes, respectively, revealing an extremely large A value.

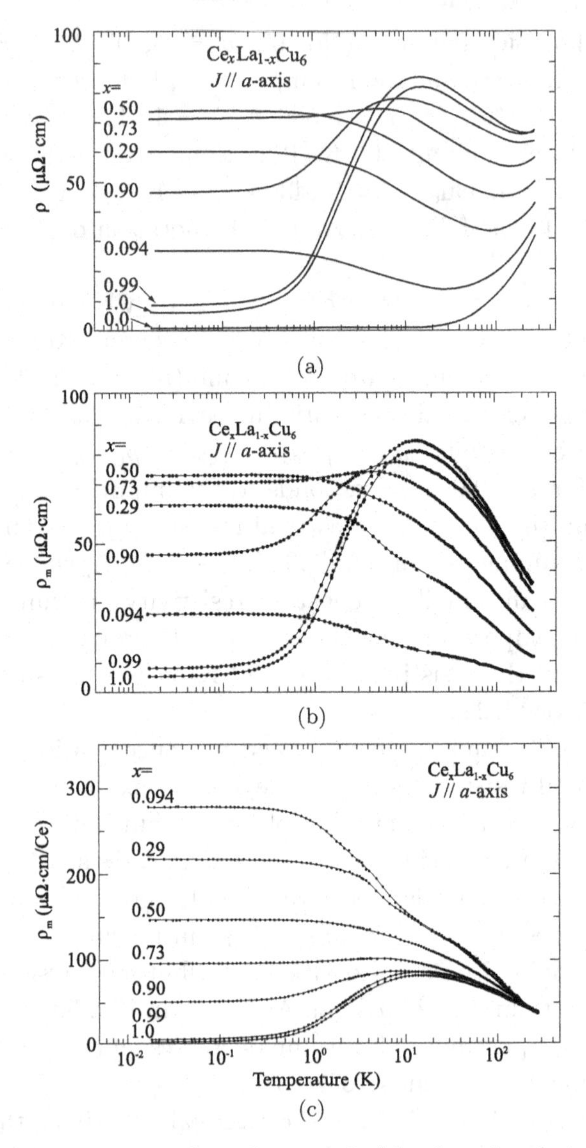

Figure 5.5: Temperature dependences of (a) the electrical resistivities, (b) magnetic resistivities, and (c) magnetic resistivities per molar cerium in $Ce_xLa_{1-x}Cu_6$, where $\rho_m = \rho_{Ce_xLa_{1-x}Cu_6} - \rho_{LaCu_6}$, cited from Ref. [119].

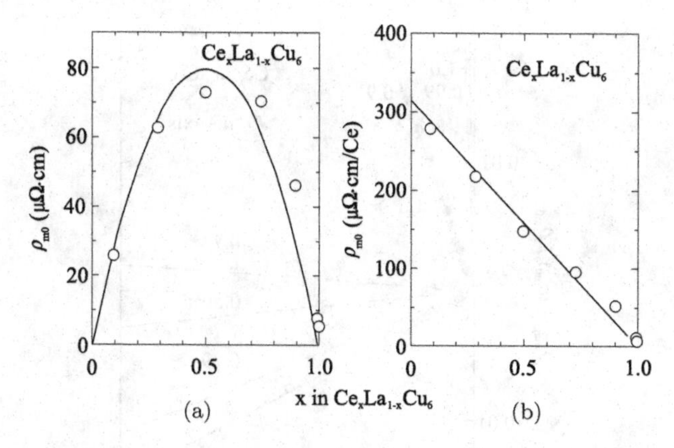

Figure 5.6: Cerium concentration dependences of (a) the residual resistivities and (b) the molar residual resistivities. Solid lines in (a) and (b) are $320x(1-x)$ $\mu\Omega$·cm and $320(1-x)$ $\mu\Omega$·cm, respectively, cited from Ref. [119].

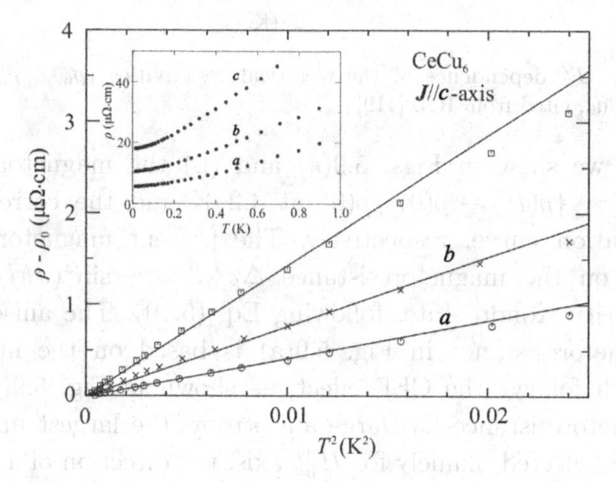

Figure 5.7: T^2 dependence of the electrical resistivities $\rho - \rho_0$ in $CeCu_6$. Inset shows the T dependence of ρ, cited from Ref. [119].

In a dilute Kondo substance, the resistivity approaches a constant value of the unitarity limit in proportion to T^2, hence the data of $Ce_xLa_{1-x}Cu_6$ can be fitted to $\rho_m = \rho_{m0}\{1-(T/T_0)^2\}$, as shown in Fig. 5.8. The slope varies from negative to positive as a function of the concentration x, reflecting a consecutive change from a dilute Kondo system to a Kondo lattice system in $Ce_xLa_{1-x}Cu_6$.

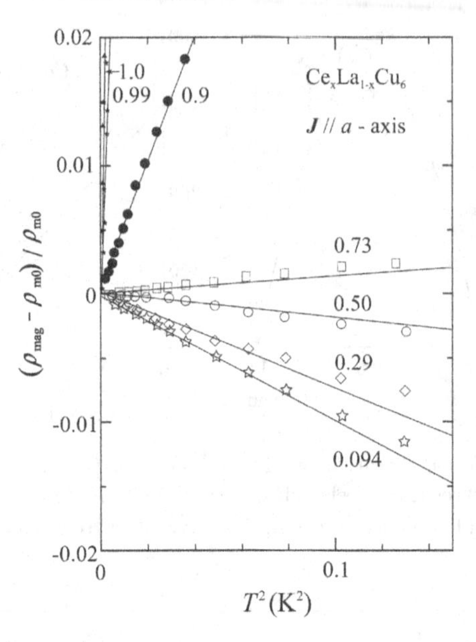

Figure 5.8: T^2 dependence of the electrical resistivities $(\rho_{\rm m} - \rho_{\rm m0})/\rho_{\rm m0}$ in $Ce_xLa_{1-x}Cu_6$, cited from Ref. [119].

Next, we show in Figs. 5.9(a) and (b) the magnetoresistance of $\Delta\rho/\rho = \{\rho(H) - \rho(0)\}/\rho(0)$ at 4.2 K and the corresponding magnetization curve, respectively. The present magnetoresistance is based on the magnetoresistance $\Delta\rho/\rho = -\sin^2(\pi M/2M_0)$ in the impurity Kondo state, following Eq. (5.10). The anisotropy of the magnetoresistance in Fig. 5.9(a) is based on the magnetization which follows the CEF effect, as shown in Fig. 5.9(b). Since the magnetoresistance has large anisotropy, the largest magnetoresistance is selected, namely for $H\|b$-axis, the direction of a magnetic easy-axis.

To compare the magnetoresistance in the dilute Kondo system with that in the Kondo lattice system, we show in Figs. 5.10(a) and (b) the temperature dependence of the magnetoresistance in $Ce_{0.094}La_{0.906}Cu_6$ and $CeCu_6$ under several magnetic fields, respectively. The magnetoresistance of $Ce_{0.094}La_{0.906}Cu_6$ is negative and becomes constant below 0.2 K in high magnetic fields, which is characteristic of a dilute Kondo system. On the other hand, the negative magnetoresistance in $CeCu_6$ possesses a minimum around 1.5 K and

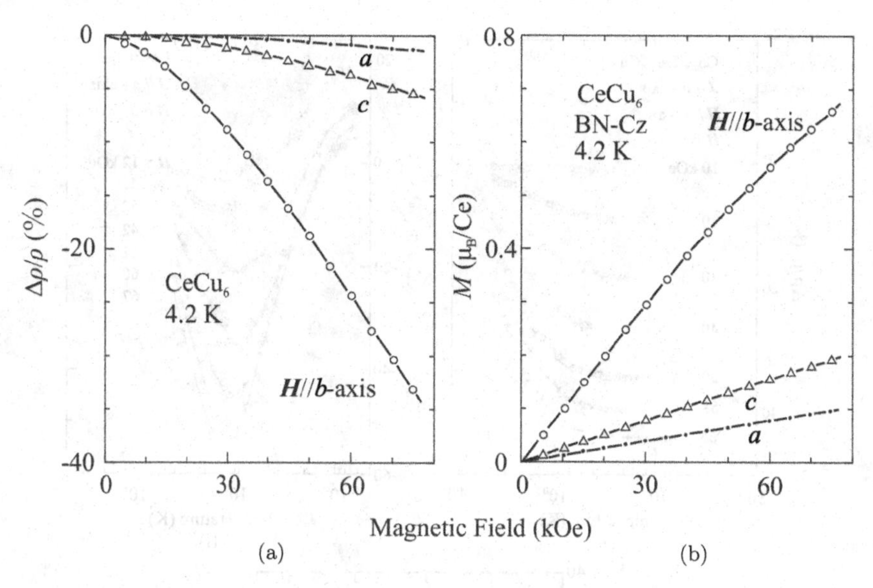

Figure 5.9: (a) Magnetoresistance and (b) the magnetization curve at 4.2 K for $H \| a$-, b-, and c-axes in $CeCu_6$, cited from Ref. [119].

changes from negative to positive sign at lower temperatures. To clarify the magnetoresistance in $CeCu_6$, we show in Fig. 5.10(c) the field dependence of the magnetoresistance under several temperatures. The negative magnetoresistance at 1.5 K follows a $-H^2$ dependence, which is characteristic of the dilute Kondo effect, as mentioned above. At lower temperatures, the magnetoresistance indicates a somewhat complicated behavior: a positive magnetoresistance appears at low fields, reduces in magnitude around 20 kOe, but increases again with further increasing field. A change of the magnetization around 20 kOe is related to the metamagnetic behavior in magnetization. The positive magnetoresistance is attributed to the usual magnetoresistance due to the Lorenz force for the conduction electrons with the heavy cyclotron masses. The singlet ground state in each cerium site forms a $4f$-derived band in the Kondo lattice system. The formation of the corresponding band is reflected in the positive magnetoresistance, as shown in Fig. 5.10(c), which occurs below about 1 K in $CeCu_6$.

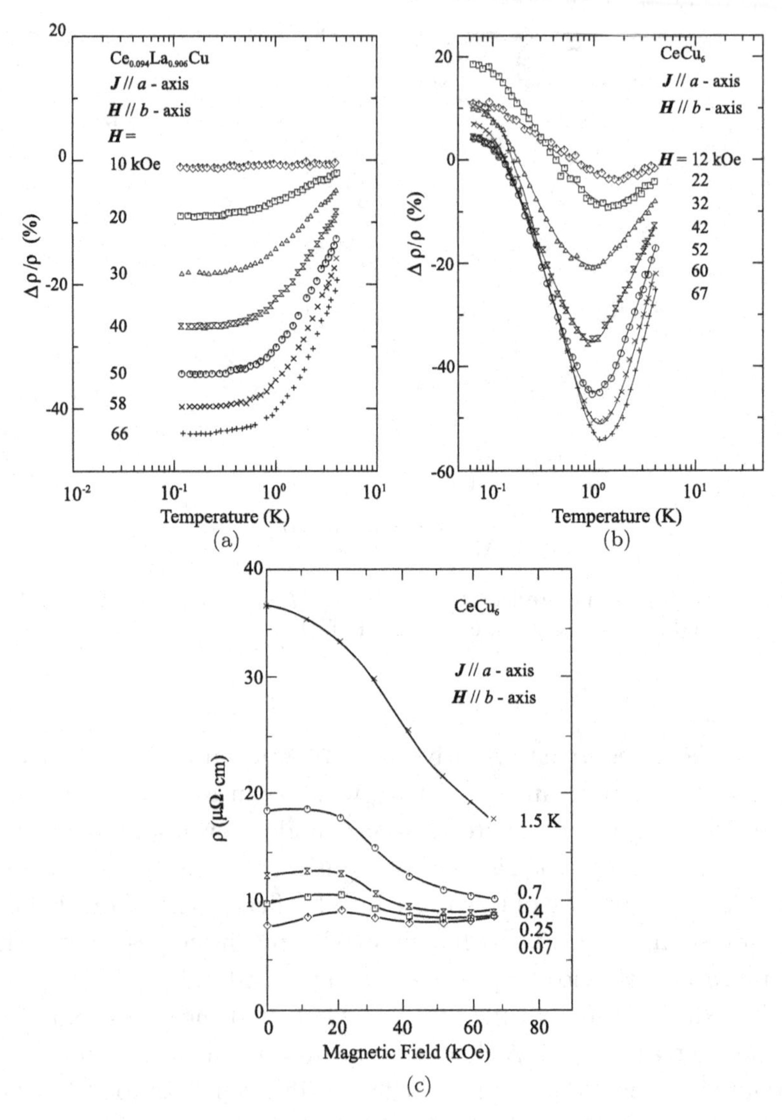

Figure 5.10: Temperature dependence of magnetoresistances of (a) $Ce_{0.094}$ $La_{0.906}Cu_6$ and (b) $CeCu_6$ under several magnetic fields, (c) the field dependence of magnetoresistances of $CeCu_6$ under several temperatures for $J\|a$-axis and $H\|b$-axis, cited from Ref. [119].

Figures 5.11(a) and (b) show the temperature dependence of the magnetic susceptibility χ and the magnetic specific heat in the form of $C_{\rm m}/T$ in $\rm Ce_xLa_{1-x}Cu_6$. The magnetic susceptibility follows the Curie-Weiss law in the temperature range from 40 K to room temperature. Since the effective Bohr magneton of 2.5 $\mu_{\rm B}$/Ce and the paramagnetic Curie temperature $\theta_{\rm p} = -8$ K are approximately independent of the cerium concentration, it is concluded that each cerium ion behaves independently at high temperatures. With decreasing temperature, the magnetic susceptibility deviates from the Curie-Weiss law and tends to reach a constant, χ_0, reflecting the disappearance of the local moment due to the Kondo effect – the Kondo lattice or the heavy fermion state is formed. A large χ_0 value corresponds to a large γ value or a large A value in $\rho = \rho_0 + AT^2$.

The magnetic specific heat in the form of $C_{\rm m}/T = \{C_{\rm Ce_{1-x}La_{1-x}Cu_6} - C_{\rm LaCu_6}\}/T$ increases steeply with decreasing temperature and becomes constant below 0.2 K: $C_{\rm m}/T = 2.1$ J/(K^2·mol) for $x = 0.29$. The Kondo temperature $T_{\rm K}$ is estimated to be 2.7 K, following a solid line obtained from name theory [124, 125]. $\rm CeCu_6$ also possesses an extremely large value of $C_{\rm m}/T = 1.5$–1.6 J/(K^2·mol).

This is simply understood as follows. At low temperatures, the magnetic entropy of the ground-state doublet in the CEF scheme, $R\log 2$, is obtained by integrating the magnetic specific heat $C_{\rm m}$ in the form of $C_{\rm m}/T$. If the magnetic specific heat is changed into the electronic specific heat γT via the Kondo effect, the following equations are obtained:

$$R\log 2 = \int_0^{T_{\rm K}} \frac{C_{\rm m}}{T} dT \tag{5.14}$$

and

$$C_{\rm m} = \gamma T. \tag{5.15}$$

The electronic specific heat coefficient γ can be obtained from Eqs. (5.14) and (5.15) as

$$\gamma = \frac{R\log 2}{T_{\rm K}} \simeq \frac{10^4}{T_{\rm K}} \quad [\text{mJ/K}^2 \cdot \text{mol}]. \tag{5.16}$$

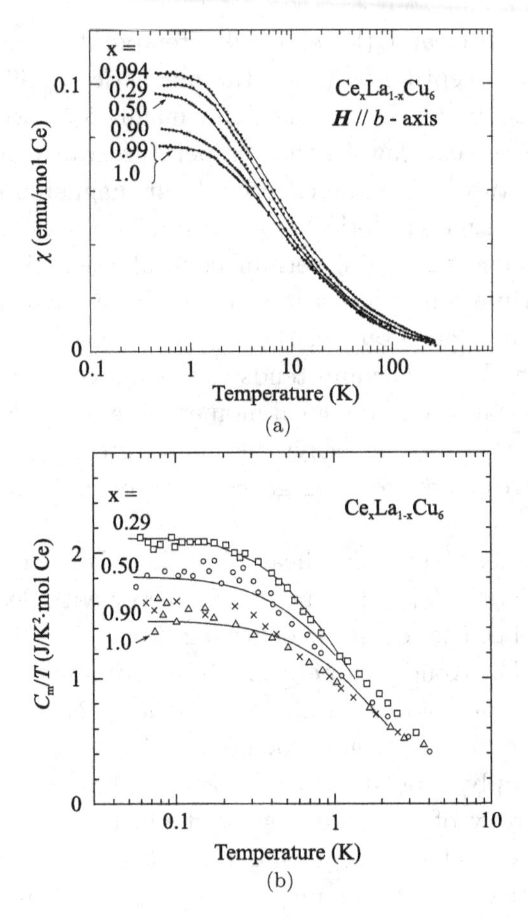

Figure 5.11: Temperature dependence of the magnetic susceptibility χ and magnetic specific heat in the form of C_{m}/T in $\mathrm{Ce}_x\mathrm{La}_{1-x}\mathrm{Cu}_6$, cited from Refs. [120, 121].

For example, the γ value is roughly estimated to be 2000 mJ/(K^2·mol) for $T_{\mathrm{K}} = 5$ K, which is consistent with the expected $\gamma = 1500$–1600 mJ/(K^2·mol) in $\mathrm{CeCu_6}$. In the s and p electron systems, the usual value of $\gamma = 1$ mJ/(K^2·mol) is obtained under $T_{\mathrm{K}} = T_{\mathrm{F}} = 10^4$ K. In fact, an extremely large cyclotron mass of $120m_0$ (m_0: rest mass of an electron) is detected in the similar heavy fermion compound $\mathrm{CeRu_2Si_2}$ with $\gamma = 350$ mJ/(K^2·mol) and $T_{\mathrm{K}} = 20$ K, as shown in Fig. 3.39. The localized $4f$-electronic states at high temperatures in these compounds are thus changed into a $4f$-derived band with a flat energy vs momentum dispersion, possessing an extremely large effective mass.

The dHvA experiment for $H \| b$-axis was carried out to confirm the presence of conduction electrons with extremely large cyclotron masses in $CeCu_6$ [126, 127]. For three dHvA branches, the corresponding cyclotron masses were determined, indicating a marked decrease of the cyclotron mass with increasing magnetic field. From the field dependence of the $\sqrt{A}$ value of the electrical resistivity $\rho = \rho_0 + AT^2$ and the γ value [128], the cyclotron masses for three dHvA branches at $H = 0$ were determined, as shown in Figs. 5.12(a) and (b). Here, the $\sqrt{A}$ values were determined for three current directions (indicated with circles for $J \| c$-axis, triangles for $J \| a$, and squares for $J \| b$) in the magnetic field along the b-axis. The cyclotron mass at $H = 0$ is estimated to be $23m_0$ for the branch with $F = 1.22 \times 10^6$ Oe, $160m_0$ for $F = 9.75 \times 10^6$ Oe, and $140m_0$ for $F = 1.24 \times 10^7$ Oe (not shown in Fig. 5.12), revealing extremely large cyclotron masses at $H = 0$. The cyclotron masses are extremely reduced in magnitude in a field range from 0 to about 5T or 50 kOe, which corresponds to the metamagnetic field region. The heavy fermion state is destroyed in magnetic fields.

Here, we reconsider the heavy fermion state of $CeCu_6$ from another viewpoint of magnetic moments obtained from the neutron scattering experiment. It is understood that localized-$4f$ electrons hybridize with conduction electrons and are changed into itinerant-$4f$ electrons at low temperatures, forming heavy fermions. In fact, $4f$-CEF excitations, which are explained in Sec. 3.2.8 for $PrIr_2Zn_{20}$, are not observed in $CeCu_6$. However, the magnetic moments are not completely reduced to zero in magnitude. New magnetic excitations are produced below 3 K in $CeCu_6$, which were observed in inelastic neutron scattering experiments with an energy resolution of 0.1 meV [129, 130]. In these experiments, it is important to consider that $CeCu_6$ has Ising-type magnetic properties, as shown in Fig. 5.9(b), revealing that an easy-axis of magnetization corresponds to the b-axis. This means that energy scans performed for scattering vectors either around $\boldsymbol{Q} = (0, 0, 1)$ or $\boldsymbol{Q} = (1, 0, 0)$ will give evidence for the magnetic scattering.

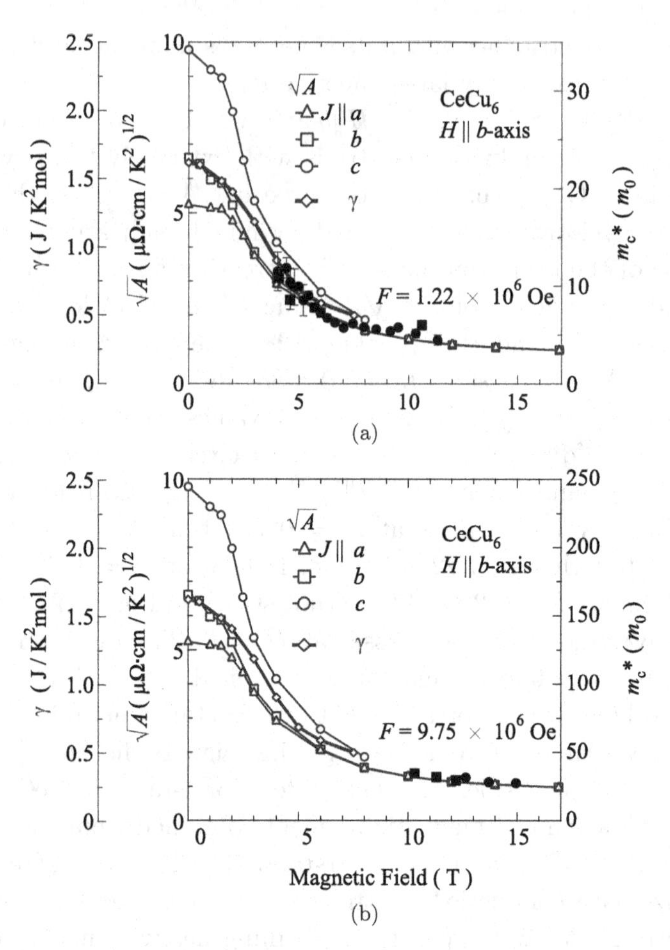

Figure 5.12: Field dependence of the cyclotron masses m_c^* for two dHvA branches with (a) $F = 1.22 \times 10^6$ Oe and (b) $F = 9.75 \times 10^6$ Oe (solid squares from Ref. [126] and solid circles from Ref. [127]), together with $\sqrt{A}$ cited from Ref. [127] and γ from Ref. [128] for $H \| b$-axis in $CeCu_6$.

Figure 5.13 shows energy scans carried for $\boldsymbol{Q} = (0, 0.9, 0)$ and $\boldsymbol{Q} = (0, 0, 0.9)$ at 0.1 K. No signal is observed for the scattering vector $\boldsymbol{Q} = (0, 0.9, 0)$, indicating longitudinal magnetic fluctuations along the b-axis. On the other hand, a signal is observed for $\boldsymbol{Q} = (0, 0, 0.9)$, which can be clearly separated from the incoherent nuclear scattering. The magnetic contribution is maximum at an energy of $\omega_0 = 0.25$ meV and extends up to about 0.9 meV.

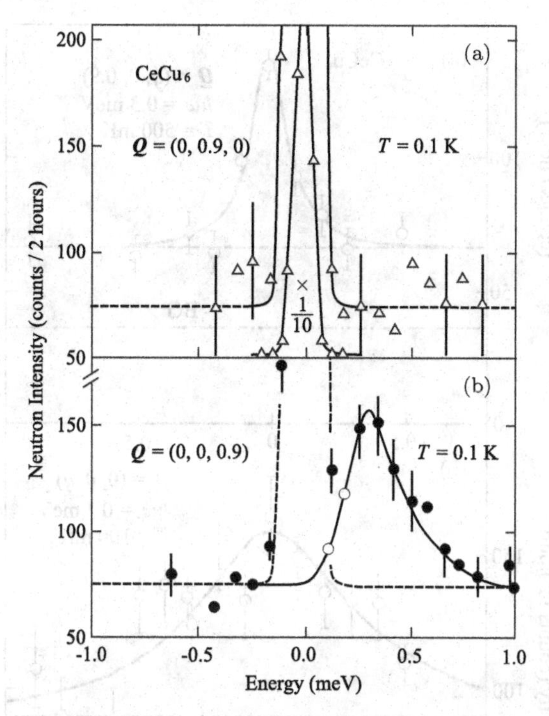

Figure 5.13: Energy scans for $CeCu_6$ at (a) $\boldsymbol{Q} = (0, 0.9, 0)$ and (b) $\boldsymbol{Q} = (0, 0, 0.9)$, cited from Ref. [129].

The present magnetic excitation was analyzed on the basis of Eq. (3.98) as follows:

$$\chi(\boldsymbol{Q}, \omega) = \chi_0 \frac{1}{1 + i\frac{\omega - \omega_0}{\Gamma}} \tag{5.17}$$

$$S(\boldsymbol{Q}, \omega) \propto \frac{1}{1 - e^{-\hbar\omega/k_B T}} \frac{\omega}{\Gamma} \frac{1}{1 + \left(\frac{\omega - \omega_0}{\Gamma}\right)^2}. \tag{5.18}$$

A solid line in Fig. 5.13(b) represents the best fit with $\omega_0 = 0.25$ meV and $\Gamma = 0.19$ meV.

Figures 5.14(a) and (b) show $\boldsymbol{Q}$-scans performed around (0,0,1) along the [100] and [001] directions, respectively. Two types of magnetic contributions can be identified from these data, especially, along [100], as shown in Fig. 5.14(a): a $\boldsymbol{Q}$-independent contribution and, superimposed to it, a more or less peaked contribution. The

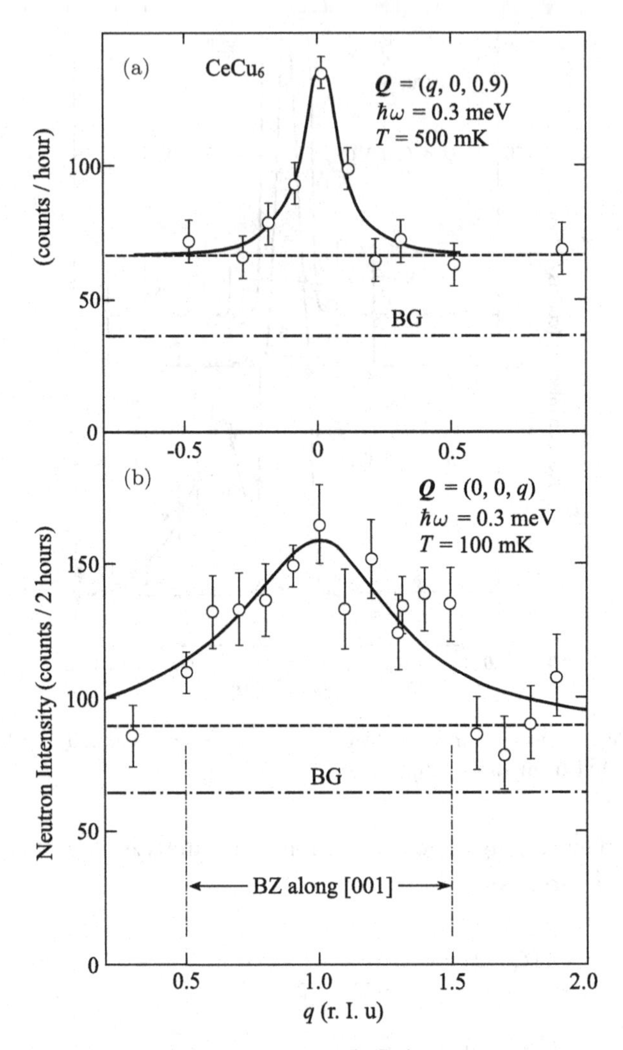

Figure 5.14: (a) and (b) $\boldsymbol{Q}$-scans around (0,0,1) in $CeCu_6$ along the [100] and [001] directions, respectively, cited from Ref. [130]. BG refers to the background.

former one indicates that there exist localized excitations arising from single-site Kondo-spin fluctuations. The latter contribution shows evidence for inter-site antiferromagnetic correlations between Ce^{3+} magnetic moments in adjacent (001) planes, with a wave vector $\boldsymbol{K} = (0, 0, 1)$. Both contributions consist of 90 % of the single-site and 10 % of the inter-site.

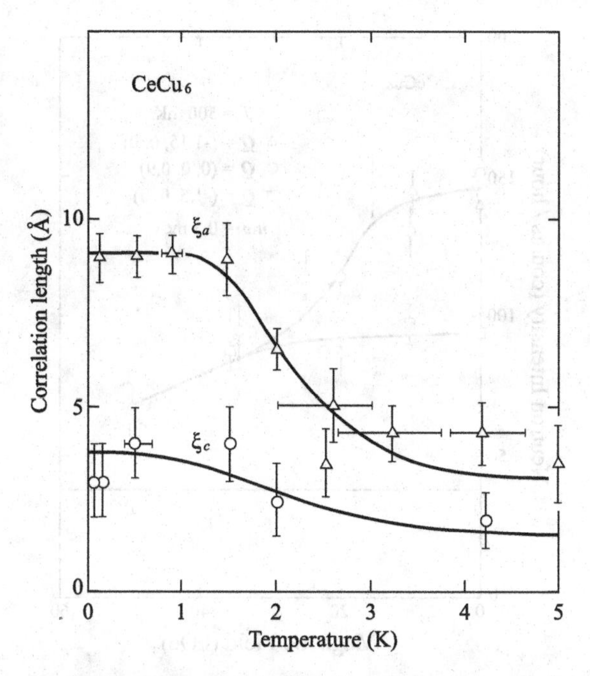

Figure 5.15: Temperature dependence of the magnetic correlation length along the a- and c-directions in $CeCu_6$, cited from Ref. [130].

Among both [100] and [001], the peaked contribution can be accounted for by a Lorentzian function, which allows us to deduce the correlation lengths $\xi_a = (9 \pm 1)$ Å and $\xi_c = (3.8 \pm 0.5)$ Å from the width of the peaked structure Δq using the relation $\xi = 2/\Delta q$. The short-range antiferromagnetic correlations extend up to the second neighbors along the a-axis and the first neighbors along the c-axis.

Furthermore, the magnetic scattering was studied as a function of temperature from 0.025 to 5 K. The $\boldsymbol{Q}$-independent contribution does not change in this temperature range, whereas the peaked contribution starts to broaden above about 1 K. The corresponding temperature dependence of the correlation length is shown in Fig. 5.15. Note that the correlation length becomes infinite when antiferromagnetic ordering is realized in $CeCu_6$. Quantum fluctuations based on the Kondo effect are strong enough to suppress the divergence of the correlation length.

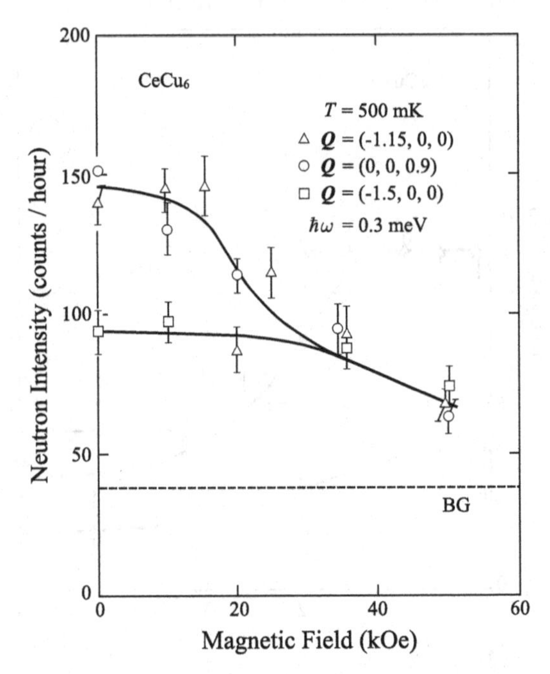

Figure 5.16: Magnetic field dependence of intensities at $\boldsymbol{Q} = (-1.15, 0, 0)$, $(0, 0, 0.9)$, and $(-1.5, 0, 0)$ for magnetic field along the b-axis of $CeCu_6$, cited from Ref. [130].

The metamagnetic behavior in $CeCu_6$, which was observed in the magnetoresistance at about 20 kOe, as shown in Fig. 5.10(c), was also studied by the inelastic neutron scattering experiment. This occurs below about 1 K only for the magnetic field along the b-axis, namely along the magnetic easy-axis. Figure 5.16 shows the field dependence of the magnetic intensities, measured at a finite energy $\hbar\omega = 0.3$ meV. The intensity measured at $\boldsymbol{Q} = (-1.5, 0, 0)$ corresponds to the single-site contribution, whereas at $\boldsymbol{Q} = (-1.15, 0, 0)$ and $\boldsymbol{Q} = (0, 0, 0.9)$, both the single-site and inter-site contributions are observed. Therefore, the difference between these two curves directly gives the inter-site contribution. Note that an inflection point for a decrease of the intensity for $\boldsymbol{Q} = (-1.15, 0, 0)$ and $\boldsymbol{Q} = (0, 0, 0.9)$ corresponds to the metamagnetic field $H_{\mathrm{m}} = 25$ kOe. It is thus concluded that the steep decrease of the intensity in the field range from 0 to 30 kOe is mainly attributed to a decay of the inter-site contribution. The field dependence of the intensity in Fig. 5.16 is very similar to the field dependences of $\sqrt{A}$, γ, and m_c^* in Fig. 5.12.

The similar metamagnetic behavior was precisely studied for $CeRu_2Si_2$ with the tetragonal structure. This is also of the Ising type [130], which occurs at 70 kOe for the magnetic field along the tetragonal [001] direction. Single-site contributions make up 60 % of the total contribution, while 40 % comes from the inter-site in $CeRu_2Si_2$. The metamagnetic behaviors are therefore strongly observed in the magnetization, specific heat, magnetoresistance, and cyclotron mass, together with the topology of the Fermi surface [6].

5.3 Doniach phase diagram and metamagnetism

The competition between the RKKY interaction and the Kondo effect was discussed by Doniach as a function of $|J_{cf}|D(\varepsilon_F)$, as shown in Fig. 5.17(a) [131, 132]. The RKKY interaction enhances

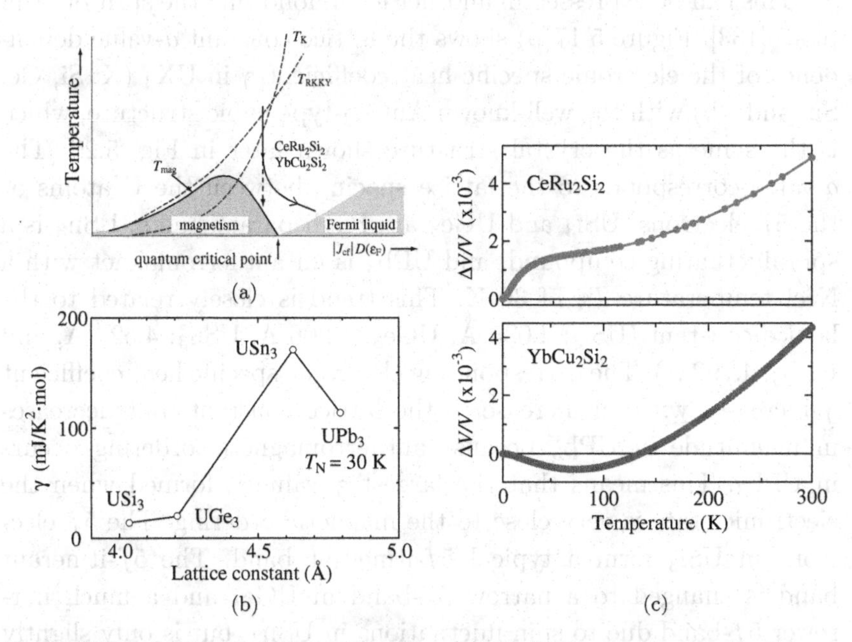

Figure 5.17: (a) Doniach phase diagram, (b) the lattice constant a dependence of the electronic specific heat coefficient γ in UX_3 (X: Si, Ge, Sn, and Pb) with the $AuCu_3$-type cubic structure, and (c) the temperature dependence of the volume in $CeRu_2Si_2$ and $YbCu_2Si_2$, cited from Ref. [131].

the long-range magnetic order, where the f electrons are treated as localized electrons, and the indirect f-f interaction mediated by the spin polarization of the conduction electrons. The intensity of the RKKY interaction is proportional to the square of the magnetic exchange interaction J_{cf} and the electronic density of states $D(\varepsilon_{\rm F})$ at the Fermi level $\varepsilon_{\rm F}$. The corresponding ordering temperature $T_{\rm RKKY}$ is characterized by

$$T_{\rm RKKY} \propto D(\varepsilon_{\rm F})|J_{cf}|^2.$$

On the other hand, the Kondo effect quenches the magnetic moment of the localized f-electrons by the spin polarization of the conduction electrons, producing a spin singlet state with a binding energy $k_{\rm B}T_{\rm K}$ so that

$$T_{\rm K} \propto e^{-\frac{1}{|J_{cf}|D(\varepsilon_{\rm F})}}.$$

This can be expressed in another form, following the sprit of "Hill plot" [133]. Figure 5.17(b) shows the lattice constant a-value dependence of the electronic specific heat coefficient γ in UX_3 (X: Si, Ge, Sn, and Pb) with the well-known $AuCu_3$-type cubic structure, which is the same as the crystal structure shown later in Fig. 5.20. The a-value corresponds to the lattice spacing between the U atoms or the $5f$ electrons. USi_3 and UGe_3 are Pauli paramagnets, USn_3 is a spin-fluctuating compound, and UPb_3 is an antiferromagnet with a Néel temperature $T_{\rm N}$ of 30 K. This trend is closely related to the lattice constant (USi_3: 4.035 Å, UGe_3: 4.206 Å, USn_3: 4.626 Å, and UPb_3: 4.792 Å). The corresponding electronic specific heat coefficient γ increases with an increase in the lattice constant, but decreases in magnitude in UPb_3 because antiferromagnetic ordering occurs in UPb_3. This means that the largest γ value is formed when the electronic state is very close to the magnetic ordering. The $5f$ electrons in USi_3 form a typical $5f$-itinerant band. The $5f$-itinerant band is changed to a narrow $5f$-band in UGe_3 and a much narrower $5f$-band due to spin fluctuations in USn_3, but is only slightly changed in UPb_3 due to the antiferromagnetic ordering. Similar plots of $T_{\rm N}$ and γ vs $\sqrt[3]{a^2c}$ are observed also in $CeTX_3$ (T: Co, Rh, Ir; X: Si, Ge) compound with the tetragonal structure [6]. It is necessary in these plots that the number of valence electrons is the same as in UX_3 or $CeTX_3$. We also note that the $5f$ electrons in U compounds are slightly different from the $4f$ electrons in Ce(Yb) compounds in nature. At low temperatures, the $5f$ electrons are itinerant, even if the corresponding U compounds are in the magnetically ordered state. On the other hand, $4f$ electrons are mainly localized in the magnetically ordered state, but changes into the $4f$-itinerant heavy fermion state under pressure. A change of the $4f$ electronic states from "localized" to "itinerant" is one of the main phenomena in Ce and Yb compounds. Nevertheless, both $5f$ and $4f$ electrons are localized in magnetism at temperatures higher than room temperature [134].

Furthermore, we will consider the electronic states of $CeRu_2Si_2$ and $YbCu_2Si_2$ with the $ThCr_2Si_2$-type tetragonal structure, focusing on the quantum critical point and change of volume in the crystal, as shown in Fig. 5.17(c) [131]. The 6-fold $4f$ electronic states based on the total angular momentum $J = 5/2$ in $CeRu_2Si_2$ (8-fold $4f$ states based on $J = 7/2$ in $YbCu_2Si_2$) split into three Kramers doublets (four doublets) in the crystalline electric field. The CEF-splitting energies are denoted as Δ_1 and Δ_2 in $CeRu_2Si_2$. Here, the electronic configuration is $4f^1$ in Ce^{3+} and $4f^{13}$ in Yb^{3+}. The ground state in $CeRu_2Si_2$ (and $YbCu_2Si_2$) is thus the Kramers doublet. This doublet induces the magnetically ordered state at low temperatures if the system only follows the RKKY interaction, as shown by a dotted line in Fig. 5.17(a). A change of the $4f$-localized state in the CEF scheme to a non-magnetic $4f$-itinerant state, as shown by a solid line in Fig. 5.17(a), corresponds to intensive shrinkage of the volume below 20 K in $CeRu_2Si_2$ (expansion of the volume below 80 K in $YbCu_2Si_2$), as shown in Fig. 5.17(c). These characteristic temperatures correspond to the Kondo temperatures: $T_K = 20$ K in $CeRu_2Si_2$ and $T_K = 80$ K in $YbCu_2Si_2$. At low temperatures, both compounds are not in the magnetically ordered state but in the $4f$-itinerant heavy fermion state.

The experimental results suggest that T_K increases as a function of pressure P in cerium compounds including $CeRu_2Si_2$ and $CeCu_6$, discussed later. Correspondingly, the magnetically ordered electronic states of the cerium compounds such as $CeRhIn_5$ with the Néel temperature T_N = 3.8 K and $CeIrSi_3$ with T_N = 5 K, shown later, can be tuned to the non-magnetic 4f-itinerant state, crossing the quantum critical point: $T_N \to 0$ for $P \to P_c$. The quantum critical point is experimentally defined as an electronic state for $T_N \to 0$ when $P \to P_c$. On the contrary, the non-magnetic 4f-itinerant state of $YbCu_2Si_2$ can be changed into the ferromagnetic state above P_c = 7 GPa [135]. Later shown through $YbIr_2Zn_{20}$, the quantum critical point is experimentally defined as an electronic state for the metamagnetic field $H_m \to 0$ for $P \to P_c$. In the quantum critical region, the heavy fermion state is realized.

These strongly correlated electrons follow the Fermi liquid relation at low temperatures. The electrical resistivity ρ varies as $\rho = \rho_0 + AT^2$, where ρ_0 is the residual resistivity. The coefficient $\sqrt{A}$ correlates with an enhanced Pauli susceptibility $\chi(T \to 0) \simeq \chi_0$ and with a large electronic specific heat coefficient $C/T(T \to 0) \simeq \gamma$. Note that the present Fermi liquid relation is not applicable to the electronic state at the quantum critical point. For example, the T^2 dependence of the electrical resistivity is changed into the T-linear dependence, as in $CeCoIn_5$ [6]. In Chap. 4, we noted the characteristic electronic properties in the quantum critical region based on the SCR theory for the 3d-electrons (see Table 4.2). These properties might be applicable to the f electrons. The nature of the non-Fermi liquid is, however, different from compound to compound. Theoretical approaches are proposed and described in Refs. [136, 137].

In this text, we mainly present the characteristic properties of the heavy fermion state, where the Fermi liquid relation is applicable. Furthermore, note that the Fermi liquid relation is applicable in magnetic fields even if it is not applicable to the electronic state at zero magnetic field. One of the characteristic properties in heavy fermion compounds is metamagnetic behavior or an abrupt nonlinear increase of magnetization at the magnetic field H_m below the characteristic temperature $T_{\chi\mathrm{max}}$ at which the magnetic susceptibility becomes maximum. Figures 5.18(a) and (b) show the logarithmic scale of temperature dependence of the magnetic susceptibility in $CeCu_6$, $CeRu_2Si_2$, UPd_2Al_3, UPt_3, USn_3, $CeSn_3$, $NpGe_3$, $YbCo_2Zn_2$,

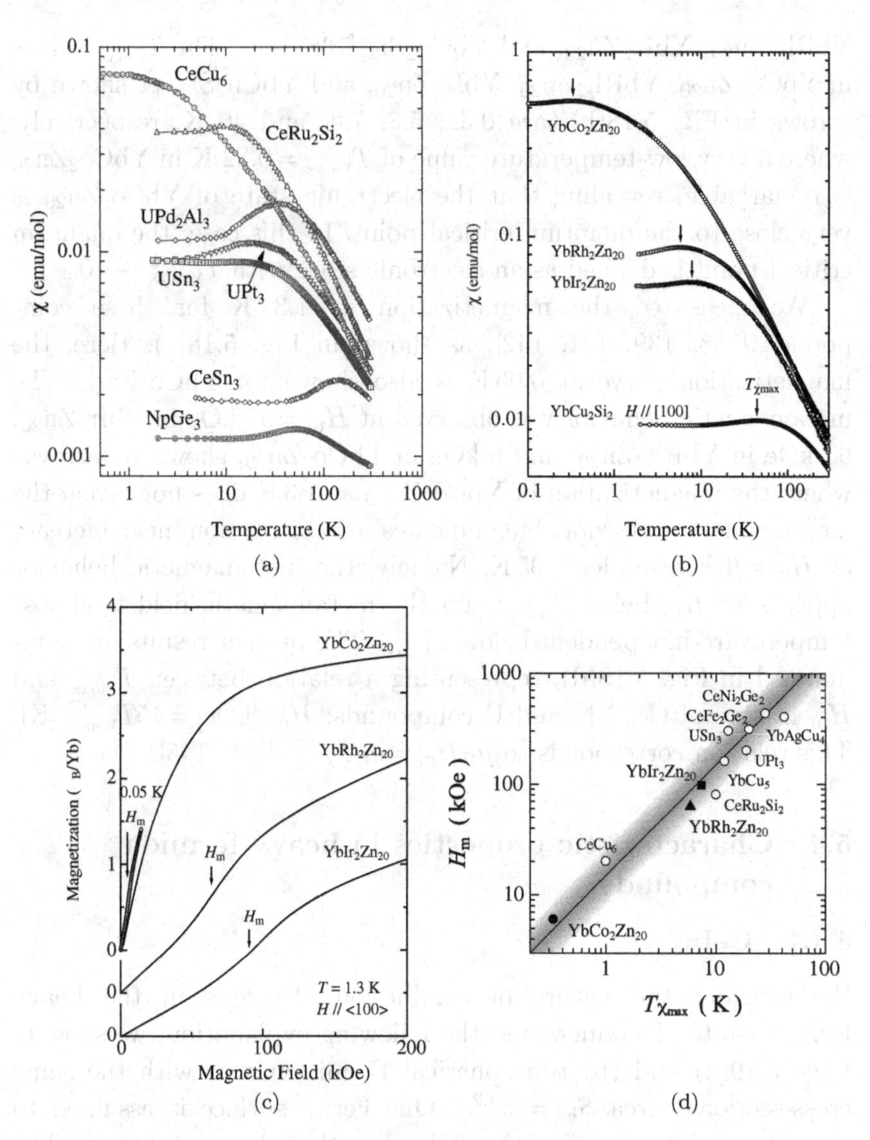

Figure 5.18: Temperature dependence of magnetic susceptibilities χ showing maxima at $T_{\chi\max}$ in (a) Ce and U compounds, and (b) $YbCu_2Si_2$ and YbT_2Zn_{20} (T: Co, Rh, In), (c) magnetization curves in YbT_2Zn_{20}, and (d) a relation between $T_{\chi\max}$ vs H_m, cited from Ref. [131].

$YbRh_2Zn_{20}$, $YbIr_2Zn_{20}$, and $YbCu_2Si_2$ [138–144]. The $T_{\chi_{max}}$ values in $YbCo_2Zn_{20}$, $YbRh_2Zn_{20}$, $YbIr_2Zn_{20}$, and $YbCu_2Si_2$, as shown by arrows in Fig. 5.18(b), are 0.32, 5.8, 7.4, and 40 K, respectively, where a very low-temperature value of $T_{\chi_{max}} = 0.32$ K in $YbCo_2Zn_{20}$ is remarkable, revealing that the electronic state of $YbCo_2Zn_{20}$ is very close to the quantum critical point. In this case, the quantum critical point is defined as an electronic state with $T_{\chi max} \rightarrow 0$.

We measured the magnetization at 1.3 K for these compounds [138, 139, 141, 142], as shown in Fig. 5.18(c). Here, the magnetization curve at 0.05 K is also shown for $YbCo_2Zn_{20}$. The metamagnetic behavior was observed at $H_m = 97$ kOe in $YbIr_2Zn_{20}$, 63 kOe in $YbRh_2Zn_{20}$, and 6 kOe in $YbCo_2Zn_{20}$, shown by arrows, where the magnetization of $YbCo_2Zn_{20}$ at 1.3 K does not reveal the metamagnetic behavior, but indicates an abrupt nonlinear increase at $H_m = 6$ kOe under 0.05 K. Namely, the metamagnetic behavior appears at H_m below $T_{\chi_{max}}$ and the metamagnetic field is almost temperature-independent below $T_{\chi_{max}}$. The present results are summarized in Fig. 5.18(d), representing a relation between $T_{\chi_{max}}$ and H_m in several Ce, Yb, and U compounds: H_m [kOe] $= 15T_{\chi_{max}}$ [K]. This relation corresponds to $\mu_B H_m = k_B T_{\chi_{max}}$ [138, 145].

5.4 Characteristic properties in heavy fermion compounds

5.4.1 $CeIn_3$

We consider the nature of conduction electrons in the heavy fermion state. To understand the following explanation, we show in Figs. 5.19(a) and (b) two spherical Fermi surfaces with the same cross-sectional area $S_F (= \pi k_F^2)$. One Fermi surface is assumed to posses a cyclotron mass m_c^*, while the other does $m_c^*(1 + \lambda)$. The Fermi velocity v_F and the scattering lifetime τ are discussed.

A conduction electron with a large cyclotron mass is believed to move slowly in the crystal [146]. It was confirmed experimentally that the product of the cyclotron mass and the inverse scattering lifetime or the Dingle temperature $m_c^* \tau^{-1}$ or $m_c^* T_D$ is constant by using two

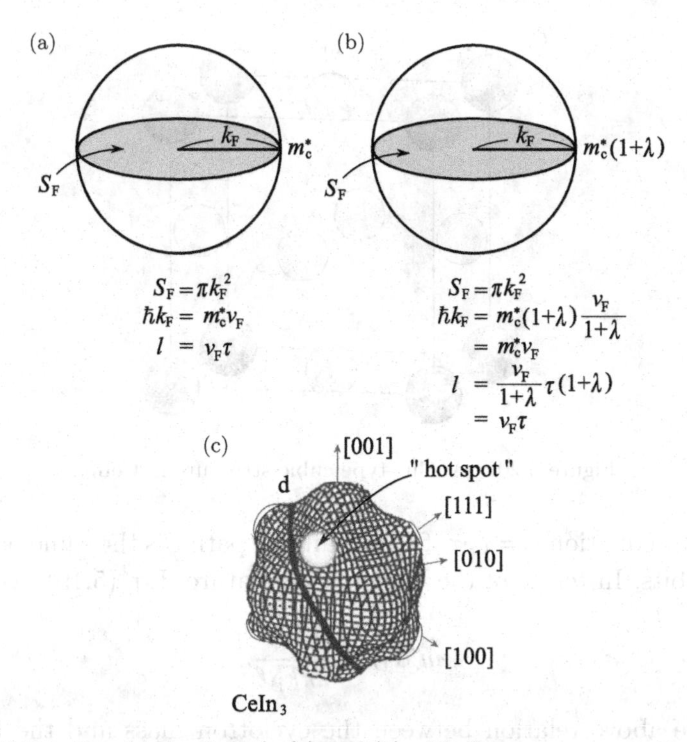

Figure 5.19: Two kinds of orbits (a) and (b) with the same cross-sectional area S_F in k-space, where the cyclotron mass is m_c^* in (a) and $m_c^*(1+\lambda)$ in (b). (c) A nearly spherical Fermi surface named "d" in $CeIn_3$, revealing "hot spots" along the $\langle 111 \rangle$ direction.

orbits of a nearly spherical Fermi surface, denoted d, in $CeIn_3$. Note that $T_D = (\hbar/2\pi k_B)\tau^{-1}$, following Eq. (3.147). Using relations $\hbar k_F = m_c^* v_F$ and $\ell = v_F \tau$, the following equation is obtained:

$$\frac{m_c^*}{\tau} = \frac{\hbar k_F}{\ell}. \tag{5.19}$$

Thus, m_c^*/τ is expected to be the same for two similar orbits with the same mean free path ℓ. When the effective mass m_c^* is enhanced by a factor of $(1+\lambda)$ due to the many-body Kondo effect, τ should also be enhanced by a factor of $(1+\lambda)$. This occurs because a large cyclotron mass is translated to a small velocity from the relation $\hbar k_F = m_c^* v_F$ so that the scattering lifetime is enhanced by $(1+\lambda)$

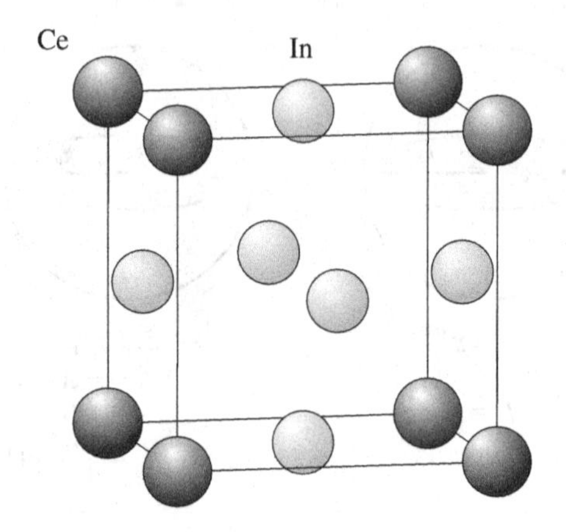

Figure 5.20: $AuCu_3$-type cubic structure in $CeIn_3$.

from the equation $\ell = v_F \tau$ if the mean free path ℓ is the same between two orbits. In terms of the Dingle temperature, Eq. (5.19) becomes

$$m_c^* T_D = \frac{\hbar^2 k_F}{2\pi k_B \ell}. \tag{5.20}$$

The above relation between the cyclotron mass and the Dingle temperature was applied to the nearly spherical Fermi surface denoted d in $CeIn_3$ with the $AuCu_3$-type cubic structure (No. 221, $a = 4.691$ Å, $Z = 1$), as shown in Fig. 5.20. The cyclotron mass m_c^* was determined from the so-called mass plot shown in Fig. 5.21(a), where a slope of the mass plot corresponds to m_c^*. The Dingle temperature T_D was determined from the Dingle plot in Fig. 5.21(b), where the slope corresponds to $m_c^* T_D$. The dHvA frequency F, the cyclotron mass m_c^* and the Dingle temperature T_D are 3.15×10^7 Oe, $2.44\, m_0$ and 0.19 K for $H \parallel \langle 111 \rangle$, and 3.29×10^7 Oe, $12\, m_0$ and 0.04 K for the field direction close to the $\langle 110 \rangle$ direction, indicating that the product of $m_c^* T_D$ is constant for the two orbits. Note that the slope of the Dingle plot is the same between two field directions, as shown in Fig. 5.21(b). We can thus conclude that the larger the cyclotron mass the longer the scattering lifetime, so that the value of $m_c^* \tau^{-1}$ or $m_c^* T_D$ is constant. It has been verified experimentally that the cyclotron mass in $CeIn_3$ is not uniform over the nearly spherical Fermi surface.

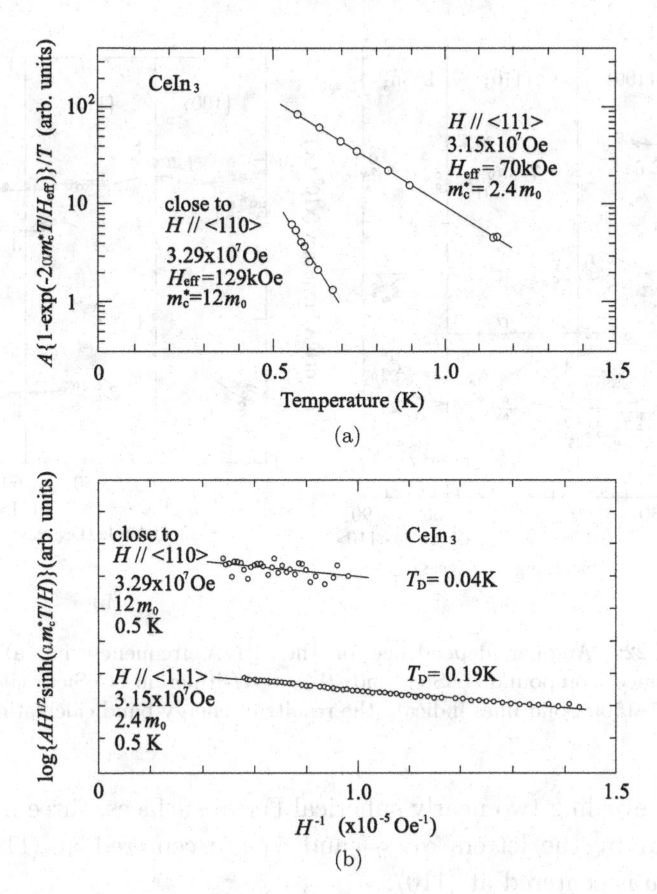

Figure 5.21: (a) Mass plot and (b) the Dingle plot on a nearly spherical Fermi surface named d in $CeIn_3$ for two field directions, cited from Ref. [146].

Instead, "hot spots" with extremely large cyclotron masses exist in the $\langle 111 \rangle$ region, as shown in Fig. 5.19(c). The cyclotron orbit for $H \| \langle 111 \rangle$ does not circulate along the hot spots, while that for $H \| \langle 110 \rangle$ does.

5.4.2 $LaSn_3$ vs $CeSn_3$

Next, we show in Fig. 5.22(a) the angular dependence of the dHvA frequency in $LaSn_3$ with the $AuCu_3$-type cubic structure [147]. The main characteristic dHvA branches in $LaSn_3$ are as follows: Two branches denoted by α and β exist in the whole range of

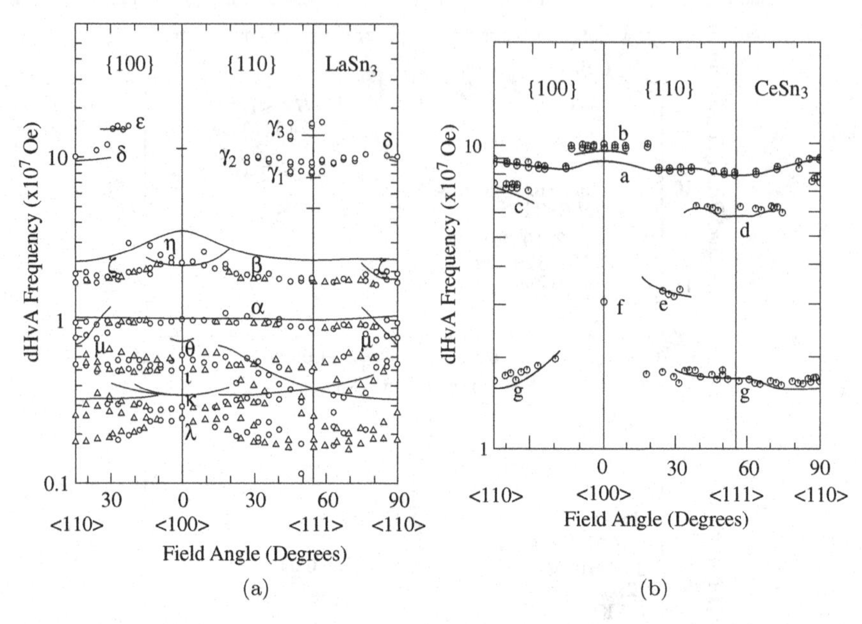

Figure 5.22: Angular dependence of the dHvA frequency in (a) a non-4f reference compound LaSn$_3$ and (b) a 4f-itinerant CeSn$_3$, cited from Refs. [147–150]. Solid lines indicate the results of energy band calculation.

angles, revealing two nearly spherical Fermi surfaces, three branches indicated by the letters γ_1, γ_2 and γ_3 are centered at $\langle 111 \rangle$, and branch δ is centered at $\langle 110 \rangle$.

LaSn$_3$ is an uncompensated metal, as one can tell from the number of valence electrons. In LaSn$_3$, band 8 yields a main Fermi surface centered at R, and two small hole closed Fermi surfaces exist at Γ in bands 7 and 8, as shown in Fig. 5.23(a). The solid lines in Fig. 5.22(a) are the result of energy band calculations [149]. The detected dHvA branches are well explained on the basis of the Fermi surface in Fig. 5.23(a).

CeSn$_3$ at low temperatures is called a valence-fluctuation compound, with a Kondo temperature of about 150 K where the magnetic susceptibility possesses a maximum at about 150 K, as shown in Fig. 5.18(a). Figure 5.22(b) shows the angular dependence of the dHvA frequency in CeSn$_3$ [148]. Some dHvA branches are similar to those of LaSn$_3$ while the other branches are considerably different.

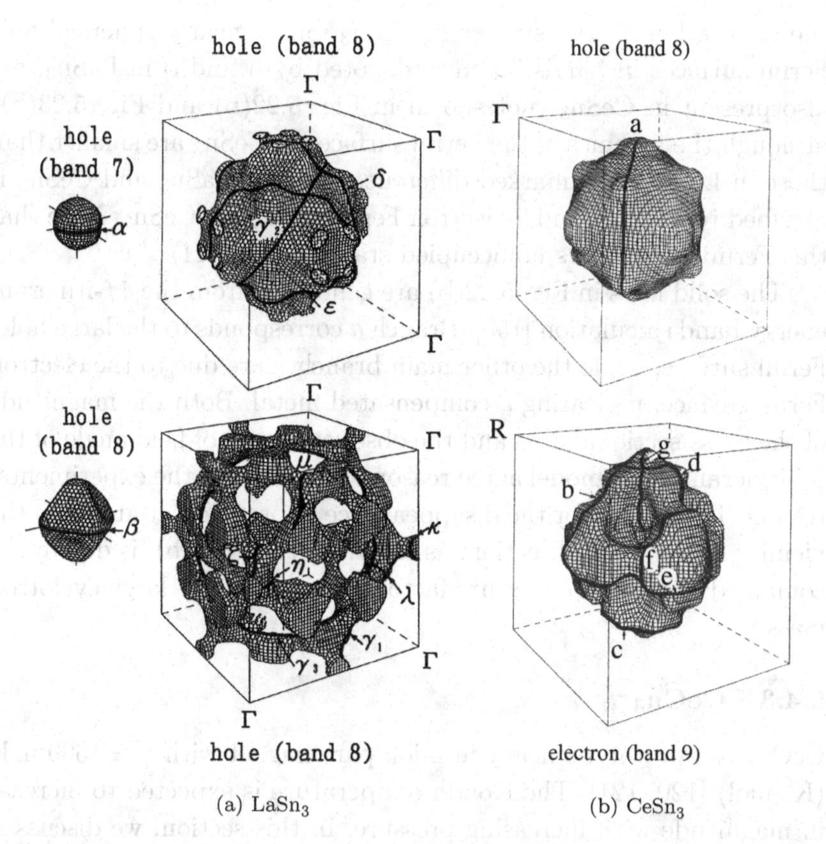

Figure 5.23: Fermi surfaces of (a) an uncompensated metal of $LaSn_3$ and (b) a compensated metal of $CeSn_3$, cited from Refs. [149, 150].

The detected cyclotron masses are roughly five times larger than those of $LaSn_3$: 4.2 m_0 for branch a in $CeSn_3$ and 0.91 m_0 for branch γ_2 in $LaSn_3$, in agreement with γ values: 53 $mJ/K^2{\cdot}mol$ in $CeSn_3$ and 11 $mJ/K^2{\cdot}mol$ in $LaSn_3$.

The energy band calculation was carried out under the assumption that the $4f$ electron is itinerant for $CeSn_3$ [150]. The calculated Fermi surfaces mainly consist of a large band 8-hole Fermi surface centered at R and a large band 9-electron Fermi surface centered at Γ. $CeSn_3$ is a compensated metal. The origin of the analogy in Fermi surfaces between $CeSn_3$ and $LaSn_3$ is as follows. The large distorted spherical band 8-hole Fermi surface in $LaSn_3$ is similar to the band 8-hole Fermi surface in $CeSn_3$. Small nearly spherical hole Fermi surfaces in bands 7 and 8, denoted by α and β in $LaSn_3$, are also present in $CeSn_3$ (not shown in Fig. 5.22(b) and Fig. 5.23(b)) although the volumes of the Fermi surfaces in $CeSn_3$ are smaller than those in $LaSn_3$. The marked difference between $LaSn_3$ and $CeSn_3$ is ascribed to a large band 9-electron Fermi surface in $CeSn_3$. Note that this Fermi surface has no occupied states along $\langle 111 \rangle$.

The solid lines in Fig. 5.22(b) are generated from the $4f$-itinerant energy band calculation [150]. Branch a corresponds to the large hole-Fermi surface, while the other main branches are due to the electron Fermi surface, indicating a compensated metal. Both the magnitude of the cross-sectional area and the observed range of field angle of the $4f$-itinerant band model agree reasonably well with the experimental results. The reason for the disappearance of branch a at angles in the vicinity of the $\langle 100 \rangle$ direction, as shown in Fig. 5.22(b), is due to the combined effect of a curvature factor and a relatively large cyclotron mass.

5.4.3 $CeCu_6$

$CeCu_6$ is a prototype heavy fermion paramagnet with $\gamma \simeq 1600$ mJ/ (K^2·mol) [120, 121]. The Kondo temperature is expected to increase in magnitude with increasing pressure. In this section, we discuss A and ρ_0 values of the Fermi liquid relation $\rho = \rho_o + AT^2$ as a function of pressure.

The temperature dependence of the electrical resistivity under various pressures is shown in Figs. 5.24(a) and (b) [127]. Below a characteristic temperature $T_{\rm max}$ where the resistivity is maximum, the resistivity decreases and finally follows the Fermi liquid relation $\rho = \rho_0 + AT^2$ at lower temperatures. With increasing pressure, $T_{\rm max}$ is shifted to higher temperatures, and the corresponding A value decreases steeply, as shown in Fig. 5.24(c). Here, the $\sqrt{A}$ value is proportional to the electronic specific heat coefficient γ, and then the $1/\sqrt{A}$ value is proportional to the Kondo temperature $T_{\rm K}$, as indicated in Eq. (5.16). $T_{\rm max}$, mentioned above, is also closely related to $T_{\rm K}$. From the low-temperature resistivities, the residual resistivities ρ_0 are found to be reduced in magnitude with increasing pressure,

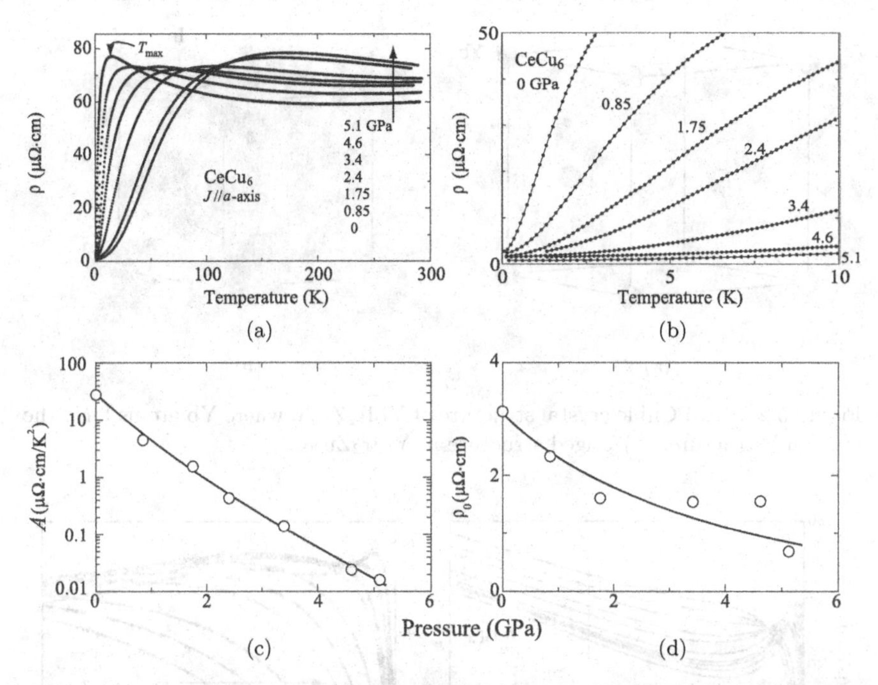

Figure 5.24: Temperature dependence of the electrical resistivity under several pressures (a) below room temperature and (b) below 10 K, and pressure dependence of (c) the A values and (d) ρ_0 values in the Fermi liquid relation of the resistivity $\rho = \rho_0 + AT^2$ in $CeCu_6$, cited from Ref. [127].

as shown in Fig. 5.24(d). The present residual resistivity $\rho_0 = 0.68$ $\mu\Omega$·cm/K^2 under 5.1 GPa is very close to the residual resistivity $\rho_0 = 0.6$ $\mu\Omega$·cm in $LaCu_6$, as shown in Figs. 5.5 and 5.24. The residual resistivity ρ_0 is thus correlated with the A value [151].

5.4.4 $YbIr_2Zn_{20}$ and $YbCo_2Zn_2$

YbT_2Zn_{20} (T: Co, Rh, Ir) crystallize in the $CeCr_2Al_{20}$-type cubic structure, as shown in Fig. 5.25. It is remarkable that the lattice constant a = 14.187 Å in $YbIr_2Zn_{20}$ is very large, and thus the distance between Yb–Yb atoms is considerably large at 6.14 Å, compared with a = 4.20 Å in $YbAl_3$ which has the $AuCu_3$-type cubic structure. Note that the Yb atom, which forms the diamond structure, shown in Fig. 5.25(a), is coordinated by 16 Zn atoms, while

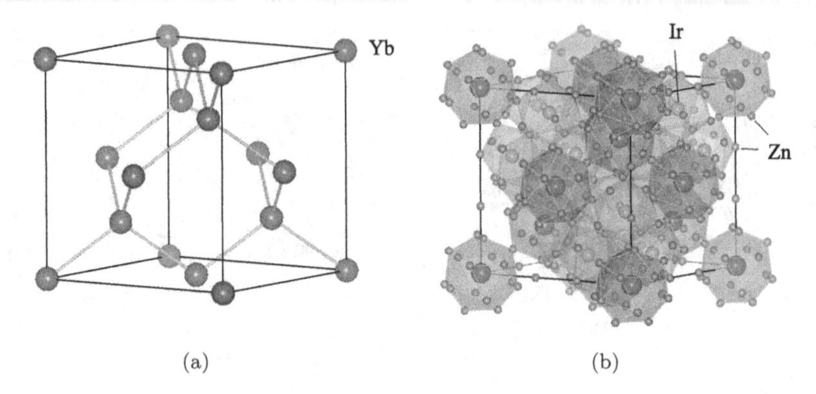

Figure 5.25: (a) Cubic crystal structure of $YbIr_2Zn_{20}$, where Yb atoms form the diamond structure. (b) Caged structure in $YbIr_2Zn_{20}$.

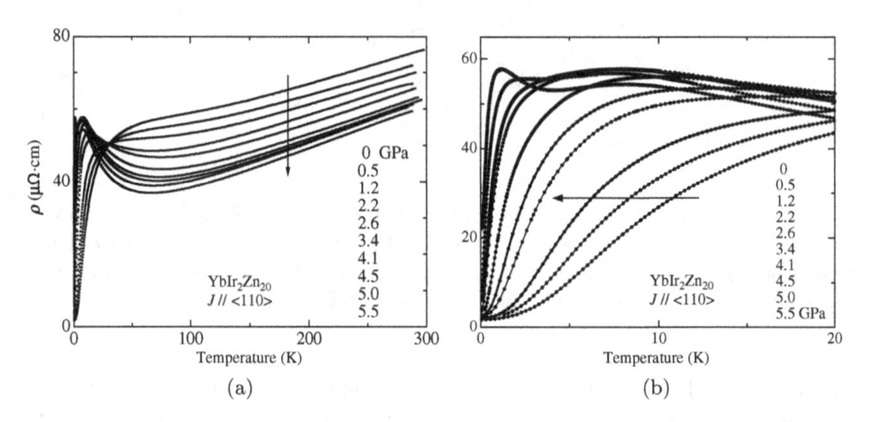

Figure 5.26: Temperature dependence of the electrical resistivity (a) below room temperature and (b) below 20 K under several pressures in $YbIr_2Zn_{20}$, cited from Ref. [144].

the Ir atom has an icosahedral zinc coordination, forming the caged structure shown in Fig. 5.25(b).

$YbIr_2Zn_{20}$ is a heavy fermion paramagnet with $\gamma = 540$ mJ/(K^2·mol) [140]. It is expected that the Kondo temperature is reduced with increasing pressure, becomes zero at $P = P_c$, namely, when crossing the quantum critical point, and the electronic state is changed into the magnetically ordered state at $P > P_c$.

Figure 5.26(a) shows the temperature dependence of the electrical resistivity ρ under several pressures in $YbIr_2Zn_{20}$ [144]. The resistivity in the temperature range from room temperature to about 50 K is found to decrease in magnitude with increasing pressure, but low-temperature resistivity increases in magnitude with increasing pressure, possessing a peak structure above 2 GPa. This is characteristic of a typical heavy fermion state. Surprisingly, above 5.0 GPa, the single peak structure is changed into a two-peak structure, revealing peaks at 2.0 and 7.5 K at 5.0 GPa, and 1.1 and 7.8 K at 5.5 GPa, as shown in Fig. 5.26(b). This is reminiscent of the temperature dependence of the cerium compounds, corresponding to two Kondo temperatures, T_K and T_K^h, respectively, as described in Sec. 5.2 [117].

Here, the low-temperature resistivity under pressure follows the Fermi liquid relation $\rho = \rho_0 + AT^2$ (ρ_0: residual resistivity) below T_{FL}. The slope of the ρ vs T^2 relation, namely, the coefficient A becomes remarkably large with increasing pressure, as shown in Fig. 5.27(a). It is noted that the A value at 5.0 GPa, namely 380 $\mu\Omega\cdot$cm/K^2, is larger than that at 5.5 GPa. From the pressure dependences of T_{FL} and A, we estimated the critical pressure to be $P_c \simeq 5.2$ GPa. The electronic state at $P > P_c$ is a magnetically ordered state. Here, the A and γ values at ambient pressure are $A = 0.29$ $\mu\Omega\cdot$cm/K^2 and $\gamma = 540$ mJ/(K$^2\cdot$mol) in $YbIr_2Zn_{20}$. The A value of 380 $\mu\Omega\cdot$cm/K^2 at 5.0 GPa in $YbIr_2Zn_{20}$ is larger than the value of A which is 150 $\mu\Omega\cdot$cm/K^2 in $YbCo_2Zn_{20}$ with $\gamma = 8$ J/(K$^2\cdot$mol) [140]. The super-heavy fermion

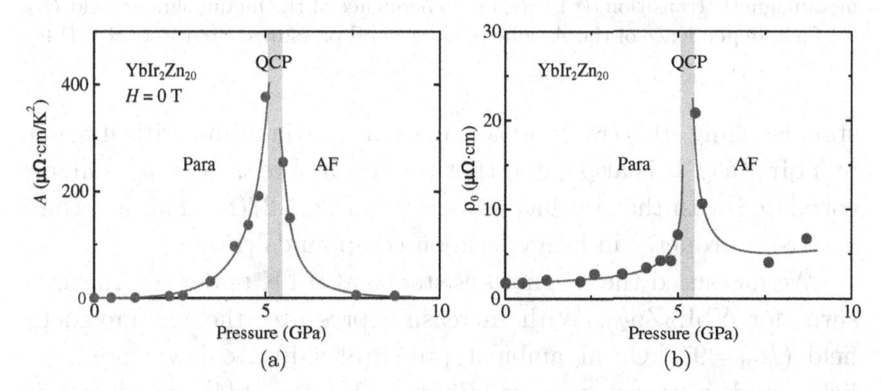

Figure 5.27: Pressure dependence of (a) A values and (b) residual resistivity ρ_0 in $YbIr_2Zn_{20}$. Para, QCP, and AF refer to the paramagnetic state, quantum critical point, and the antiferromagnetic state respectively, cited from Ref. [144].

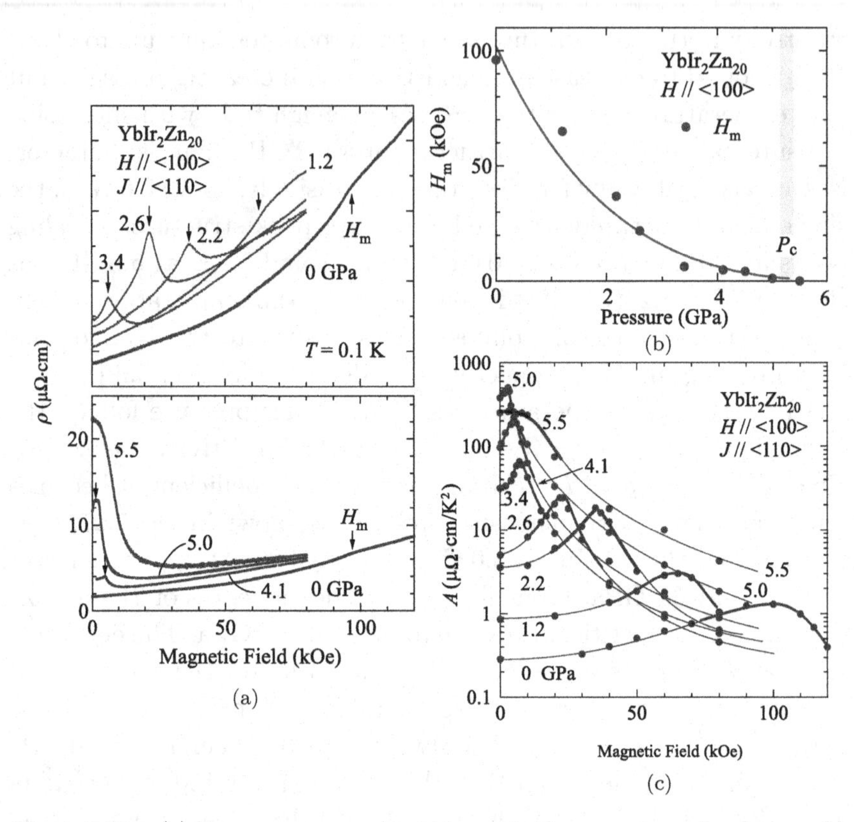

Figure 5.28: (a) Transverse magnetoresistance in the pressure region from ambient pressure to 3.4 GPa and from 4.1 to 5.5 GPa in $YbIr_2Zn_{20}$. Arrows indicate the metamagnetic transition. (b) Pressure dependence of the metamagnetic field H_m. (c) Field dependence of the A value under several pressures, cited from Ref. [144].

state reaching 10 J/(K^2·mol) is realized at the quantum critical region in $YbIr_2Zn_{20}$. It is also noted that the residual resistivity ρ_0 is highly correlated with the A value, as shown in Fig. 5.27(b). This is a characteristic property in heavy fermion compounds [151].

We measured the magnetoresistance at 0.1 K under several pressures for $YbIr_2Zn_{20}$. With increasing pressure, the metamagnetic field ($H_m = 97$ kOe at ambient pressure) shifts to lower magnetic fields and becomes zero at $P_c \simeq 5.2$ GPa [144], as shown in Fig. 5.28(a) by arrows. It is remarkable that the shoulder-like feature of metamagnetic behavior at 0 and 1.2 GPa is changed into a sharp peak at higher pressures. The present peak, associated with the metamagnetic behavior, is a guiding parameter to reach the quantum critical point: $P \to P_c$ for $H_m \to 0$, as shown in Fig. 5.28(b). Here, the A value is extremely enhanced with increasing pressure and has a maximum at $P_c \simeq 5.2$ GPa, with $A = 380\,\mu\Omega\cdot\text{cm/K}^2$ at 5.0 GPa, as shown in Fig. 5.27(a).

The relation between A and γ is known as the Kadowaki-Woods plot [42]. The A and γ values in YbT_2Zn_{20} (T: Co, Rh, Ir) is found to belong to the generalized Kadowaki-Woods relation for $N = 4$, as shown in Fig. 3.6 [43, 44, 140]. The value of $A = 380\,\mu\Omega\cdot\text{cm/K}^2$ at 0 kOe under 5.0 GPa exceeds $\gamma = 10\,\text{J/(K}^2\cdot\text{mol)}$ if we follow this relation. The super-heavy fermion state is realized in $YbIr_2Zn_{20}$ under pressure. Surprisingly, $YbCo_2Zn_{20}$ is very close to this state: $A = 160\,\mu\Omega\cdot\text{cm/K}^2$ and $\gamma = 8000\,\text{mJ/(K}^2\cdot\text{mol)}$ at ambient pressure.

Furthermore, the temperature dependence of the electrical resistivity below 0.8 K under magnetic fields and pressures was measured to determine the A value. The Fermi liquid relation of $\rho = \rho_0 + AT^2$ is satisfied in these experimental conditions, and the obtained A value is shown in Fig. 5.28(c) as a function of magnetic field. A broad peak at $H_m = 97$ kOe with ambient pressure in the A value is changed into a distinct peak at 2.2 and 3.4 GPa, together with an anomalous enhancement of the value of A at higher pressures: $A = 380\,\mu\Omega\cdot\text{cm/K}^2$ at 0 kOe under 5.0 GPa, as mentioned above, which is strongly reduced with increasing magnetic field: $A = 1.45\,\mu\Omega\cdot\text{cm/K}^2$ at 80 kOe under 5.0 GPa.

The mass reduction is obtained in the dHvA experiment, carried out at ambient pressure for a super-heavy fermion compound $YbCo_2Zn_{20}$. Figure 5.29(a) shows the FFT spectrum for $H \parallel \langle 100 \rangle$ in $YbCo_2Zn_{20}$. The dHvA frequency ranges from 1×10^6 to 4×10^7 Oe. From the size of the small Brillouin zone, these dHvA branches correspond to main Fermi surfaces in $YbCo_2Zn_{20}$ [138, 139, 142]. The cyclotron effective mass is found to be highly field-dependent, decreasing steeply with increasing magnetic field. Figures 5.29(b)–(d) show the field dependence of the cyclotron mass for the dHvA frequency $F = 3.15 \times 10^7$, 1.15×10^7, and 0.45×10^7 Oe, respectively,

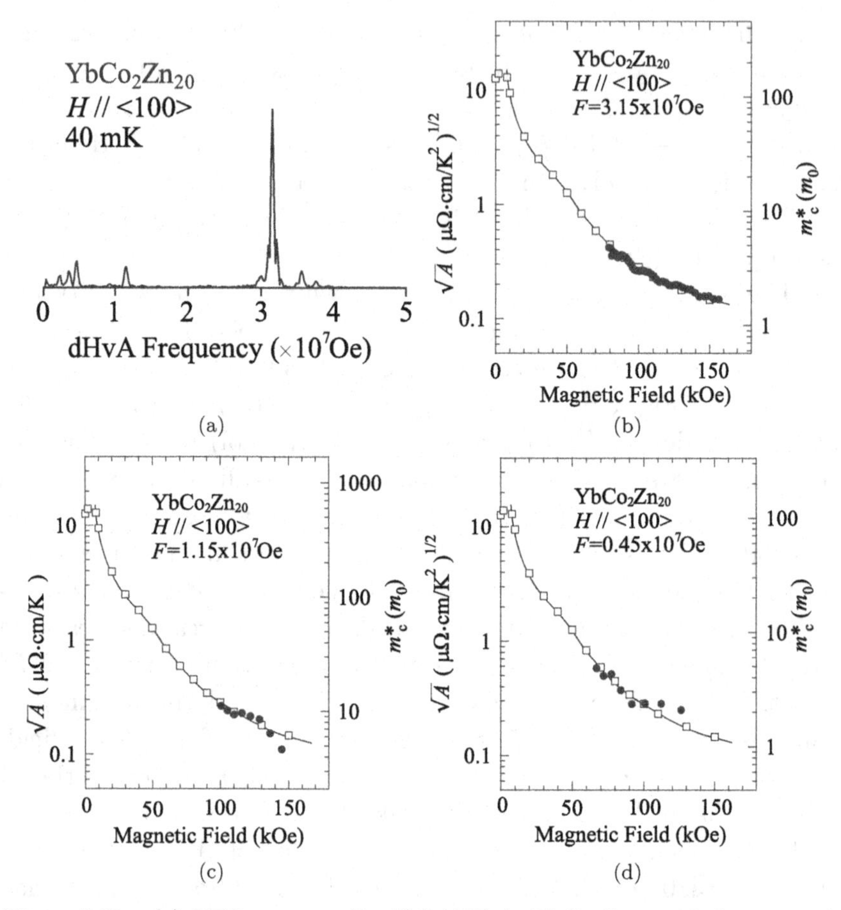

Figure 5.29: (a) FFT spectrum for $H \parallel \langle 100 \rangle$ in $YbCo_2Zn_{20}$ and the magnetic field dependence of the $\sqrt{A}$ value for $J \parallel H \parallel \langle 100 \rangle$, shown by squares, together with the field dependence of the cyclotron effective mass m_c^*, shown by closed circles for $H \parallel \langle 100 \rangle$ in $YbCo_2Zn_{20}$, for (b) $F = 3.15 \times 10^7$, (c) 1.15×10^7, and (d) 0.45×10^7 Oe, cited from Ref. [138]. The solid lines connecting the data are guide lines.

together with the $\sqrt{A}$ value obtained from the electrical resistivity under magnetic field. Following the Fermi liquid relation, the $\sqrt{A}$ value correlates well with the cyclotron effective mass. The cyclotron masses at 0 kOe are thus estimated to be 150, 500, and $110 m_0$, respectively. The present experimental results indicate, however, that the super-heavy cyclotron effective masses at 0 kOe are strongly reduced in high magnetic fields, as shown in Fig. 5.12.

5.4.5 $CeCoIn_5$, $CeRhIn_5$, $NpPd_5Al_2$, $PuCoGa_5$, and $PuRhGa_5$

The electronic state is tuned to the quantum critical point by applying pressure in Ce-based antiferromagnets. At the quantum critical point, the electrical and magnetic properties do not follow the Fermi liquid relations but instead, the electronic state is characterised by non-Fermi liquid properties. $CeCoIn_5$ corresponds to a typical quantum critical point compound at ambient pressure.

Here, the electronic states of a superconductor of $CeCoIn_5$ and related compounds of $CeRhIn_5$, $PuCoGa_5$ and $PuRhGa_5$ are presented in this section. These compounds crystallize in the $HoCoGa_5$-type tetragonal structure (No. 123, a = 4.601 Å, c = 7.540 Å, $Z = 1$ in $CeCoIn_5$), as shown in Fig. 5.30(a). This is closely related to the electronic state. The uniaxially distorted $AuCu_3$-type layers of $CeIn_3$ and the $CoIn_2$ layers are stacked sequentially along the tetragonal [001] direction (c-axis). For example, the $3d$ electrons of Co in $CeCoIn_5$ hybridize with the $5p$ electrons of In, which has a small density of states around the Fermi energy. This means that

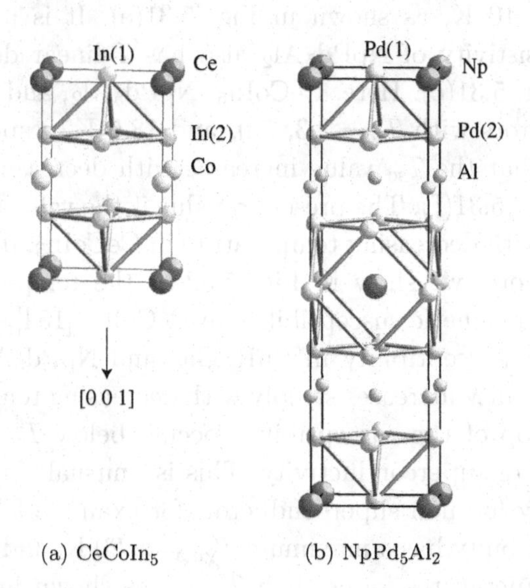

Figure 5.30: Tetragonal crystal structures of (a) $CeCoIn_5$ and (b) $NpPd_5Al_2$, cited from Ref. [152].

there are very few conduction electrons in the $CoIn_2$ layer and the $CeIn_3$ layer is conductive, producing nearly cylindrical Fermi surfaces along the [001] direction. The crystal structure of $NpPd_5Al_2$ shown in Fig. 5.30(b) is very similar to the $HoCoGa_5$-type structure [152].

First, we show in Figs. 5.31(a) and (b) the temperature dependence of the electrical resistivity in the pressure range from 0 to 2.08 GPa for an antiferromagnet $CeRhIn_5$ with T_N = 3.8 K. The tetragonal crystal structure of $CeRhIn_5$ is the same as that of $CeCoIn_5$, and similar to that of $NpPd_5Al_2$, as shown in Fig. 5.30. With increasing pressure, the Néel temperature of $CeRhIn_5$, shown by arrows, increases slightly up to about 1 GPa, but decreases gradually to 1.8 GPa. Superconductivity appears above 1.60 GPa. Overall the resistivity of $CeRhIn_5$ at 2.08 GPa is typical of the heavy fermion state. This is very similar to that of $CeCoIn_5$ in Fig. 5.31(d). The temperature vs pressure phase diagram for $CeRhIn_5$ is shown in Fig. 5.31(c), revealing three phases of the antiferromagnetic (AF) state, coexistence of the antiferromagnetic and superconducting states, and the superconducting (SC) state.

The electrical resistivity of $CeCoIn_5$ follows a T-linear dependence below 10 K, as shown in Fig. 5.31(e). It is noted that the electrical resistivity of $NpPd_5Al_2$ also has T-linear dependence, as shown in Fig. 5.31(e). Here, $CeCoIn_5$, $NpPd_5Al_2$, and $PuCoGa_5$ are superconductors with $T_{sc} = 2.3$, 5.0, and 18.5 K, respectively [153]. It is found that the T_{sc} value increases with decreasing γ value, as shown in Fig. 5.31(f). The present γ value is the value at T_{sc} because γ increases with decreasing temperature in $CeCoIn_5$, discussed later.

Furthermore, we show in Fig. 5.32(a) the temperature dependence of the magnetic susceptibility in $CeCoIn_5$ [154], together with the magnetic susceptibility in $PuRhGa_5$ and $NpPd_5Al_2$ [155, 156]. The susceptibility increases steeply with decreasing temperature, but a sudden drop of the susceptibility occurs below T_{sc} = 2.3 K due to the onset of superconductivity. This is unusual. In the case of a typical heavy fermion superconductor, for example UPt_3, the magnetic susceptibility has a maximum $T_{\chi max} \simeq 20$ K, and becomes constant at temperatures lower than $T_{\chi max}$, as shown in Fig. 5.18(a). Superconductivity occurs below T_{sc} = 0.5 K in UPt_3. This means

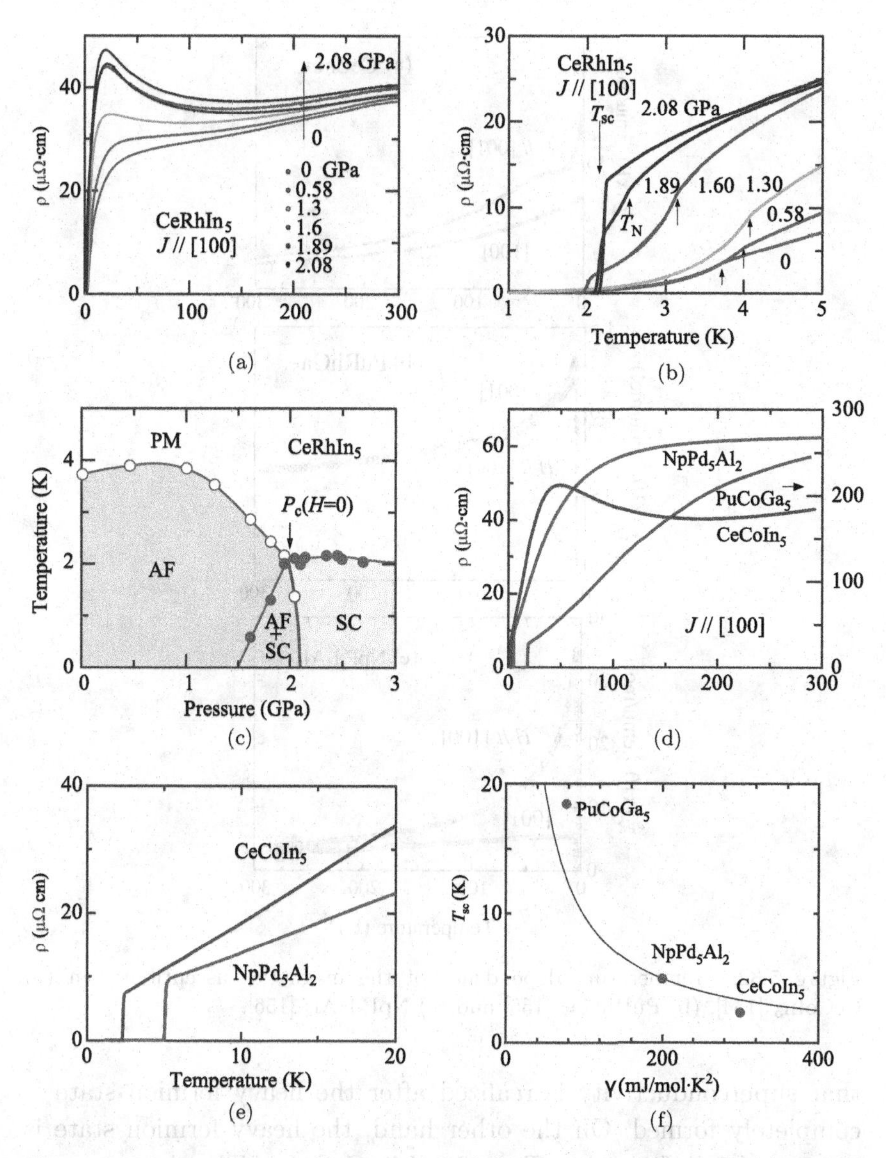

Figure 5.31: Temperature dependence of the electrical resistivity in (a) and (b) $CeRhIn_5$ under pressure, together with (c) the T-P phase diagram, and (d) and (e) $CeCoIn_5$, $PuCoGa_5$ and $NpPd_5Al_2$, together with (f) their γ values at T_{sc} vs T_{sc} phase diagram, cited from Refs. [152, 154–156]. Only $PuCoGa_5$ is a polycrystal sample, and the scale is shown on the right in (d), cited from Ref. [153].

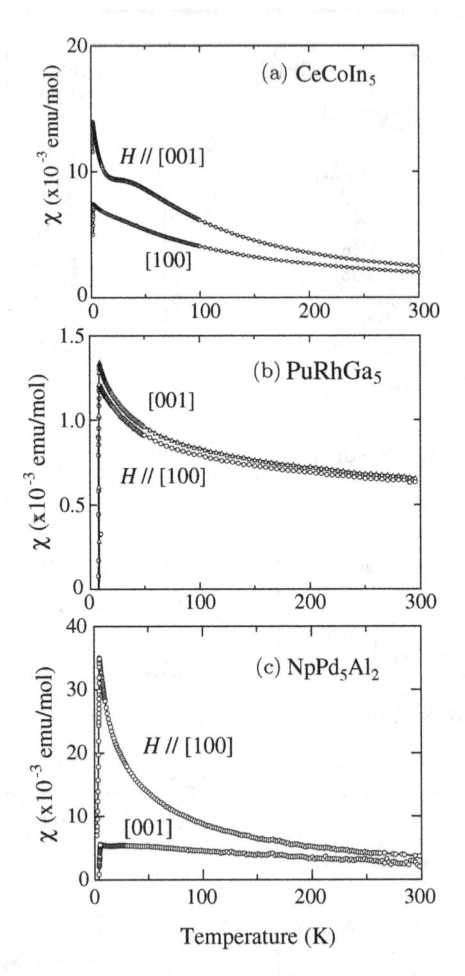

Figure 5.32: Temperature dependence of the magnetic susceptibility in (a) $CeCoIn_5$ [154], (b) $PuRhGa_5$ [155] and (c) $NpPd_5Al_2$ [156].

that superconductivity is realized after the heavy fermion state is completely formed. On the other hand, the heavy fermion state is not completely formed at $T_{sc} = 2.3\,K$ in $CeCoIn_5$. That is, superconductivity of $CeCoIn_5$ is realized in the non-Fermi liquid state. Similar behavior is observed at least in $NpPd_5Al_2$, as shown in Fig. 5.32(c), which is described later.

Superconductivity in the non-Fermi liquid state of $CeCoIn_5$ can be also confirmed through the specific heat data [157, 158].

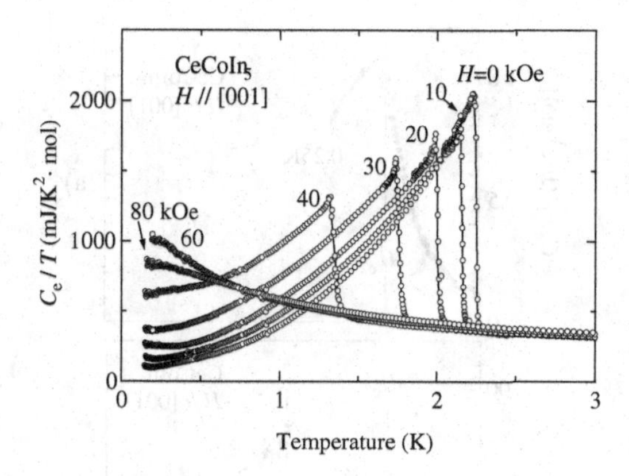

Figure 5.33: Temperature dependence of the electronic specific heat in the form of C_e/T in $CeCoIn_5$, cited from Ref. [157].

Figure 5.33 shows the temperature dependence of the specific heat in the form C/T under several magnetic fields. The value of C/T increases with decreasing temperature under magnetic fields. The γ value is $380\,\mathrm{mJ/K^2{\cdot}mol}$ at $T_{sc} = 2.3\,\mathrm{K}$, but is extremely enhanced to $1070\,\mathrm{mJ/K^2{\cdot}mol}$ at 0.2 K. Note that the upper critical field $H_{c2} = 49.5\,\mathrm{kOe}$ for $H\|$ [001]. The specific heat in the form of C/T in the normal state (under H = 60 kOe) does not saturate at low temperatures, but increases with decreasing temperature. The non-Fermi liquid state is thus observed in the specific heat, together with the magnetic susceptibility shown in Fig. 5.32(a).

We also show in Fig. 5.34(a) the magnetic field dependence of the electronic specific heat coefficient C_e/T at 0.25 K. A large value of γ = 1070 mJ/(K^2·mol) at H_{c2} = 49.5 kOe decreases steeply with increasing field. Correspondingly, the large cyclotron mass in $CeCoIn_5$ is field-dependent, and decreases as a function of the magnetic field, as shown in Fig. 5.34(b). This behavior is characteristic in heavy fermions. These cyclotron masses in $CeCoIn_5$ are large compared with $0.87m_0$ in branch β_2 and 0.89–$0.95m_0$ in branches α_i in $ThRhIn_5$, where the topology of the Fermi surface is similar between $CeCoIn_5$ and $ThRhIn_5$ because both compounds have the same number of valence electrons. The large cyclotron mass in $CeCoIn_5$ is mainly due to the many-body Kondo effect. From the energy band calculation, it has been verified that the $4f$-electron contribution to

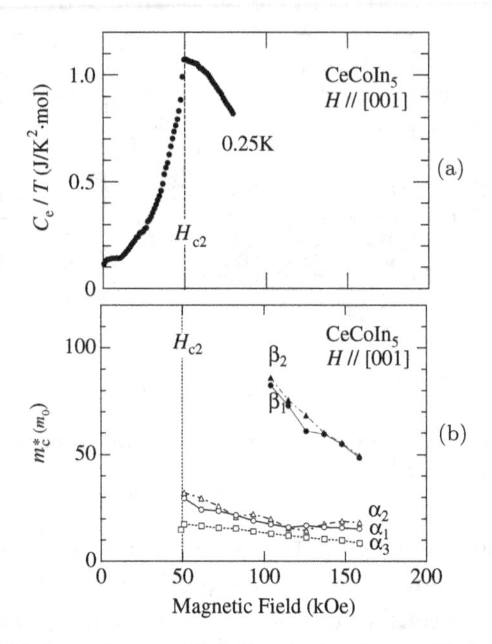

Figure 5.34: Field dependence of (a) the electronic specific heat C_e/T and (b) cyclotron mass in $CeCoIn_5$, cited from Ref. [159].

the density of states at the Fermi energy in $CeCoIn_5$ is about 70% in magnitude. Note that the $4f$ electron does not contribute to the electronic state in $ThRhIn_5$ because of the lack of $4f$ electrons, revealing a small cyclotron mass.

The characteristic features of superconductivity in $CeCoIn_5$ are as follows. Superconductivity is of the $d_{x^2-y^2}$ type, with line nodes for the [110] direction, elongating along the tetragonal [001] direction on the quasi-two dimensional Fermi surfaces, which was confirmed from the angular dependence of the thermal conductivity and specific heat in magnetic fields [160]. Note that the $d_{x^2-y^2}$-type gap in $CeCoIn_5$ is discussed later in Sec. 6.4.2. In Fig. 5.35(b) we schematically show the gap with the line nodes. The usual uniform gap is shown in Fig. 5.35(a) for comparison.

The first-order phase transition from the superconducting mixed state to the normal state occurs at the upper critical field H_{c2} in $CeCoIn_5$ in the magnetization curve [161], as shown in Fig. 5.36(a), together with the similar step-like magnetization curves of $NpPd_5Al_2$

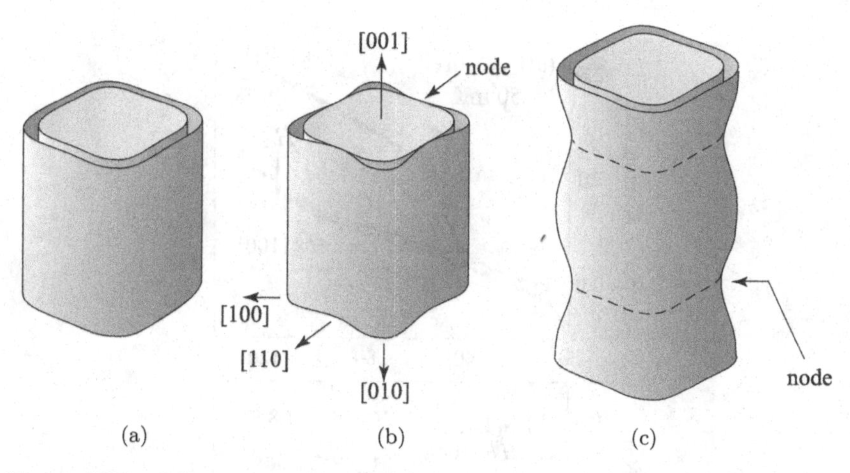

Figure 5.35: Schematic pictures of the two dimensional superconducting gap with (a) an isotropic gap, and (b) vertical and (c) horizontal line node gaps.

in Fig. 5.36(b) [156]. This persists up to about $0.3\,T_{sc}$ ($T_{sc} = 2.3\,\text{K}$) for $H \parallel [001]$ and $0.4\,T_{sc}$ for $H \parallel [100]$ in $CeCoIn_5$. Note that the phase transition at H_{c2} is of second order in the usual superconductors. H_{c2} is shown in Fig. 5.37(b) as a function of temperature. A strong Pauli paramagnetic effect is realized in the present superconductive state of $CeCoIn_5$, which is described in the next chapter.

Here, we note the characteristic superconducting property in $CeCoIn_5$, shown by the dark-gray regions in the mixed state close to H_{c2} (see Fig. 5.37(b)). These regions were discussed with an approach stemming from the view that the Fulde, Ferrel, Larkin, Ovchinikov (FFLO) state most likely exists in the superconducting mixed state close to H_{c2} at temperatures far lower than $T_{sc} = 2.3\,\text{K}$ [162, 163]. The FFLO state is a novel superconducting state in strong magnetic fields characterized by the formation of Cooper pairs with a nonzero total momentum ($\boldsymbol{k}\uparrow,\ -\boldsymbol{k}+\boldsymbol{q}\downarrow$), instead of the ordinary BCS pairs ($\boldsymbol{k}\uparrow,\ -\boldsymbol{k}\downarrow$). Similar characteristic phenomena were once observed in $CeRu_2$ and UPd_2Al_3. The magnetization in $CeRu_2$ and UPd_2Al_3 exhibited a hysteresis loop with a sharp peak in the narrow field region close to H_{c2}. However, the FFLO state was denied in $CeRu_2$ and UPd_2Al_3. The peak effect in $CeRu_2$ and UPd_2Al_3 is

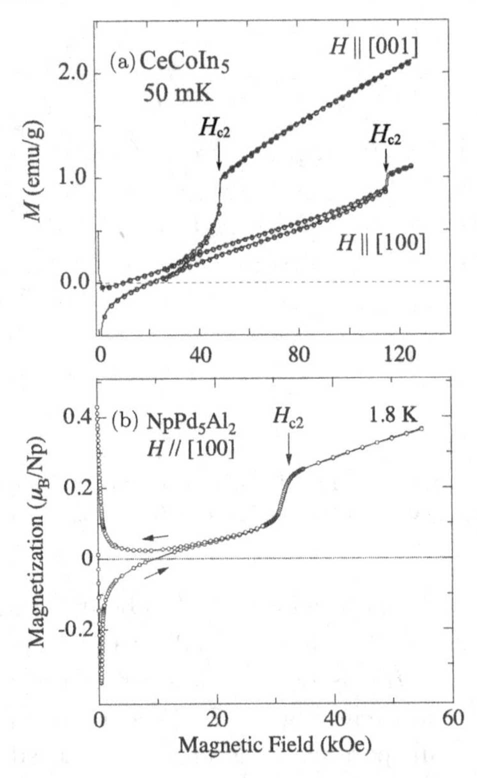

Figure 5.36: Magnetization curves of (a) $CeCoIn_5$ [161] and (a) $NpPd_5Al_2$ [156], revealing a step-like increase of the magnetization at H_{c2}.

now considered to be due to a weak vortex pinning effect. On the other hand, $CeCoIn_5$ is a good candidate for the FFLO state. This is because the sample is extremely clean and strong Pauli paramagnetic suppression is realized for the $H_{c2}(0)$ value. Moreover, $CeCoIn_5$ is considered to be situated in an antiferromagnetic quantum critical electronic state. Nevertheless, the existence of the FFLO state in the dark-gray regions was rejected because the neutron scattering experiment revealed the existence of the field-induced antiferromagnetic phase in these regions [164]. Strangely, this phase disappears in the normal state when the magnetic field crosses H_{c2} with increasing magnetic fields. The dark-gray regions are different from the antiferromagnetic phase in $CeRhIn_5$, as later shown in Fig. 6.44(b). The FFLO state for $CeCoIn_5$ is discussed in detail in Ref. [165].

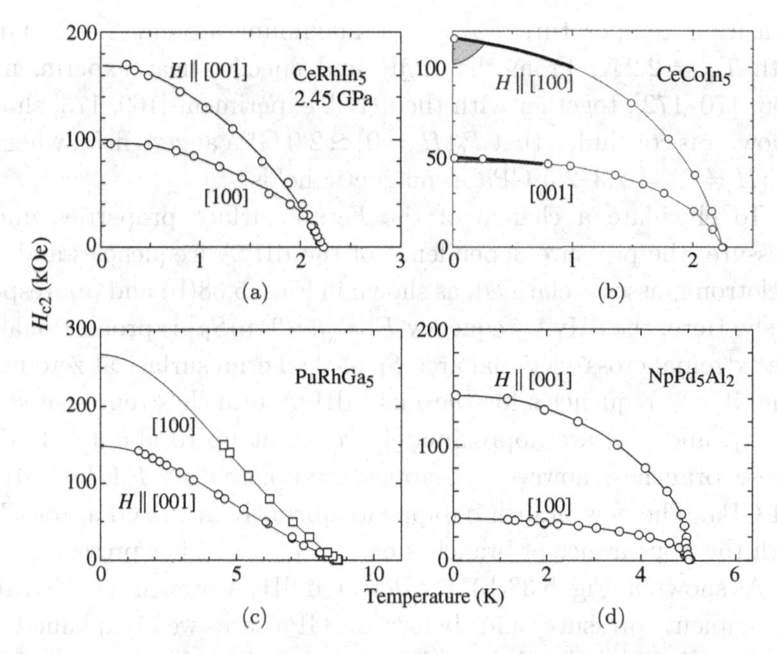

Figure 5.37: Temperature dependence of the upper critical field H_{c2} in (a) $CeRhIn_5$ [166], (b) $CeCoIn_5$ [167], (c) $PuRhGa_5$ [155] and (d) $NpPd_5Al_2$ [156]. Thick solid lines in $CeCoIn_5$ indicate the first-order phase transition at H_{c2}, and the dark-gray regions in the mixed state were discussed from an approach involving the FFLO state, cited from Refs. [162, 163].

The Fermi surface properties were studied as a function of pressure, especially at P_c, by the dHvA experiment for some compounds such as antiferromagnets of $CeRh_2Si_2$, $CeIn_3$ and $CeRhIn_5$ [168]. The following discuss the experimental results of $CeRhIn_5$.

The dHvA results for an antiferromagnet $CeRhIn_5$ were studied under pressure in comparison with those of a non-$4f$ reference compound $LaRhIn_5$ and a heavy fermion superconductor $CeCoIn_5$ without magnetic ordering [154, 166, 167, 169]. With increasing pressure P, the Néel temperature $T_N = 3.8$ K in $CeRhIn_5$ increases, reaches a maximum around 1 GPa, and decreases with further increasing pressure, as shown in Fig. 5.31(c). The critical pressure P_c is 2.0 GPa, while a smooth extrapolation indicates $T_N \rightarrow 0$ at a pressure of 2.3–2.5 GPa. $CeRhIn_5$ reveals superconductivity over a wide pressure region from 1.6–5.2 GPa. The superconducting transition temperature T_{sc} has a maximum around 2.3–2.5 GPa, with $T_{sc} = 2.2$ K. From the NQR and specific heat experiments [166, 170–172], together with the dHvA experiment [169, 173] shown below, it is concluded that $P_c(H = 0) \simeq 2.0$ GPa at zero field, whereas P_c $(H \neq 0) = 2.3$–2.5 GPa in magnetic fields.

To elucidate a change of the Fermi surface properties under pressure, the pressure dependence of the dHvA frequencies and the cyclotron masses is clarified, as shown in Figs. 5.38(b) and (e), respectively. Here, the dHvA frequency $F[= (\hbar c/2\pi e)S_F]$ is proportional to the extremal cross-sectional area S_F of the Fermi surface at zero field. The dHvA frequencies for the main dHvA branches represented by β_2, α_1, and $\alpha_{2,3}$ are approximately constant up to about 2.3 GPa. These branches, however, disappear completely at $P_c(H \neq 0) \simeq$ 2.4 GPa. The new branch α_3 appears abruptly at 2.4 GPa, together with the appearance of branches α_1 and α_2 at higher pressures.

As shown in Fig. 5.38(b), the detected dHvA branches of $CeRhIn_5$ at ambient pressure and below 2.3 GPa are well explained by the experimental dHvA branches in a non-4f reference compound $LaRhIn_5$ and the result of energy band calculations for $LaRhIn_5$, as shown in Figs. 5.38(a) and (d), respectively. The corresponding topologies of the main Fermi surfaces in $CeRhIn_5$ are nearly cylindrical and are found to be approximately the same as those in $LaRhIn_5$, indicating that the 4f electron in $CeRhIn_5$ is almost localized and scarcely contributes to the volume of the Fermi surfaces.

Above $P_c(H \neq 0) \simeq 2.4$ GPa, the detected dHvA frequencies change abruptly in magnitude, but correspond to those in $CeCoIn_5$, as shown in Fig. 5.38(c). The main Fermi surfaces in $CeCoIn_5$, as shown in Fig. 5.38(f), are also nearly cylindrical. The topologies of two kinds of cylindrical electron Fermi surfaces in $CeCoIn_5$ are similar to those in $LaRhIn_5$($CeRhIn_5$), but the Fermi surfaces of $CeCoIn_5$ are larger than those of $LaRhIn_5$ in volume. The detected dHvA branches in $CeCoIn_5$ are consistent with the 4f-itinerant band model [167]. This is because one 4f electron in each Ce site in $CeCoIn_5$ becomes a valence electron and contributes to the conduction electrons. The contribution of the 4f electron to the density of states at the Fermi

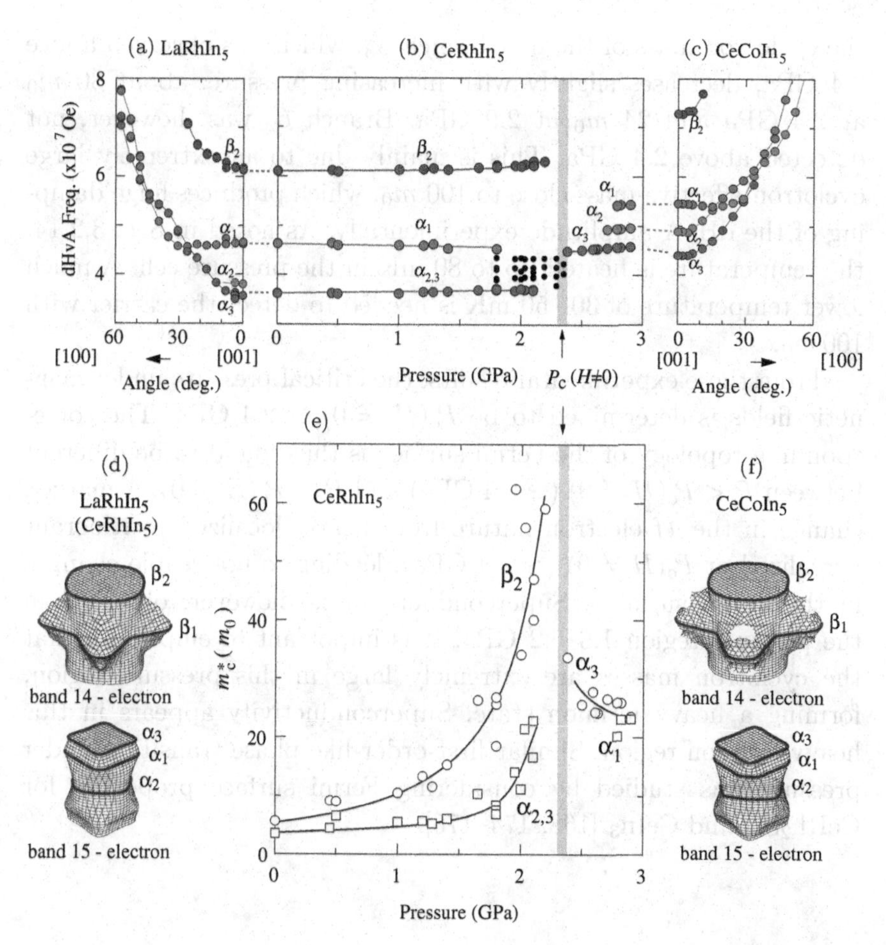

Figure 5.38: (a) Angular dependence of the dHvA frequency in $LaRhIn_5$, (b) pressure dependence of the dHvA frequency for $H \parallel [001]$ in $CeRhIn_5$, and (c) angular dependence of the dHvA frequency in $CeCoIn_5$, (d) Fermi surfaces of $LaRhIn_5$, (e) pressure dependence of the cyclotron mass m_c^* for $H \parallel [001]$ in $CeRhIn_5$, and (f) Fermi surfaces of $CeCoIn_5$, cited from Refs. [154, 166, 167, 169].

energy is about 70% theoretically. The detected cyclotron masses of 5–87 m_0 in $CeCoIn_5$ are extremely large.

As shown in Fig. 5.38(e), the cyclotron masses of the main branches β_2 and $\alpha_{2,3}$ in $CeRhIn_5$ increase rapidly above 1.6 GPa, where superconductivity sets in: 5.5 m_0 at ambient pressure, 20 m_0 at 1.6 GPa, and 60 m_0 at 2.2 GPa for branch β_2. On the other hand, the cyclotron mass of the new branch α_3, which was observed above 2.4 GPa, decreases slightly with increasing pressure: about 30 m_0 at 2.4 GPa and 24 m_0 at 2.9 GPa. Branch β_2 was, however, not detected above 2.4 GPa. This is mainly due to an extremely large cyclotron effective mass close to 100 m_0, which produces huge damping of the dHvA amplitude experimentally. As noted in Sec. 3.2.14, the temperature is heated up to 80 mK in the pressure cell. A much lower temperature of 30−50 mK is needed to detect the carrier with 100 m_0.

From these experimental results, the critical pressure under magnetic fields is determined to be $P_c(H \neq 0) \simeq 2.4$ GPa. The corresponding topology of the Fermi surface is thus found to be different between $P < P_c(H \neq 0)$ ($\simeq 2.4$ GPa) and $P > P_c(H \neq 0)$. A marked change in the $4f$-electron nature from nearly localized to itinerant is realized at $P_c(H \neq 0)$ ($\simeq 2.4$ GPa), leading to noticeable changes in the Fermi surfaces. Superconductivity is, however, observed in the pressure region 1.6–5.2 GPa. It is important to emphasize that the cyclotron masses are extremely large in this pressure region, forming a heavy fermion state. Superconductivity appears in this heavy fermion region. Similar first-order-like phase transition under pressure was studied by considering Fermi surface properties for $CeRh_2Si_2$ and $CeIn_3$ [168, 174–176].

Chapter 6

Superconductivity

The microscopic BCS theory of superconductivity, which was proposed by Bardeen, Cooper and Schrieffer in 1957, is based on the following idea [177]. When an attractive interaction between fermions is present, the stable ground state is no longer the degenerate Fermi gas but a coherent state in which the electrons are combined into pairs of spin-singlets with zero total momentum ($\boldsymbol{k}\uparrow, -\boldsymbol{k}\downarrow$: Cooper pairs). A conduction electron distorts the lattice by moving in the lattice and attracting the positive ion. This distortion attracts another conduction electron. Namely, the interaction between two electrons mediated by the phonon forms the Cooper pair of the two electrons. A BCS-type superconductor has an isotropic superconducting gap which is opened over the entire Fermi surface. Superconductivity was regarded as one of the more well-understood many-body problems. The BCS-superconductivity has been, however, altered in heavy fermion superconductors. The pairing mechanism is not due to phonons. These alternative mechanisms are as follows: pairing mechanism correlated with antiferromagnetism (or antiferromagnetic fluctuations), ferromagnetism vs superconductivity, quadrupole interaction vs superconductivity, and a novel pairing state in the non-centrosymmetric crystal structure.

6.1 BCS superconductivity

First, we explain the basic concept of the Cooper pair [178]. We add two electrons of $(\boldsymbol{k}_1, \boldsymbol{\sigma}_1)$ and $(\boldsymbol{k}_2, \boldsymbol{\sigma}_2)$ on a spherical Fermi surface

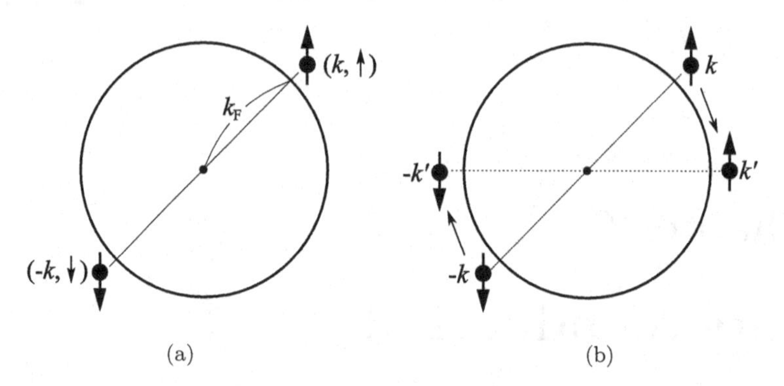

Figure 6.1: (a) Two electrons of $(\boldsymbol{k}, \uparrow)$ and $(-\boldsymbol{k}, \downarrow)$ on a spherical Fermi surface of which the total momentum is conserved, with a spin singlet state. (b) Scattering of two electrons from $(\boldsymbol{k}, -\boldsymbol{k})$ to $(\boldsymbol{k}', -\boldsymbol{k}')$.

at 0 K, which is fully occupied by electrons up to a Fermi energy $\varepsilon_{\rm F}(=\hbar^2 k_{\rm F}^2/2m)$, and then k must be larger than $k_{\rm F}$ in magnitude, as shown in Fig. 6.1. The two-electron wave function $\Psi(\boldsymbol{r}_1, \boldsymbol{\sigma}_1, \boldsymbol{r}_2, \boldsymbol{\sigma}_2)$ is a product of the orbital part $\psi(\boldsymbol{r}_1, \boldsymbol{r}_2)$ and spin part $\chi(\boldsymbol{\sigma}_1, \boldsymbol{\sigma}_2)$. The spin is either $\uparrow$ or $\downarrow$, and the spin part $\chi(\boldsymbol{\sigma}_1, \boldsymbol{\sigma}_2)$ is either the spin singlet or spin triplet. Here, we consider the spin singlet state, namely, one electron is in the $\uparrow$ spin state and the other is in the $\downarrow$ spin state.

The Schrödinger equation of the two-electron wave function $\psi(\boldsymbol{r}_1, \boldsymbol{r}_2)$ is

$$\left\{-\frac{\hbar^2}{2m}\left(\boldsymbol{\nabla}_1^2 + \boldsymbol{\nabla}_2^2\right) + V(\boldsymbol{r}_1, \boldsymbol{r}_2)\right\}\psi(\boldsymbol{r}_1, \boldsymbol{r}_2) = (\varepsilon + 2\varepsilon_{\rm F})\psi(\boldsymbol{r}_1, \boldsymbol{r}_2) \tag{6.1a}$$

where ε is the energy of the two electrons relative to the Fermi level $2\varepsilon_{\rm F}$. The wave function $\psi(\boldsymbol{r}_1, \boldsymbol{r}_2)$ is

$$\begin{aligned}\psi(\boldsymbol{r}_1, \boldsymbol{r}_2) &= \sum_{\boldsymbol{k}_1, \boldsymbol{k}_2} C_{\boldsymbol{k}_1, \boldsymbol{k}_2} e^{i\boldsymbol{k}_1\cdot\boldsymbol{r}_1} e^{i\boldsymbol{k}_2\cdot\boldsymbol{r}_2} \\ &= \sum_{\boldsymbol{k}_1, \boldsymbol{k}_2} C_{\boldsymbol{k}_1, \boldsymbol{k}_2} e^{i(\boldsymbol{k}\cdot\boldsymbol{r} + \boldsymbol{K}\cdot\boldsymbol{R})} \end{aligned} \tag{6.2a}$$

where $\boldsymbol{r} = \boldsymbol{r}_1 - \boldsymbol{r}_2$, $\boldsymbol{R} = (\boldsymbol{r}_1 + \boldsymbol{r}_2)/2$, $\boldsymbol{k} = (\boldsymbol{k}_1 - \boldsymbol{k}_2)/2$, and $\boldsymbol{K} = \boldsymbol{k}_1 + \boldsymbol{k}_2$. We assume the center of a wave vector $\boldsymbol{K} = 0$. Correspondingly, the two electrons are ($\boldsymbol{k}$, ↑) and ($-\boldsymbol{k}$, ↓), as shown in Fig. 6.1. The wave function depends only on the relative coordinate $\boldsymbol{r} = \boldsymbol{r}_1 - \boldsymbol{r}_2$, and the electron interaction $V(\boldsymbol{r}_1, \boldsymbol{r}_2)$ is also expressed as $V(\boldsymbol{r})$. The Schrödinger equation and the wave function are

$$\left\{-\frac{\hbar^2}{2m}\nabla^2 + V(\boldsymbol{r})\right\}\psi(\boldsymbol{r}) = (\varepsilon + 2\varepsilon_\mathrm{F})\psi(\boldsymbol{r}) \tag{6.1b}$$

and

$$\psi(\boldsymbol{r}) = \sum_{\boldsymbol{k}} C_{\boldsymbol{k}} e^{i\boldsymbol{k}\cdot\boldsymbol{r}} \tag{6.2b}$$

On inserting Eq. (6.2b) into Eq. (6.1b) and integrating over the volume ($= L^3$), we obtain the following equations

$$L^3\frac{\hbar^2\boldsymbol{k}^2}{m}C_{\boldsymbol{k}} - V\sum_{\boldsymbol{k}'} C_{\boldsymbol{k}'} = L^3(\varepsilon + 2\varepsilon_\mathrm{F})C_{\boldsymbol{k}} \tag{6.3}$$

$$\int e^{-i\boldsymbol{k}\cdot\boldsymbol{r}}V(\boldsymbol{r})e^{i\boldsymbol{k}'\cdot\boldsymbol{r}}d\boldsymbol{r} = V_{\boldsymbol{k},\boldsymbol{k}'} = -V \ (V > 0). \tag{6.4}$$

Here, we assumed that the two electrons are scattered from ($\boldsymbol{k}$, $-\boldsymbol{k}$) to ($\boldsymbol{k}'$, $-\boldsymbol{k}'$) via a phonon with a wave vector $\boldsymbol{q}$ and the phonon's cutoff energy or a maximum energy $\hbar\omega_\mathrm{D}$ ($= k_\mathrm{B}\theta_\mathrm{D}$, θ_D: Debye temperature), as shown in Figs. 6.1(b) and 6.2. The interaction of the two electrons is attractive ($V > 0$).

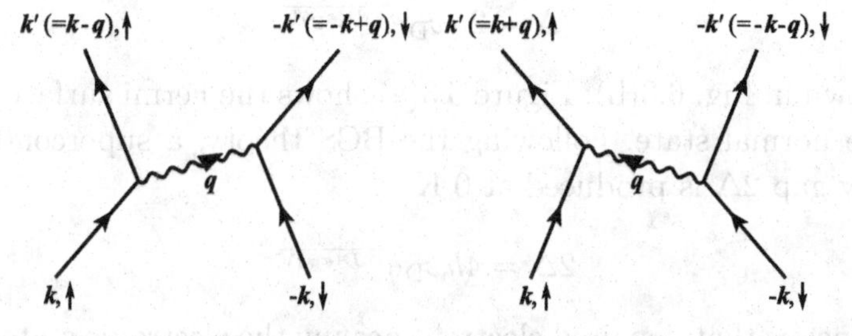

Figure 6.2: Electron-electron interaction via a phonon, when the electron $\boldsymbol{k}$ emits a phonon $\boldsymbol{q}$, and the electron $-\boldsymbol{k}$ absorbs the phonon $\boldsymbol{q}$, together with the opposite process.

From Eq. (6.3), we obtain

$$1 = \frac{V}{L^3} \sum_{\boldsymbol{k}'} \frac{1}{\dfrac{\hbar^2 \boldsymbol{k}'^2}{m} - (\varepsilon + 2\varepsilon_{\mathrm{F}})}. \tag{6.5}$$

By using the relation $L^{-3} \sum \to (2\pi)^{-3} \int d\boldsymbol{k}$ and $\varepsilon' = \hbar^2 \boldsymbol{k}'^2 / 2m - \varepsilon_{\mathrm{F}}$, Eq. (6.5) is written as

$$\begin{aligned} 1 &= V \int_0^{\hbar\omega_{\mathrm{D}}} 2D(\varepsilon' + \varepsilon_{\mathrm{F}}) \frac{1}{2\varepsilon' - \varepsilon} d\varepsilon' \\ &= 2D(\varepsilon_{\mathrm{F}})V \int_0^{\hbar\omega_{\mathrm{D}}} \frac{1}{2\varepsilon' - \varepsilon} d\varepsilon' \end{aligned} \tag{6.6}$$

where $D(\varepsilon + \varepsilon_{\mathrm{F}}) \simeq D(\varepsilon_{\mathrm{F}})$ because $\varepsilon_{\mathrm{F}} \sim 10^4$ K and $\hbar\omega_{\mathrm{D}} \sim 10^2$ K. From Eq. (6.6), we obtain

$$\begin{aligned} \varepsilon &= -\frac{2\hbar\omega_{\mathrm{D}}}{e^{1/D(\varepsilon_{\mathrm{F}})V} - 1} \\ &\simeq -2\hbar\omega_{\mathrm{D}} e^{-\frac{1}{D(\varepsilon_{\mathrm{F}})V}} \quad \text{if} \quad D(\varepsilon_{\mathrm{F}})V \ll 1. \end{aligned} \tag{6.7}$$

The energy ε of the two electrons relative to $2\varepsilon_{\mathrm{F}}$ is reduced, $\varepsilon < 0$, via the phonon-mediated interaction. The two electrons are called a Cooper pair.

The formation of Cooper pairs is applied to electrons in the energy range from ε_{F} to $\varepsilon_{\mathrm{F}} - \hbar\omega_{\mathrm{D}}$. If the phonon-mediated attractive interaction overcomes Coulomb repulsion, the superconducting state is realized. In the superconducting state, the Fermi surface shrinks by

$$\Delta = 2\hbar\omega_{\mathrm{D}} e^{-\frac{1}{D(\varepsilon_{\mathrm{F}})V}} \tag{6.8}$$

as shown in Fig. 6.3(b). Figure 6.3(a) shows the Fermi surface at 0 K in the normal state. Following the BCS theory, a superconducting energy gap 2Δ is produced at 0 K

$$2\Delta = 4\hbar\omega_{\mathrm{D}} e^{-\frac{1}{D(\varepsilon_{\mathrm{F}})V}}. \tag{6.9}$$

This means that depaired electrons occupy the electronic states separated by Δ from the Fermi energy, or by 2Δ from the highest energy level of the Fermi surface, as shown in Fig. 6.3(c). The depaired

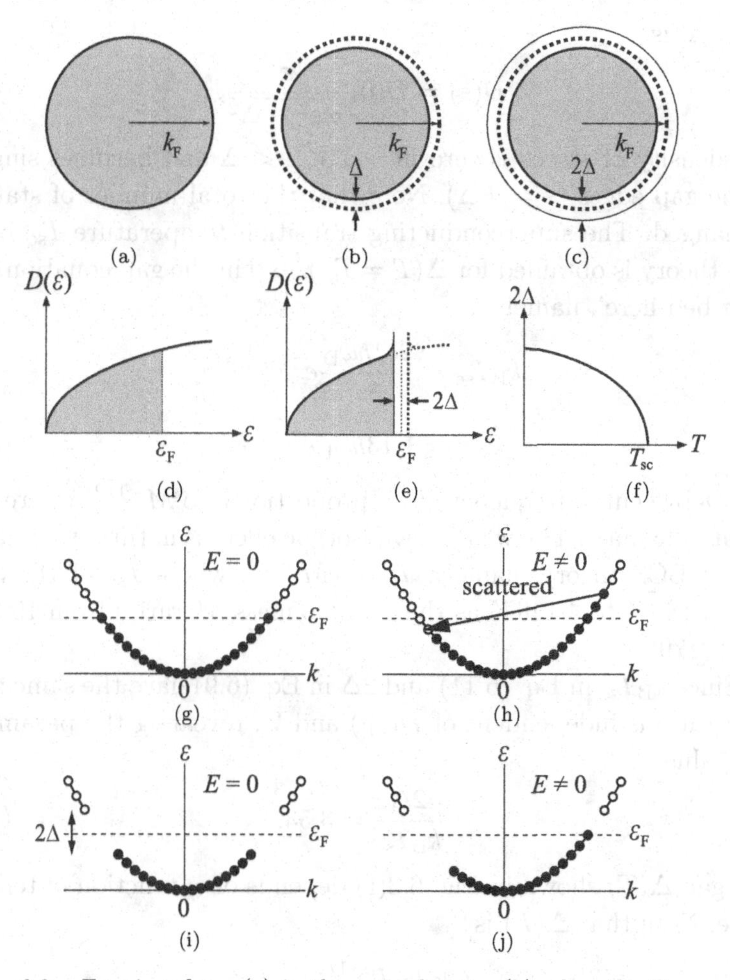

Figure 6.3: Fermi surfaces (a) in the normal state, (b) when the Cooper pairs are formed, and (c) in the BCS superconducting state at 0 K. The densities of states $D(\varepsilon)$ in (d) the normal state and (e) the BCS superconducting state at 0 K. (f) Temperature dependence of the superconducting energy gap 2Δ. Energy bands at (g) an electric field $E = 0$ and (h) $E \neq 0$ in the normal state. Energy bands at (i) $E = 0$ and (j) $E \neq 0$ in the superconducting state.

electrons are often called quasiparticles. The density of states in the normal state, as shown in Fig. 6.3(d), is changed in the superconducting state, as shown in Fig. 6.3(e). If the density of states at the Fermi energy ε_F, $D(\varepsilon_F)$, is expressed as $D(0)$ by shifting the energy, the superconducting density of states $D(\varepsilon)$, for $\varepsilon > \Delta$ and

$\varepsilon < -\Delta$, is

$$D(\varepsilon) = D(0)\frac{\varepsilon}{\sqrt{\varepsilon^2 - \Delta^2}}. \tag{6.10}$$

The density of states is zero in $-\Delta < \varepsilon < \Delta$ and becomes singular at the gap edges ($\varepsilon = \pm\Delta$). Note that the total number of states is unchanged. The superconducting transition temperature $T_{\rm sc}$ in the BCS theory is obtained for $\Delta(T = T_{\rm sc}) \to 0$ in the gap equation, 'not described here', namely

$$\begin{aligned} k_{\rm B}T_{\rm sc} &= \frac{2e^{\gamma}\hbar\omega_{\rm D}}{\pi} e^{-\frac{1}{D(\varepsilon_{\rm F})V}} \\ &= 1.13\hbar\omega_{\rm D} e^{-\frac{1}{D(\varepsilon_{\rm F})V}}. \end{aligned} \tag{6.11}$$

The Debye cutoff frequency $\omega_{\rm D}$ is proportional to $M^{-1/2}$, where M is the atomic mass. The well-known isotope effect is naturally explained by the BCS theory, namely, $T_{\rm sc} \sim M^{-1/2}$, where $T_{\rm sc}$ of Hg varies from 4.185 K to 4.146 K as the isotopic mass M varies from 199.5 to 203.4 [179].

Since $k_{\rm B}T_{\rm sc}$ in Eq. (6.11) and 2Δ in Eq. (6.9) have the same form, their ratio is independent of $D(\varepsilon_{\rm F})$ and V, revealing the parameter-free value

$$\frac{2\Delta}{k_{\rm B}T_{\rm sc}} = 3.53. \tag{6.12}$$

The gap $\Delta(T)$ shown in Fig. 6.3(f) depends on a function of temperature. Note that $\Delta(T)$ is

$$\frac{\Delta(T)}{\Delta(0)} \simeq 1 - \left(\frac{2\pi k_{\rm B}T}{\Delta}\right)^{1/2} e^{-\frac{\Delta}{k_{\rm B}T}} \quad (T \ll T_{\rm sc}) \tag{6.13a}$$

$$\frac{\Delta(T)}{\Delta(0)} \simeq 1.74\left(1 - \frac{T}{T_{\rm sc}}\right)^{1/2} \quad (\text{close to } T_{\rm sc}) \tag{6.13b}$$

where $\Delta = \Delta(T = 0)$.

It is easy to understand the zero resistivity of $\rho(T) = 0$ in the superconducting state because the gap 2Δ is opened in the conduction band at the Fermi energy $\varepsilon_{\rm F}$. Figure 6.3(g) represents the energy band in the normal state, revealing that the energy band is fully occupied up to the Fermi energy $\varepsilon_{\rm F}(= \hbar^2\boldsymbol{k}_{\rm F}^2/2m)$ under an electric field

$E = 0$. Under $E \neq 0$, an electron is scattered from the state k to an unoccupied state k', as shown in Fig. 6.3(h), which is caused by structural imperfections in the lattice, leading to Ohm's law. On the other hand, the superconducting gap 2Δ is opened in the superconducting state, as shown in Fig. 6.3(i). Therefore, no unoccupied state exists in the scattering process under $E \neq 0$, as shown in Fig. 6.3(j). The current flows without scattering, revealing the zero resistivity. This is realized even at $T \neq 0$, specifically below T_{sc}.

The temperature dependence of physical quantities such as the specific heat C and the nuclear spin-lattice relaxation rate $1/T_1$ obeys the exponential law. The electronic specific heat based on $C_{\text{e}} = \gamma T$ in the normal state possesses a jump at T_{sc}, ΔC

$$\frac{\Delta C}{C(T = T_{\text{sc}})} = \frac{\Delta C}{\gamma T_{\text{sc}}} = 1.43. \tag{6.14}$$

This is a result of the weak coupling BCS theory. Note that the BCS superconductivity is based on the weak coupling of the electron-phonon interaction, which produces relations of $2\Delta/k_{\text{B}}T_{\text{sc}} = 3.53$ and $\Delta C/\gamma T_{\text{sc}} = 1.43$. In the exact superconductors, the electron-phonon coupling is not always weak. Strong-coupling superconductors exist: Pb, for example, or $CeCoIn_5$ and $CeIrSi_3$ shown later. The temperature dependence of the electronic specific heat C_{e} below T_{sc} is

$$C_{\text{e}} \sim T^{-\frac{3}{2}} e^{-\Delta/k_{\text{B}}T} \quad (T < 0.1T_{\text{sc}}) \tag{6.15a}$$

$$C_{\text{e}} \sim e^{-\Delta'/k_{\text{B}}T} \quad (0.2 < T < 0.5T_{\text{sc}}) \tag{6.15b}$$

where Eq. (6.15a) is an expression based on the BCS theory at $T \ll T_{\text{sc}}$, but Eq. (6.15b) is an empirical one. The gap Δ' in Eq. (6.15b) is not the same as Δ but is close to Δ.

The nuclear spin-lattice relaxation rate $1/T_1$ is

$$\frac{1}{T_1} = \pi\gamma_{\text{n}}^2\hbar \langle A_q A_{-q} \rangle \int_0^\infty \int_0^\infty \left(1 + \frac{\Delta^2}{\varepsilon\varepsilon'}\right) D(\varepsilon) n(\varepsilon)$$
$$\times D(\varepsilon') \left(1 - n(\varepsilon')\right) \delta(\varepsilon - \varepsilon') d\varepsilon d\varepsilon'$$

$$= 2\pi\gamma_{\rm n}^2\hbar \langle A_q A_{-q}\rangle \int_0^\infty \left[D^2(\varepsilon) + M^2(\varepsilon)\right] n(\varepsilon)\,(1 - n(\varepsilon))\,d\varepsilon \tag{6.16}$$

$$M(\varepsilon) = \frac{D(0)}{4\pi} \int_0^{2\pi} \int_0^{\pi} \frac{\Delta(\theta,\phi)}{\sqrt{\varepsilon^2 - \Delta^2(\theta,\phi)}} \sin\theta d\theta d\phi \tag{6.17}$$

where $D(\varepsilon)$ corresponds to Eq. (6.10). $M(\varepsilon)$ is called an anomalous density of states and originates from the coherent effect. The density of states $D(\varepsilon)$ in Eq. (6.10) diverges at $\varepsilon = \Delta$ correspondingly, revealing a divergence of $1/T_1$. The lifetime effect of quasiparticles by the electron-phonon and the electron-electron interactions, however, broadens the quasiparticle density of states. This results in suppression of the divergence of $1/T_1$, but the production of a peak at $T_{\rm sc}$. This is called the coherence peak or a Hebel-Slichter peak, after they first observed this peak in Al experimentally [180]. Note that the coherent peak is not always observed, especially in the case of strong-coupling superconductors and the anisotropic superconductors described below. In any case, $1/T_1$ in the BCS superconductors is evidenced by the exponential law below $T_{\rm sc}$:

$$\frac{1}{T_1} \sim e^{-\frac{\Delta}{k_{\rm B}T}}.$$

Moreover, the spin susceptibility $\chi_{\rm s}$, which is obtained by measuring the Knight shift K in the NMR experiment as described in Sec. 3.2.7, obeys the exponential law. This is because the spin susceptibility, which is due to quasiparticles, decreases exponentially as a Yosida function and becomes zero at $T = 0$ [181]: Therefore,

$$\begin{aligned} \chi_{\rm s} &= -4\mu_{\rm B}^2 \int_0^\infty D(\varepsilon)\frac{\partial n}{\partial \varepsilon} d\varepsilon \\ &= \chi_0 \left(\frac{2\pi\Delta}{k_{\rm B}T}\right)^{\frac{1}{2}} e^{-\frac{\Delta}{k_{\rm B}T}} \quad (T \ll T_{\rm sc}). \end{aligned} \tag{6.18}$$

This means that the number of quasiparticles becomes zero at 0 K because the quasiparticles completely condense into the Cooper pairs.

6.2 Anisotropic superconductivity

Superconductivity is now a fundamental phenomenon in metallic compounds, observed in the s- and p-electron systems as well as in the strongly correlated d- and f-electron systems. Even if the phenomenon of superconductivity is commonly attributed to be of the BCS-type, careful measurements revealed unconventional properties, for example in $CeRu_2$. A reduced coherent peak was observed at $T_{sc} = 6.1\,K$ in the Ru-NQR experiment for $CeRu_2$ [182], revealing that a superconducting gap is opened in the whole momentum k-space but is anisotropic. The laser photoemission spectroscopic experiment also claimed that the gap is not homogeneously opened [183]. The specific heat experiment also suggested that the electronic specific heat coefficient C_e/T below T_{sc} does not increase linearly as a function of applied magnetic field H but indicates a $\sqrt{H}$-dependence [184]. From the field-angle-resolved specific heat measurement, it was verified that the gap minimum exists in the cubic $\langle 110 \rangle$ direction [185].

Similarly, multigap/multiband superconductivity is another concept of BCS-type superconductivity, which was precisely studied in MgB_2, for example. Two different Fermi surfaces based on the σ and π bands possess two different gaps [186, 187].

The pairing mechanism is based on phonons in these BCS-type superconductors, but is due to magnetic spin fluctuations in strongly correlated electron systems. The typical d-electron systems are high-T_c cuprates [188]. Heavy fermion superconductors such as $CeCoIn_5$, UPt_3, $NpPd_5Al_2$ and $PuCoGa_5$ belong to the f-electron systems [189]. In these strongly correlated electron systems, the superconducting pairing favors the d-wave spin singlet and/or p-(or f-)wave spin triplet states in nature [190]. These superconductors do not obey the exponential law of temperature but the power law. Next, we explain the unconventional (anisotropic) superconductivity.

Heavy fermion superconductors are known to follow the power law in physical properties such as the electronic specific heat C_e and the nuclear spin-lattice relaxation rate $1/T_1$. They do not display an exponential dependence as predicted by the BCS theory [191], namely

$$C_e = \begin{cases} \sim T^3 & \text{(axial type, point node)} \\ \sim T^2 & \text{(polar type, line node)} \\ \sim T & \text{(gapless)} \end{cases} \tag{6.19}$$

$$\frac{1}{T_1} = \begin{cases} \sim T^5 & \text{(axial type, point node)} \\ \sim T^3 & \text{(polar type, line node)} \\ \sim T & \text{(gapless).} \end{cases} \tag{6.20}$$

This indicates the existence of an anisotropic gap, namely the existence of a node in the energy gap. The corresponding density of states is changed on the basis of the anisotropic gap, specifically,

1) for the polar-type line node ($\Delta(\theta, \phi) = \Delta \cos\theta$)

$$D(\varepsilon) = \begin{cases} D(0)\dfrac{\varepsilon}{\Delta}\sin^{-1}\dfrac{\varepsilon}{\Delta} & |\varepsilon| \geq \Delta \quad (6.21\text{a}) \\ \\ D(0)\dfrac{\pi}{2}\dfrac{\varepsilon}{\Delta} & |\varepsilon| < 0 \quad (6.21\text{b}) \end{cases}$$

2) for the axial-type point node ($\Delta(\theta, \phi) = \Delta \sin\theta e^{i\phi}$)

$$D(\varepsilon) = D(0)\frac{\varepsilon}{2\Delta}\ln\left|\frac{\varepsilon + \Delta}{\varepsilon - \Delta}\right| \tag{6.22}$$

When we compare the phonon-mediated attractive interaction based on the BCS theory with the strong repulsive interaction among the f-electrons, it is theoretically difficult for the former interaction to overcome the latter one. To avoid a large overlap of the wave functions of the paired particles, the heavy electron system would rather choose an anisotropic channel, like a p-wave spin triplet or a d-wave spin singlet state, to form Cooper pairs.

Figure 6.4 shows a schematic view of the superconducting parameters with the s-, d- and p-wave pairings. The order parameter $\Psi(\boldsymbol{r})$ with the even parity(s- and d-wave) is symmetric with respect to $\boldsymbol{r}$, where one electron with the up-spin state of the Cooper pair is simply considered to be located at the center of $\Psi(\boldsymbol{r})$, $\boldsymbol{r} = 0$, and the

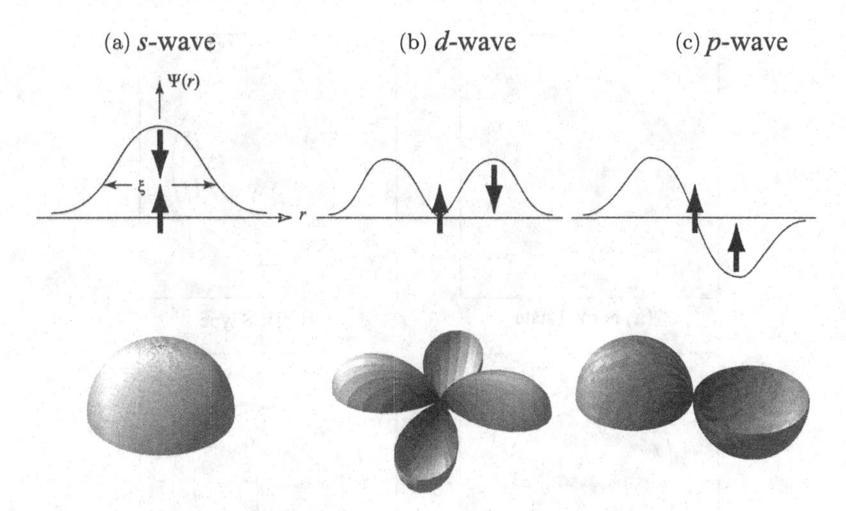

Figure 6.4: Schematic view of the superconducting parameters with the *s*-, *d*- and *p*-wave pairings, cited from Ref. [27].

other electron with the down-spin state is located at $\boldsymbol{r}$. The width of $\Psi(\boldsymbol{r})$ with respect to $\boldsymbol{r}$ is called the coherence length ξ. For example, UPd_2Al_3 is considered to be a *d*-wave superconductor from the NMR Knight shift experiment [192], which corresponds to the case (b) in Fig. 6.4. On the other hand, $\Psi(\boldsymbol{r})$ with odd parity (*p*-wave) is not symmetric with respect to $\boldsymbol{r}$, where the parallel spin state is shown in Fig. 6.4(c). From the NMR Knight shift experiment, it is concluded that UPt_3 possesses odd parity in symmetry, as described in Sec. 3.2.7 [59, 193].

For an anisotropic state, there are three kinds of gap structures, as shown in Fig. 6.5. The first gap structure indicates the isotropic superconducting gap, as shown in Fig. 6.5(b), which is the same as the *s*-wave one and is isotropic. This is called the Balian-Werthamer (BW) state. The second one shows a line node in the equator on the Fermi surface. This structure is called the polar type, as shown in Fig. 6.5(c). The third one has a point node in the pole on the Fermi surface. This condition has the Anderson-Brinkman-Morel (ABM) state. This is called the axial type, as shown in Fig. 6.5(d).

Here, we explain the framework of anisotropic superconductivity in a simplified form. The two-electron wave function $\Psi(\boldsymbol{r}_1, \boldsymbol{\sigma}_1, \boldsymbol{r}_2, \boldsymbol{\sigma}_2)$

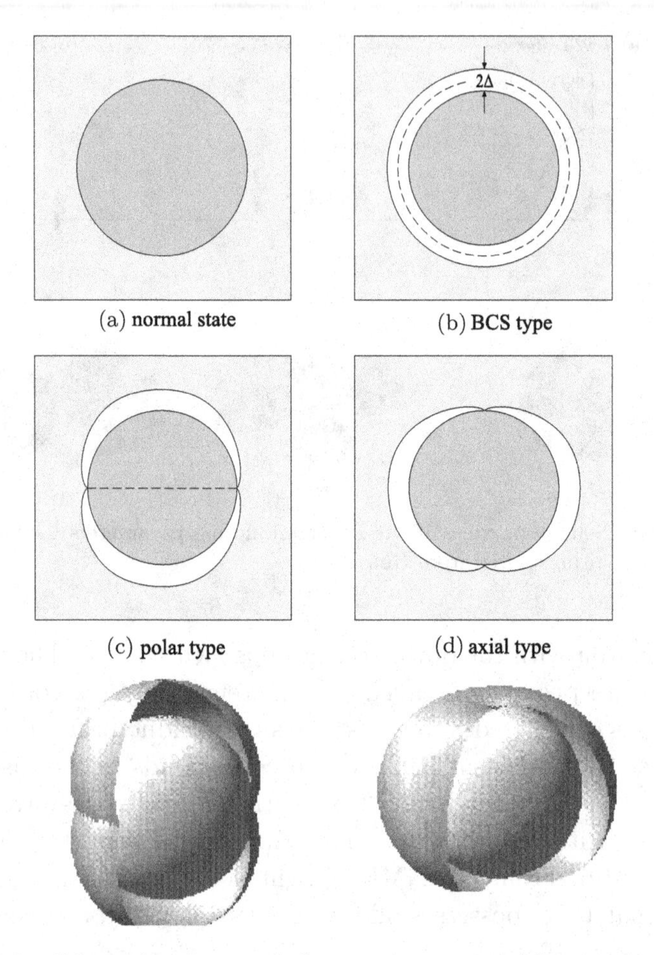

Figure 6.5: Schematic picture of the gap structures: (a) normal state, (b) BCS-type superconductor, which has an isotropic gap, (c) polar type and (d) axial type, cited from Ref. [27].

is a product of the orbital part $\psi(\boldsymbol{r}_1, \boldsymbol{r}_2)$ and the spin part $\chi(\boldsymbol{\sigma}_1, \boldsymbol{\sigma}_2)$, as mentioned above. The spin is either $\uparrow$ or $\downarrow$. The spin part is thus classified into the following two groups

1) Total spin $S = 0$ (spin singlet)

$$\chi^{S=0}(\boldsymbol{\sigma}_1, \boldsymbol{\sigma}_2) = \frac{1}{\sqrt{2}} (|\uparrow\downarrow\rangle - |\downarrow\uparrow\rangle)$$

2) Total spin $S = 1$ (spin triplet)

$$\chi^{S=1}(\boldsymbol{\sigma}_1, \boldsymbol{\sigma}_2) = \begin{cases} |\uparrow\uparrow\rangle & : S_z = 1 \\ \dfrac{1}{\sqrt{2}}(|\uparrow\downarrow\rangle + |\downarrow\uparrow\rangle) & : S_z = 0 \\ |\downarrow\downarrow\rangle & : S_z = -1. \end{cases}$$

Note that it is easier to think about the projection of the total spin $S = 1$, namely $S_z = 1,\ 0,\ -1$, as shown in Fig. 6.6(b). The spin singlet case with the total spin $S = 0$ is also shown in Fig. 6.6(a).

The orbital part $\psi(\boldsymbol{r}_1, \boldsymbol{r}_2)$ is easily understood if the electron interaction $V(r)$ is a centrifugal force, as in the hydrogen atom. In this case, $\psi(r)$ is expressed using polar coordinates as

$$\psi(r) = R_{nl}(r)Y_l^m(\theta, \phi) \tag{6.23}$$

where $Y_l^m(\theta, \phi)$ is

$$1)\ l = 0\ (s \text{ wave}):\ Y_0^0 = \frac{1}{\sqrt{4\pi}} \tag{6.24}$$

$$2)\ l = 1\ (p \text{ wave}):\ Y_1^0 = \sqrt{\frac{3}{4\pi}}\cos\theta$$

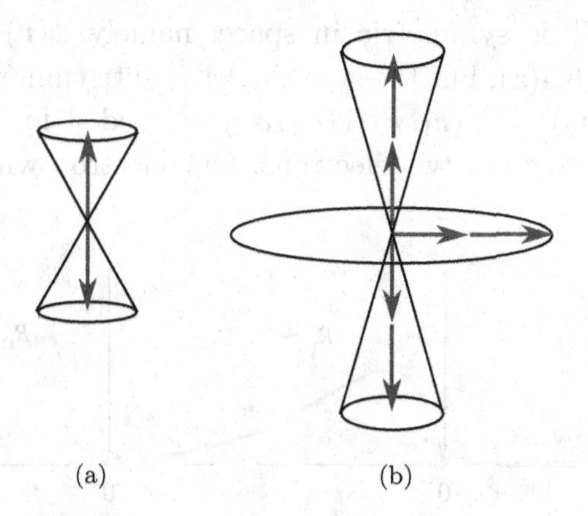

Figure 6.6: (a) Total spin $S = 0$ and (b) total spin $S = 1$ for two spins $\boldsymbol{\sigma}_1$ and $\boldsymbol{\sigma}_2$.

$$Y_1^{\pm 1} = \mp\sqrt{\frac{3}{8\pi}}\sin\theta\, e^{\pm i\phi} \tag{6.25}$$

$$3)\ l = 2\ (d\text{ wave}):\ Y_2^0 = \sqrt{\frac{5}{16\pi}}\left(3\cos^2\theta - 1\right)$$

$$Y_2^{\pm 1} = \mp\sqrt{\frac{15}{8\pi}}\sin\theta\cos\theta\, e^{\pm i\phi}$$

$$Y_2^{\pm 2} = \sqrt{\frac{15}{32\pi}}\sin^2\theta\, e^{\pm 2i\phi}, \tag{6.26}$$

and $R_{nl}(r)$ is

$$1)\ l = 0\ (s\text{ wave}):\ R_{10} = 2\left(\frac{1}{a_0}\right)^{3/2} e^{-\frac{r}{a_0}} \tag{6.27}$$

$$2)\ l = 1\ (p\text{ wave}):\ R_{21} = \frac{1}{\sqrt{3}}\left(\frac{1}{2a_0}\right)^{3/2}\frac{r}{a_0}e^{-\frac{r}{2a_0}} \tag{6.28}$$

$$3)\ l = 2\ (d\text{ wave}):\ R_{32} = \frac{4}{27\sqrt{10}}\left(\frac{1}{3a_0}\right)^{3/2}\left(\frac{r}{a_0}\right)^2 e^{-\frac{r}{3a_0}}. \tag{6.29}$$

$R_{nl}(r)$ is schematically illustrated in Fig. 6.7.

In the case of $l = 0$ (even numbers), the orbital wave function $\psi(r) = R_{10}Y_0^0$ is symmetric in space, namely $\psi(\boldsymbol{r}) = \psi(-\boldsymbol{r})$, as shown in Fig. 6.4(a), but the spin singlet is antisymmetric. Note that $\Psi(\boldsymbol{r}_1, \boldsymbol{\sigma}_1, \boldsymbol{r}_2, \boldsymbol{\sigma}_2) = \psi(\boldsymbol{r}_1, \boldsymbol{r}_2)\chi(\boldsymbol{\sigma}_1, \boldsymbol{\sigma}_2)$ is needed for antisymmetry regarding a change of two electrons. One electron with the up spin

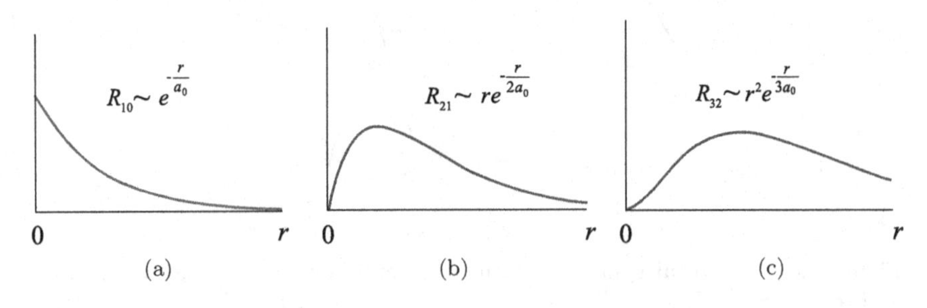

Figure 6.7: Radial wave function $R_{nl}(r)$.

state (↑) is located at $r = 0$ and the other electron with the down spin state (↓) spreads over the distance ξ. The magnitude of the Cooper pair is the coherent length ξ.

In the case of $l = 1$ (odd numbers), the orbital wave function $\psi(r) = R_{21}Y_1^0$ and $R_{21}Y_1^{\pm 1}$ is antisymmetric, namely $\psi(\boldsymbol{r}) \neq \psi(-\boldsymbol{r})$ but the spin triplet state is symmetric, as shown in Fig. 6.4(c). Note that $\boldsymbol{r} \to -\boldsymbol{r}$ corresponds to $r \to r$, $\theta \to \pi - \theta$, and $\phi \to \phi + \pi$ in the polar coordinates. When one electron with the up spin state is located at $r = 0$, the other electron with the up spin state does not exist at $r = 0$ because $R_{21} = 0$ at $r = 0$. The other electron is separated from the electron at $r = 0$.

In the case $l = 2$ (even numbers), the orbital wave function $\psi(r) = R_{32}Y_2^0$, $R_{32}Y_2^{\pm 1}$, and $R_{32}Y_2^{\pm 2}$ is symmetric in space, as shown in Fig. 6.4(b), but the spin singlet is antisymmetric. In this case, two electrons are separated from each other.

The energy gap is anisotropic and is expressed as

$$\Delta(k) = \Delta \sum_{l,m} \lambda_m Y_l^m(\theta, \phi). \tag{6.30}$$

In the case of the s wave, the gap is opened uniformly because of $Y_0^0 = 1/\sqrt{4\pi}$, as shown in Fig. 6.5(b). The gap has a node in the case of $l \neq 0$. The polar node ($\theta = \pi/2$) is realized, for example, for $l = 1$ and $m = 0$ because $Y_1^0 = \sqrt{3/4\pi}\cos\theta$. The gap is

$$\Delta(\theta) = \Delta \cos\theta \tag{6.31}$$

as shown in Fig. 6.5(c)

In the case of $l = 1$ and $m = 1$, the Cooper pair is in the parallel spin state, $(\boldsymbol{k}\uparrow, -\boldsymbol{k}\uparrow)$ and $(\boldsymbol{k}\downarrow, -\boldsymbol{k}\downarrow)$, and the gap is

$$\Delta(\theta) = \Delta_0 \sin\theta \, e^{i\phi} \tag{6.32}$$

revealing an axial node, as shown in Fig. 6.5(d).

As noted above, the node in the superconducting gap can be detected by measuring the specific heat, spin-lattice relaxation rate, and thermal conductivity. Through the recent development of techniques, specifically the field-angle-resolved specific heat and thermal conductivity measurements, it is possible to determine the

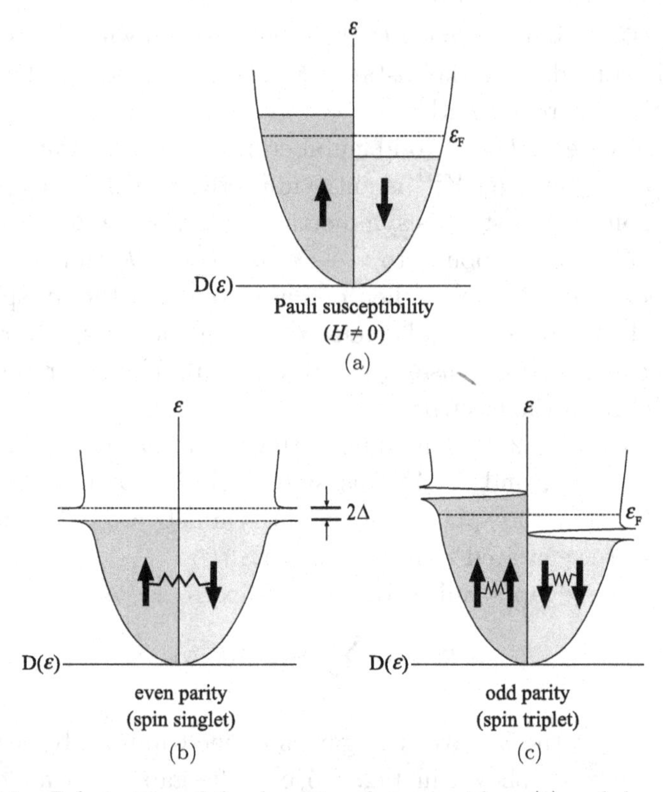

Figure 6.8: Polarization of the densities of states with up(↑) and down(↓) spin states in (a) the normal state, revealing the Pauli susceptibility, (b) the spin singlet and (c) spin triplet superconducting states, shown as in Fig. 4.2.

position of the node on the Fermi surface, as shown later for $CeCoIn_5$ and UPt_3.

Furthermore, it is possible to distinguish between the spin singlet (s and d wave types, even parity) and the spin triplet (p and f wave types, odd parity) by measuring the spin susceptibility or the NMR Knight shift. The spin susceptibility χ_s corresponds to the susceptibility based on the Zeeman shift of the up(↑) and down(↓) densities of states under magnetic field. In the normal state, it is the Pauli susceptibility $\chi_P(= \chi_n = \chi_0)$, as shown in Fig. 6.8(a). In the case of spin singlet, Cooper pairs with the up and down spin states do not contribute to the spin susceptibility, which is due to the thermally excited quasiparticles at finite temperature, as shown in Fig. 6.8(b).

χ_s decreases below T_{sc}, as follows:

$$\chi_s = \begin{cases} \chi_0 \left(\dfrac{2\pi\Delta}{k_B T}\right)^{1/2} e^{-\frac{\Delta}{k_B T}} & T \ll T_{sc} \\ \sim T & \text{polar-type } d \text{ wave} \\ \sim T^2 & \text{axial-type } d \text{ wave.} \end{cases} \tag{6.33}$$

On the other hand, the spin susceptibility of spin triplet is unchanged below T_{sc} because the Cooper pairs of $(\uparrow\uparrow)$ and $(\downarrow\downarrow)$ are polarized in magnetic fields, as shown in Fig. 6.8(c). As described in Sec. 3.2.7, no change of the Knight shift in UPt_3 is found below $T_{sc} = 0.58$ K, suggesting spin triplet superconductivity.

6.3 Ginzburg-Landau theory and related superconducting properties

First, we explain the superconducting properties from a thermodynamics viewpoint. The free energy in the superconducting state G_s is reduced by $H_c^2/8\pi$ below T_{sc} compared with the free energy in the normal state G_n, as shown in Fig. 6.9(a). H_c is the thermodynamic critical field. The experimental results can be approximately described by a quadratic temperature dependence

$$H_c(T) = H_c(0)[1 - (T/T_{sc})^2]. \tag{6.34}$$

The corresponding entropies in the superconducting and normal states S_s and S_n are also shown in Fig. 6.9(b). The superconducting transition is thus of second-order and the specific heat indicates a jump at T_{sc}, revealing $\Delta C/\gamma T_{sc} = 1.43$ in the BCS superconductivity, as shown in Fig. 6.9(c).

When the magnetic field H is applied to elemental superconductors such as Sn and Al, the Meissner current flows across the surface to shield against the applied magnetic field, shown in Fig. 6.11(a). The corresponding free energy in the superconducting state increases

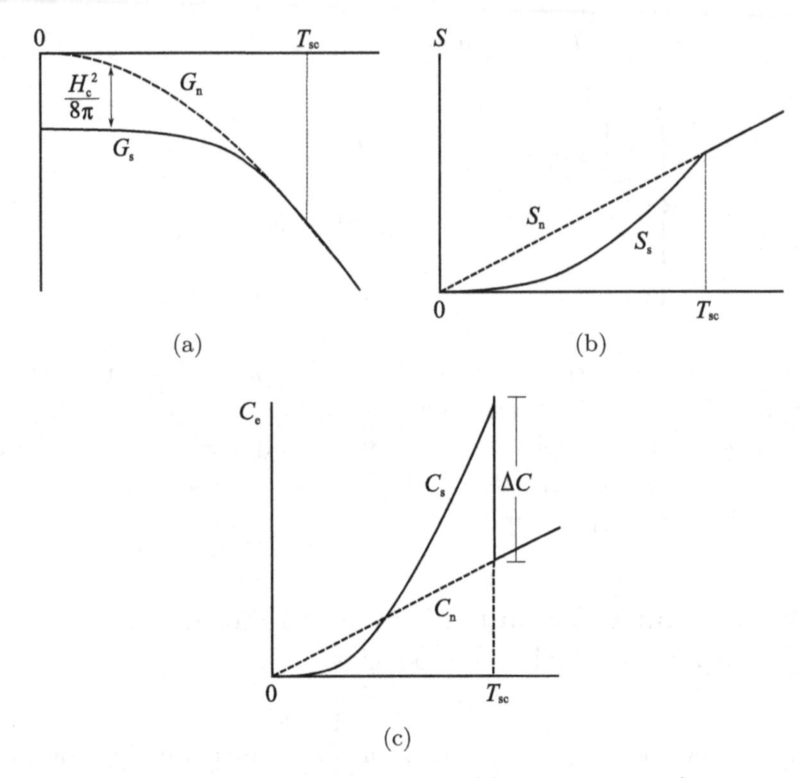

Figure 6.9: Temperature dependences of (a) free energies (G_n and G_s), (b) entropies (S_n and S_s), and (c) electronic specific heats (C_n and C_s) in the normal and superconducting states, respectively.

as a function of magnetic field:

$$G_s(H) = G_n(0) + \frac{1}{8\pi}(H^2 - H_c^2), \tag{6.35}$$

following line 0N in Fig. 6.10(a). Here we assume $G_n(H) = G_n(0)$, neglecting the magnetization $\chi_s H$ because of the small value of the magnetic susceptibility in the superconducting state χ_s in Sn and Al. These superconductors are classified as type I.

In the usual superconducting compounds, which are classified as type II, the free energy in the superconducting state reveals a line 0N′ in Fig. 6.10(a), where H_{c2} is called the upper critical field and is larger than H_c. In this text we discuss type II superconductors on the basis of the Ginzburg-Landau theory. The corresponding magnetization

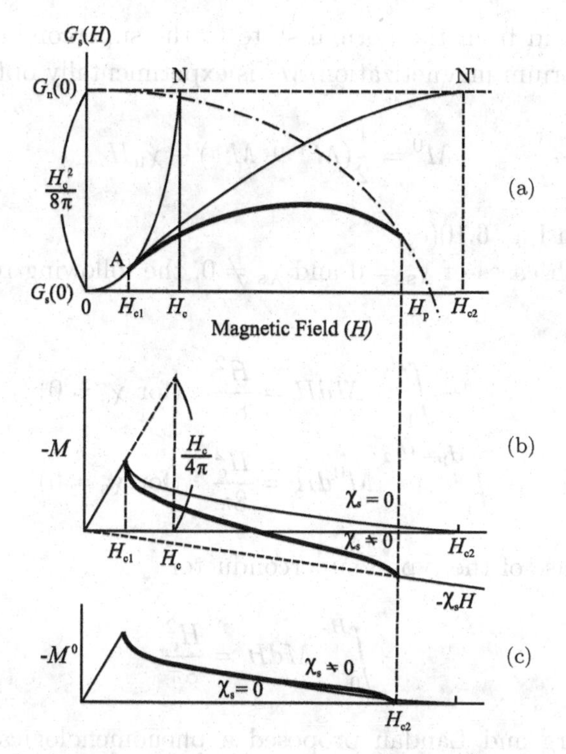

Figure 6.10: Field dependence of (a) free energies (G_{n} and G_{s}) in the normal and superconducting states respectively, (b) magnetization curves in the cases of $\chi_{\mathrm{s}} = 0$ and $\chi_{\mathrm{s}} \neq 0$, and (c) equilibrium magnetization curves in the cases of $\chi_{\mathrm{s}} = 0$ and $\chi_{\mathrm{s}} \neq 0$.

curves in the superconducting state ($-M$) are shown in Fig. 6.10(b), where two cases of $\chi_{\mathrm{s}} = 0$ and $\chi_{\mathrm{s}} \neq 0$ are shown in Fig. 6.10(b) for type II superconductors.

In the case of $\chi_{\mathrm{s}} \neq 0$, the free energy becomes

$$G_{\mathrm{n}}(H) = G_{\mathrm{n}}(0) - \frac{1}{2}\chi_{\mathrm{s}} H^2. \tag{6.36}$$

In this case, the upper critical field corresponds to H_{P} which is smaller than that in the case of $\chi_{\mathrm{s}} = 0$, as shown in Fig. 6.10(b). H_{P} is the Pauli limiting field. Experimentally, we measure the magnetization (M^+) by increasing magnetic field from the superconducting state to the normal state and the magnetization (M^-) by decreasing magnetic field from the normal state to the superconducting state. The equilibrium magnetization M^0 is experimentally obtained as

$$M^0 = \frac{1}{2}(M^+ + M^-) - \chi_n H \tag{6.37}$$

as shown in Fig. 6.10(c).

For both cases of $\chi_s = 0$ and $\chi_s \neq 0$, the following relations are satisfied:

$$-\int_0^{H_{c2}} M dH = \frac{H_c^2}{8\pi} \quad (\text{for } \chi_s = 0) \tag{6.38a}$$

$$-\int_0^{H_p = H_{c2}} M^0 dH = \frac{H_c^2}{8\pi} \quad (\text{for } \chi_s \neq 0) \tag{6.38b}$$

as in the case of the type I superconductor

$$-\int_0^{H_c} M dH = \frac{H_c^2}{8\pi}. \tag{6.38c}$$

Ginzburg and Landau proposed a phenomenological theory of superconductivity, which is obtained from a general theory on the second-order phase transition [194]. The free energy G is expanded in terms of an order parameter $|\psi|^2$, which corresponds to a local density n_s of superconducting electrons. The free energy between the superconducting and normal states in magnetic fields is

$$G_s - G_n = \alpha|\psi|^2 + \frac{\beta}{2}|\psi|^4 + \frac{1}{2m^*}\left|\left(-i\hbar\nabla + \frac{e^*}{c}\boldsymbol{A}\right)\psi\right|^2 + \frac{h^2}{8\pi} \tag{6.39}$$

where m^* and e^* are the mass and charge of a Cooper pair, respectively, giving $m^* = 2m_0$ and $e^* = -2e$.

First, we consider the case with no magnetic field. Under the condition of uniform ψ over space, Eq. (6.39) becomes

$$G_s - G_n = \alpha|\psi|^2 + \frac{\beta}{2}|\psi|^4 \tag{6.40}$$

The equilibrium value of the order parameter is obtained from $\partial(G_s - G_n)/\partial|\psi|^2 = 0$, so that

$$|\psi|^2 = -\frac{\alpha}{\beta} \equiv |\psi_\infty|^2 \tag{6.41}$$

$$G_s - G_n = -\frac{\alpha^2}{2\beta} = -\frac{H_c^2}{8\pi}. \tag{6.42}$$

Here, $\beta > 0$ and then $\alpha < 0$ are required. Note that the difference between the free energy in the normal and superconducting states per unit volume is $H_c^2/8\pi$. The superconducting transition temperature T_{sc} is defined as the temperature showing $\alpha = 0$, and near T_{sc}, α is

$$\alpha = \alpha'(T - T_{sc}). \tag{6.43}$$

Then H_c is obtained from Eq. (6.42)

$$H_c \sim T_{sc} - T \quad (\text{near } T_{sc}) \tag{6.44a}$$

where $H_c(T)$ is approximately described in the entire temperature range with a quadratic temperature dependence

$$H_c(T) = H_c(0)\left[1 - \left(\frac{T}{T_{sc}}\right)^2\right]. \tag{6.44b}$$

By setting the variation of Eq. (6.39) with respect to the order parameter equal to zero, we obtain the following two equations:

$$\alpha\psi + \beta|\psi|^2\psi + \frac{1}{2m^*}\left(-i\hbar\nabla + \frac{e^*}{c}\boldsymbol{A}\right)^2 \psi = 0 \tag{6.45a}$$

$$J(\boldsymbol{r}) = \frac{ie^*\hbar}{2m^*}(\psi^*\nabla\psi - \psi\nabla\psi^*) - \frac{e^{*2}}{m^*c}|\psi|^2\boldsymbol{A}. \tag{6.45b}$$

If the applied magnetic field is small, the change of ψ is small, revealing $\nabla\psi = \nabla\psi^* = 0$. Then, Eq. (6.45b) becomes

$$J(\boldsymbol{r}) = -\frac{e^{*2}}{m^*c}n_s^*\boldsymbol{A} \tag{6.46}$$

where $|\psi|^2 = n_s^*$. Using the relations $\mathrm{rot}h(\boldsymbol{r}) = (4\pi/c)J(\boldsymbol{r})$ and $\mathrm{rot}(\mathrm{rot}h) = -\nabla^2 h$, we obtain

$$h_z = He^{-\frac{x}{\lambda}} \tag{6.47}$$

$$J_y = \frac{c}{4\pi}He^{-\frac{x}{\lambda}} \tag{6.48}$$

$$\lambda = \sqrt{\frac{m^* c^2}{4\pi n_s^2 e^{*2}}}. \tag{6.49}$$

When a small magnetic field H is applied along the z-axis, the Meissner current J_y, which is a flow of Cooper pairs, is directed along the y-axis within a distance λ from the surface and shields the interior of the sample from the external magnetic field, as shown in Fig. 6.11(a). The magnetic field decays from the surface into a superconductor bulk exponentially. λ is the London penetration depth.

Furthermore, we consider the superconducting state in the presence of a magnetic field H along the z-axis. We neglect a term of $\beta|\psi|^2\psi$ in Eq. (6.45a) and use a convenient gauge of $\boldsymbol{A} = (0, Hx, 0)$ from $\mathrm{rot}\boldsymbol{A} = H$. Equation (6.45a) becomes

$$\frac{1}{2m^*}\left(-i\hbar\nabla + \frac{e^*}{c}\boldsymbol{A}\right)^2 \psi = -\alpha\psi \tag{6.50a}$$

or

$$-\frac{\hbar^2}{2m^*}\frac{d^2\psi}{dx^2} + \frac{1}{2}m^*\omega_c^2 x^2\psi = -\alpha\psi \tag{6.50b}$$

where $\omega_c = e^*H/m^*c$. The corresponding harmonic oscillator eigenvalues are

$$-\alpha = \left(n + \frac{1}{2}\right)\hbar\omega_c. \tag{6.51}$$

A minimum value is obtained at $n = 0$. We can define the upper critical field H_{c2} from Eq. (6.51), using the relation $\alpha = -\hbar^2/2m^*\xi^2$

$$H \equiv H_{c2} = \frac{\hbar c/e^*}{2\pi\xi^2} = \frac{\phi_0}{2\pi\xi^2} = \sqrt{2}\kappa H_{c2} \tag{6.52}$$

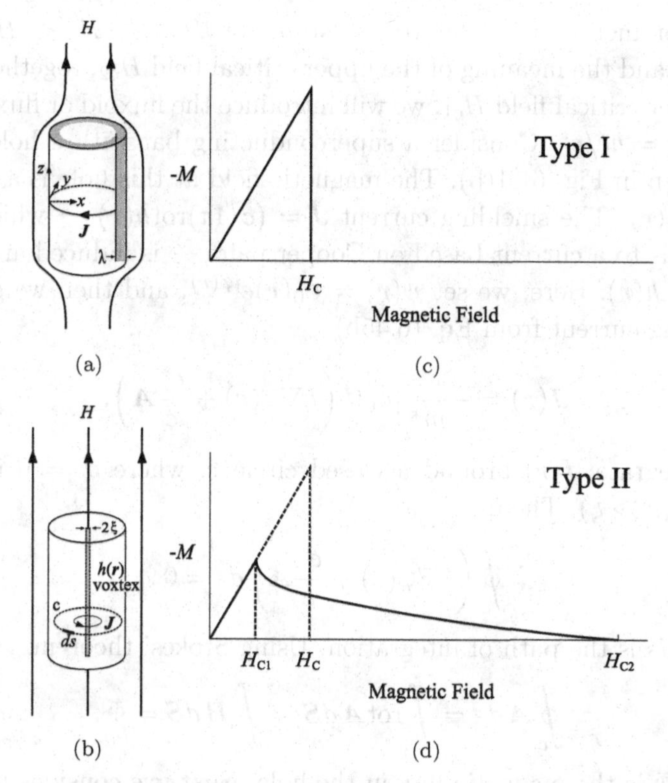

Figure 6.11: (a) Meissner current in a weak field ($H < H_{\rm c}$ or $H_{\rm c1}$) and (b) vortex in a strong field ($H_{\rm c1} < H < H_{\rm c2}$). Magnetizations in (c) type I and (d) type II superconductor.

where relations of $H_{\rm c2}^2 = 4\pi\alpha^2/\beta$ and $n_{\rm s} = -\alpha/\beta = m^*c^2/4\pi e^{*2}\lambda^2$ are used, and we set a GL parameter $\kappa = \lambda/\xi$. Elemental superconductors such as Sn and Al have $\kappa \ll 1$ because $\xi \gg \lambda$. On the other hand, typical superconductors of compounds are not pure compared to elemental superconductors. When decreasing the magnetic field for these superconductors, superconductivity will spontaneously nucleate at $H = H_{\rm c2}$. If $\kappa > 1/\sqrt{2}$, $H_{\rm c2}(> H_{\rm c})$ is realized. Two types of superconductors should be distinguished: type I with $\kappa < 1/\sqrt{2}$ and type II with $\kappa > 1/\sqrt{2}$. The type II superconducting properties are the primary subject in this text.

The field dependence of the magnetization is very different between type II and type I, as shown in Fig. 6.10(b). The type II superconductor is in the vortex state for $H_{c1} < H < H_{c2}$. To understand the meaning of the upper critical field H_{c2}, together with the lower critical field H_{c1}, we will introduce the fluxoid or flux quantum $\phi_0 = \hbar c/e^*$. Consider a superconducting bar with a hole in it, as shown in Fig. 6.11(b). The magnetic field at this hole is assumed to be $h(\boldsymbol{r})$. The shielding current $\boldsymbol{J} = (c/4\pi)\mathrm{rot}h(\boldsymbol{r})$ — which corresponds to a current based on Cooper pairs — is induced around a hole or $h(\boldsymbol{r})$. Here, we set $\psi(\boldsymbol{r}) = \psi_0(\boldsymbol{r})e^{i\varphi(\boldsymbol{r})}$, and then we get the shielding current from Eq. (6.45b)

$$\boldsymbol{J}(\boldsymbol{r}) = -\frac{e^*}{m^*}|\psi_0|^2\left(\hbar\nabla\varphi(\boldsymbol{r}) + \frac{e^*}{c}\boldsymbol{A}\right). \tag{6.53}$$

We integrate $\boldsymbol{J}(\boldsymbol{r})$ around a closed circle c, where $\boldsymbol{J} = 0$ in that region ($r \gg \xi$). Then,

$$\oint_c \left(\hbar\nabla\varphi(\boldsymbol{r}) + \frac{e^*}{c}\boldsymbol{A}\right) ds = 0 \tag{6.54}$$

where ds is the path of integration. Using Stokes' theorem,

$$\oint_c \boldsymbol{A}\,d\boldsymbol{s} = \int \mathrm{rot}\boldsymbol{A}\,d\boldsymbol{S} = \int \boldsymbol{B}\,d\boldsymbol{S} = \Phi \tag{6.55}$$

where Φ is the magnetic flux in the hole. Next, we consider integration of the phase. Since the order parameter $\psi(\boldsymbol{r})$ must be single-valued by the Bohr condition, its phase must change an integer multiple $2\pi n$ in going around the hole. Thus,

$$\oint_c \nabla\varphi d\boldsymbol{s} = 2\pi n. \tag{6.56}$$

From these results, the quantization of flux is obtained to be

$$\Phi = n\frac{hc}{e^*} = n\frac{hc}{2e} = n\phi_0 \tag{6.57}$$

where $e^* = 2e$ is inserted and $\phi_0 = 2.068\times10^{-7}$ Oe·cm^2 is the fluxoid or flux quantum.

As shown in Fig. 6.11(d), the type II sample is in the normal state above H_{c2}. For $H_{c1} < H < H_{c2}$, the superconductor is in a vortex state. Below H_{c1}, the magnetic field is expelled as in the type I superconductor.

The vortex is a normal phase with a cylindrical shape of radius ξ. The vortex energy per unit length I is

$$I = \int \frac{h^2(\boldsymbol{r})}{8\pi} d^3\boldsymbol{r} + \int \frac{1}{2} m^* \boldsymbol{v}_{\mathrm{s}}^2 n_{\mathrm{s}}^* d^3\boldsymbol{r} \simeq \left(\frac{\Phi}{4\pi\lambda}\right)^2 \log\kappa. \tag{6.58}$$

The first term in Eq. (6.58) is the energy due to magnetic field $h(\boldsymbol{r})$. The second term corresponds to the energy due to Cooper pairs induced by $h(\boldsymbol{r})$, where $\boldsymbol{v}_{\mathrm{s}}$ is the velocity of the Cooper pair in relation to $\boldsymbol{J} = (c/2\pi)\mathrm{rot}h(\boldsymbol{r}) = -e^* n_{\mathrm{s}}^* \boldsymbol{v}_{\mathrm{s}}$.

As the energy I is proportional to Φ^2, a vortex containing n flux quanta has a higher energy than n vortices, which results in a vortex with one flux quantum. The energy E of a superconductor with n vortices is

$$E = nI - \frac{n\phi_0 H}{4\pi}. \tag{6.59}$$

The lower critical field H_{c1} is obtained if $E \leq 0$:

$$H_{\mathrm{c1}} = \frac{\phi_0}{4\pi\lambda^2} \log\kappa. \tag{6.60}$$

In the vortex state, the vortices act as line magnets and repel each other. In the absence of vortex pinning centers, the net repulsion energy is minimized when the vortices form a hexagonal array or a vortex lattice.

As for the upper critical field H_{c2}, we note the conventional analyses. From the slope of the upper critical field $(-dH_{\mathrm{c2}}/dT)_{T=T_{\mathrm{sc}}}$, we can estimate the $H_{\mathrm{c2}}(0)$ value. The temperature dependence of $H_{\mathrm{c2}}(T)$ was theoretically discussed by Werthamer, Helfand, and Hohenberg (known as WHH) [195]. The upper critical field at $T =$ 0 K, $H_{\mathrm{c2}}(0)$, can be described using the WHH formula:

$$H_{\mathrm{c2}}(0) = 0.7 \left| \frac{dH_{\mathrm{c2}}}{dT} \right|_{T_{\mathrm{sc}}} \cdot T_{\mathrm{sc}}. \tag{6.61}$$

For type-II superconductors, the coherence length ξ can be estimated from $H_{\mathrm{c2}}(0)$ using Eq. (6.52). The upper critical field based on Eqs. (6.52) and (6.61) is called the orbital limiting field H_{orb}.

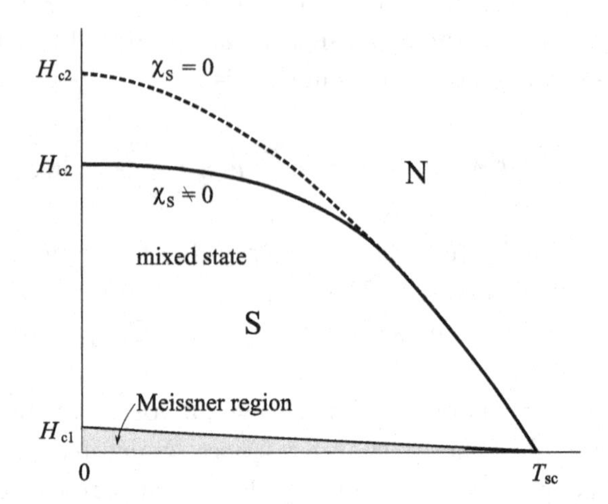

Figure 6.12: Upper and lower critical fields in the cases of $\chi_s = 0$ and $\chi_s \neq 0$.

The temperature dependence of H_{c2} in the usual type II superconductors can be modelled with the WHH theory, using appropriate parameters. The $H_{c2}(0)$ values in heavy fermion superconductors are, however, extremely reduced compared to the $H_{c2}(0)$ values estimated from Eq. (6.61). This is mainly due to the Pauli paramagnetic suppression, as shown in Fig. 6.12.

The influence of the magnetic field on the spins of the electrons in the superconducting states has first been reported by Clogston [196] and by Chandrasekhar [197]. Based on Eq. (6.36), the normal state becomes energetically more favorable for the system when the magnetic energy $\frac{1}{2}\chi_s H^2$ of the normal state reaches the condensation energy $\frac{H_c^2}{8\,\pi}$ of the superconductor

$$\frac{1}{2}\chi_s H_P^2 = \frac{H_c^2}{8\pi}. \tag{6.62}$$

The upper critical field H_{c2} should be smaller than the Pauli limiting field H_P. For example, we estimate the $H_{c2}(0)$ value of a d-wave superconductor $CeCoIn_5$, shown in Fig. 5.37(b), on the basis of Eq. (6.61): $H_{c2}(0) = 480$ kOe for $|dH_{c2}/dT|_{T_{sc}} = 305$ kOe/K and $T_{sc} = 2.24$ K in the field along the [100] direction. The exact $H_{c2}(0)$ value is, however, 117 kOe. The Pauli paramagnetic suppression is realized in $CeCoIn_5$. We can rather simply obtain H_P [kOe]$=18.6 T_{sc}$

using $\chi_s = \chi_n = 2\mu_B^2 D(\varepsilon_F)$ and BCS relations of $\frac{H_c^2}{8\pi} = \frac{1}{2}D(\varepsilon_F)\Delta^2$ and $2\Delta = 3.53k_B T_{sc}$: $H_P = 42$ kOe for $T_{sc} = 2.24$ K in $CeCoIn_5$. The simple relation H_P [kOe]$=18.6T_{sc}$ is not applicable to heavy fermion compounds. This is mainly due to the small value of spin susceptibility χ_s in Eq. (6.62). The susceptibility consists of the spin and orbital parts: $\chi(T) = \chi_s(T) + \chi_{orb}$, as mentioned in Sec. 3.2.7. χ_s is related to superconductivity and to Eq. (6.62). It is also noted that the paramagnetic suppression of H_{c2} is not realized in spin-triplet superconductors such as UPt_3. This is also due to a small χ_s value.

6.4 Heavy fermion superconductivity

6.4.1 $CeRu_2$

$CeRu_2$, with the cubic Laves-phase structure (No. 227, $a = 7.534$ Å, $Z = 8$) shown in Fig. 6.13, is also a fascinating superconducting compound [184, 198–200]. $CeRu_2$ is a valence-fluctuation compound, where the $4f$ electron is itinerant and contributes to the conduction electrons at low temperatures, as in $CeSn_3$ described in Sec. 5.4.2 [148, 150]. The magnetic susceptibility is approximately temperature-independent. The electronic specific heat coefficient γ is slightly enhanced at 27 mJ/(K^2·mol). The experimental dHvA data are convincingly explained by the results of $4f$-itinerant band calculations [200]. The theoretical Fermi surfaces with three-dimensional electronic state are shown in Fig. 6.14. The main dHvA branches

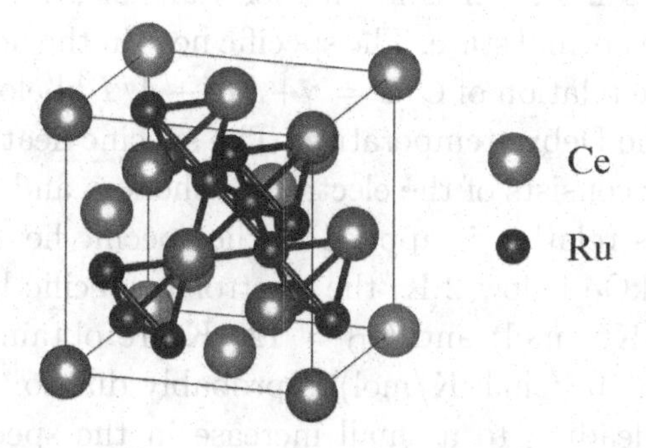

Figure 6.13: Cubic Laves-phase crystal structure in $CeRu_2$.

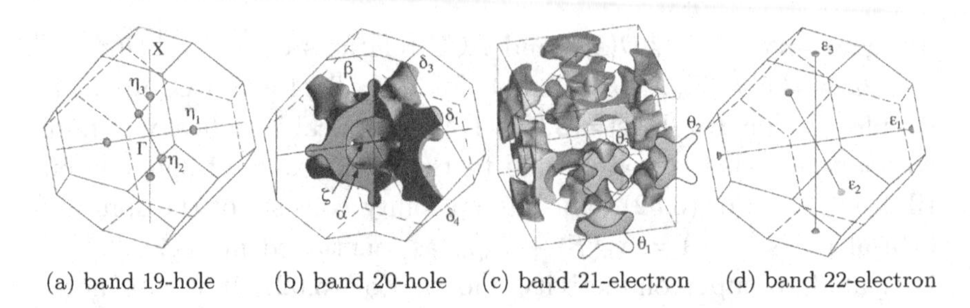

(a) band 19-hole (b) band 20-hole (c) band 21-electron (d) band 22-electron

Figure 6.14: Theoretical Fermi surfaces in $CeRu_2$, cited from Ref. [200].

detected, which correspond to orbits denoted by α, β, and δ_1 in a band 20-hole Fermi surface in Fig. 6.14, possess relatively large cyclotron masses of 2.6, 7.6, and 1.6 m_0, respectively.

Figure 6.15 shows the temperature dependence of the spin-lattice relaxation rate $1/T_1$ obtained from ^{101}Ru-NQR experiment in $CeRu_2$, together with UPt_3 and UPd_2Al_3. A coherent peak appears at $T_{sc} = 6.3$ K and follows the temperature dependence expected by the BCS theory, indicating a *s*-wave superconductor. Do note the characteristic features — no coherent peak and a T^3-dependence — in UPt_3 and UPd_2Al_3, suggesting the existence of a line node in the superconducting gap.

Figure 6.16 shows the T^2-dependence of the specific heat in the form of C/T under several constant external magnetic fields [184]. The jump in the specific heat at the zero field is sharp at the superconducting critical temperature $T_{sc} = 6.3$ K, indicating a high-quality single crystal sample. In fields higher than 55 kOe, the sample is driven into the normal state. The specific heat in the normal state follows the simple relation of $C/T = \gamma + \beta T^2 + A/T^3$ below $T < \Theta_D/50$, where Θ_D is the Debye temperature. The specific heat C at low temperatures thus consists of the electronic, phonon, and nuclear contributions. If this relation is applied to the specific heat data under a field of 55–80 kOe below 2 K, the electronic specific heat coefficient $\gamma = 27$ mJ/ $(K^2 \cdot \text{mol})$ and $\Theta_D = 120$ K are obtained. The A/T^2 term ($A = 2 \times 10^{-2}$ mJ $\cdot$ K/mol) is probably due to the Ru nuclear specific heat, leading to a small increase in the specific heat data below 0.3 K.

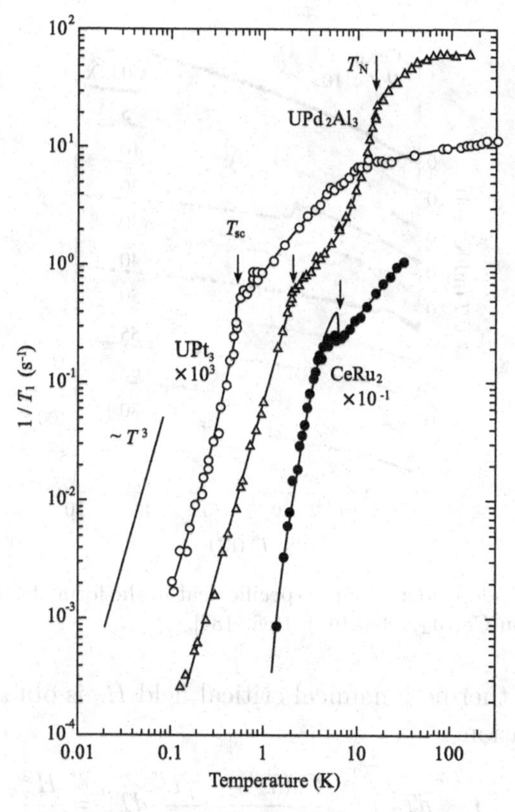

Figure 6.15: Temperature dependence of the spin-lattice relaxation rate $1/T_1$ in $CeRu_2$, together with UPt_3 and UPd_2Al_3, courtesy of Y. Kitaoka.

The electronic specific heat C_e in the superconducting state, which is obtained by subtracting the phonon and nuclear contributions from the total specific heat at the zero field, is convincingly explained by the BCS theory. If the BCS relation $C_e \sim \exp(-\Delta(0)/k_B T)$ is applied to the specific heat data in the temperature range 4.2–1.6 K, the superconducting parameters $2\Delta(0)/k_B T_{sc} = 3.7$ and $\Delta(0) = 12$ K are obtained. The value of $2\Delta(0)/k_B T_{sc}$ is slightly larger than the BCS value of 3.53. Values of $2\Delta(0)/k_B T_{sc} = 3.8$ and 4.0 are also obtained in NQR experiments [201, 202]. In the present experiment, $\Delta C/\gamma T_{sc}$ is obtained as 2.0, where ΔC is the jump in the specific heat at T_{sc}. This value is also larger than the BCS value of 1.43.

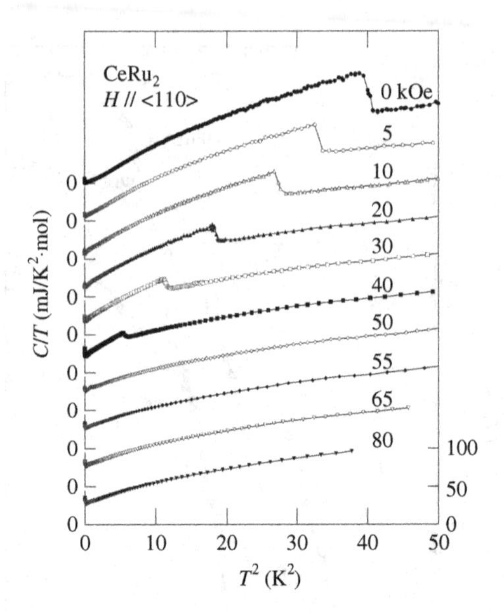

Figure 6.16: T^2-dependence of the specific heat in the form of C/T under several constant fields in $CeRu_2$, cited from Ref. [184].

Next, the thermodynamical critical field H_c is obtained from the following relation

$$\int_{T}^{T_{sc}} dT' \int_{T'}^{T_{sc}} \frac{C_e(T'') - \gamma T''}{T''} dT'' = \frac{H_c^2}{8\pi} \tag{6.63}$$

where C_e and $\gamma T''$ are the electronic specific heat in the superconducting and normal states, respectively. H_c thus obtained is plotted in Fig. 6.17(a) as a function of temperature. Note that the temperature dependence of H_c slightly deviates from the simple relation $H_c(0)(1-t^2)$, where $t = T/T_{sc}$ and $H_c(0)$ at 0 K is 1.49 kOe.

Next, the upper critical field H_{c2} is obtained from the specific heat data in Fig. 6.16. The data are plotted in Fig. 6.17(b) as a function of temperature. $H_{c2}(0)$ at 0 K is estimated as 52.3 kOe. From the slope of H_{c2} against the temperature, $\mathrm{d}H_{c2}/\mathrm{d}T$ at $T = T_{sc}$ and the jump of the specific heat ΔC at T_{sc}, the Maki parameter κ_2 is obtained using the following relation [203]:

$$\left(\frac{\Delta C}{T}\right)_{T_{sc}} = \frac{1}{4\pi\beta_A(2\kappa_2^2 - 1)} \left(\frac{dH_{c2}}{dT}\right)^2_{T_{sc}} \tag{6.64}$$

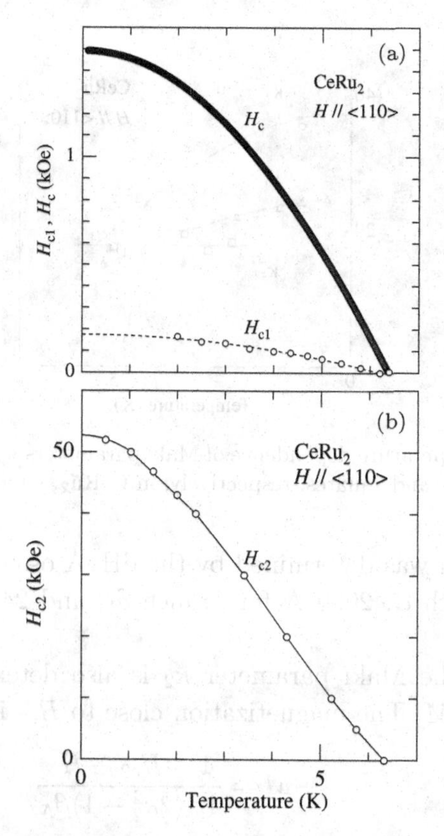

Figure 6.17: Temperature dependence of (a) the thermodynamical critical field H_c and lower critical field H_{c1} and (b) upper critical field H_{c2} in $CeRu_2$, cited from Ref. [184].

where β_A is about 1 in value. The temperature dependence of κ_2 is plotted with circles in Fig. 6.18. The κ_2 value is 12 at around T_{sc} and increases with decreasing temperature. The data do not diverge as $\sqrt{\ln(T_{sc}/T)}$ at low temperatures, but saturate at a finite value when extrapolating the temperature down to 0 K. This means that the sample is, strictly speaking, not in the pure limit [204]. It is, however, close, because the theoretical ratio of the κ_2 value at 0 K to that at T_{sc} is 2.0 for $\xi/\ell = 0.05$, where ℓ is the mean free path [204]. The corresponding ratio for the present sample is about 3. In fact, the ξ/ℓ value is approximately 0.03–0.04, because the coherence length is simply determined as 79.3 Å from $H_{c2}(0)$ $(= \phi_0/2\pi\xi^2)$ and the mean

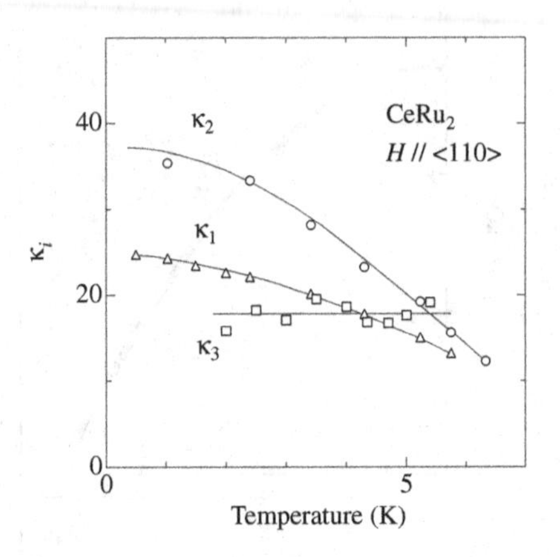

Figure 6.18: Temperature dependence of Maki parameters κ_1, κ_2, and κ_3, shown by triangles, circles and squares, respectively, in CeRu$_2$, cited from Ref. [184].

free path, which was determined by the dHvA oscillation, is 2600 Å for dHvA branch α, 2000 Å for branch δ_1, and 2400 Å for branch $\varepsilon_{1,2,3}$ [200].

Note that the Maki parameter κ_2 is also determined from the magnetization M. The magnetization close to H_{c2} is

$$-M = \frac{1}{4\pi}\frac{H_{c2} - H}{(2\kappa_2^2 - 1)\beta_A} \tag{6.65}$$

$$-\left.\frac{dM}{dH}\right|_{H=H_{c2}} = -\frac{1}{4\pi}\frac{1}{(2\kappa_2^2 - 1)\beta_A}. \tag{6.66}$$

Figure 6.19 shows the magnetization curves (M^+ and M^-) [205], where the inset shows the equilibrium magnetization M^0 or M^0_{eq}, obtained from Eq. (6.37). From the magnetization data M^0, the κ_2 values are determined, which are the same values as in Fig. 6.18. The H_c values are also obtained from Eq. (6.38b). Note that the peak effect mentioned in Sec. 5.4.5 is observed in magnetic field close to H_{c2}. This is due to weak vortex pinning in this compound.

Furthermore, the Maki parameter κ_1 is obtained from H_c and H_{c2} using the following relation

$$H_{c2} = \sqrt{2}\kappa_1 H_c \tag{6.67}$$

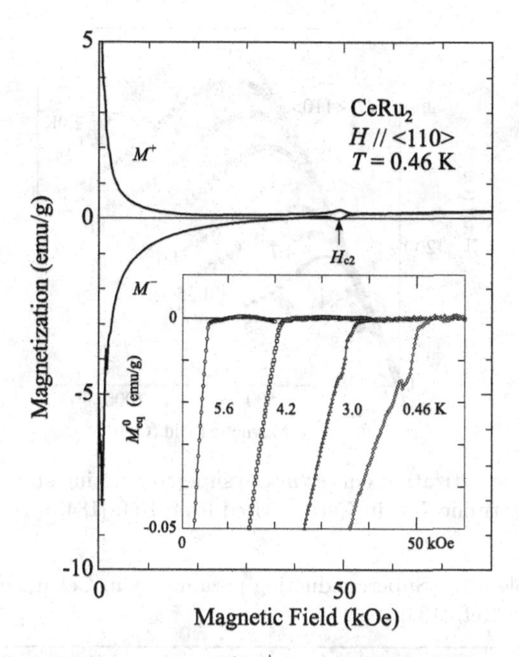

Figure 6.19: Magnetization curves (M^+ and M^-) in $CeRu_2$. Inset shows the equilibrium magnetization M^0_{eq}, courtesy of K. Tenya and T. Sakakibara, cited from Ref. [205].

The data are plotted with triangles in Fig. 6.18 as a function of temperature. The κ_1 value is almost the same as κ_2 near T_{sc}, but becomes smaller than κ_2 with decreasing temperature. This behavior is normal in type II superconductors [204].

Magnetization M at low fields was measured in the superconducting state. Figure 6.20 shows the magnetization curves at various temperatures. The data were taken after zero-field cooling. The magnetization at 2.5 K, for example, increases linearly with increasing field and has a maximum at 610 Oe, when the whole sample enters the superconducting mixed state. It is not easy to determine the lower critical field H_{c1} from the magnetization curve. The conventional method was applied to determine H_{c1} [206]. Figure 6.17(a) shows the temperature dependence of H_{c1}. The broken line in Fig. 6.17(a) is a plot of the conventional relation

$$H_{c1}(t) = H_{c1}(0)(1 - t^2)^{\gamma}$$

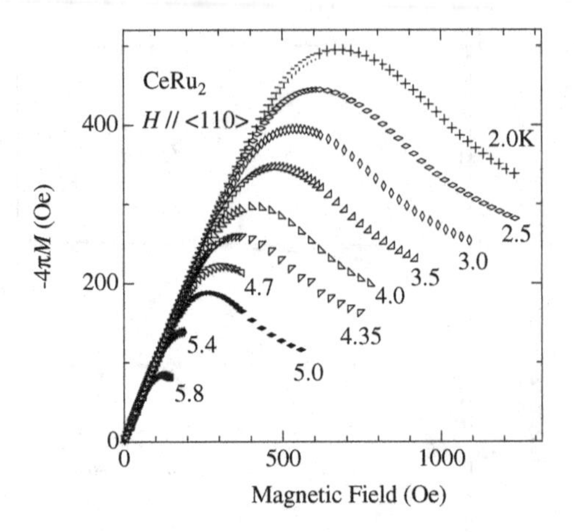

Figure 6.20: Magnetization curves in the superconducting state at various temperatures to determine H_{c1} in $CeRu_2$, cited from Ref. [184].

Table 6.1: Superconducting parameters in $CeRu_2$, cited from Ref. [184].

$H_{c2}(0)$ kOe	$H_c(0)$ kOe	$H_{c1}(0)$ kOe	ξ Å	λ Å	κ	γ mJ/(K^2·mol)
52	1.5	0.18	79	2000	25	27

where γ is 0.99 and $H_{c1}(0) = 175$ Oe. The Maki parameter κ_3 is obtained from H_{c1} and H_c using the following relation

$$H_{c1} = \frac{H_c \log \kappa_3}{\sqrt{2}\,\kappa_3}. \tag{6.68}$$

The κ_3 value is shown by squares in Fig. 6.18, and is almost constant in the measured temperature range, in contrast to the temperature dependent κ_2 and κ_1. Note that the present temperature-independent behavior of κ_3 is close to the case for the pure limit [207].

The experimentally obtained values for $H_{c2}(0)$, $H_c(0)$, and $H_{c1}(0)$ as well as the calculated values of ξ, λ, and κ in the framework of Ginzburg-Landau (GL) theory are summarized in Table 6.1. The coherence length ξ is obtained from $H_{c2}(0)$ as mentioned above. The GL parameter κ is also calculated from Eq. (6.67) using the $H_{c2}(0)$

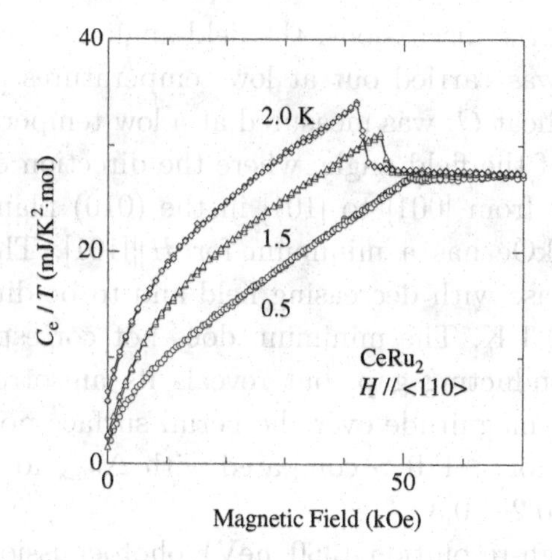

Figure 6.21: Field dependence of the electronic specific heat in the form of C_e/T for $CeRu_2$, cited from Ref. [184].

and $H_c(0)$ values. The parameter κ_1 in Eq. (6.67) becomes κ. The penetration depth λ is obtained from $\kappa(=\lambda/\xi)$ and ξ.

The field dependence of the electronic specific heat C_e at 0.5, 1.5, and 2.0 K was measured, as shown in Fig. 6.21 [184]. Note that C_e/T at 0.5 K is proportional to the field in the range from 20 to 50 kOe but deviates from the linear relation with decreasing field, indicating $\sqrt{H}$ behavior. The linear region becomes smaller with increasing temperature.

$CeRu_2$ is an s-wave superconductor which approximately follows the BCS theory, but the field dependence of the electronic specific heat in the form of C_e/T indicates the $\sqrt{H}$ dependence in the superconducting mixed state. The field dependence of C_e/T in the superconducting mixed state has been reported, for example, in the high-temperature superconductor $YBa_2CuO_{6.95}$ and the organic superconductor κ-(BEDT-TFF)$_2$Cu[N(CN)$_2$]Br, indicating the $\sqrt{H}$ dependence [208, 209]. These superconductors with two-dimensional electronic states are expected to be unconventional, because the electronic specific heat shows a T^2 dependence claiming a line node in the gap structure. The $\sqrt{H}$ dependence of the density of states $D(\varepsilon_F)$ at the Fermi level was predicted by Volovik for superconductivity with a line node in the gap [210].

Based on these discussions, the field-angle-resolved specific heat measurement was carried out at low temperatures [185]. The electronic specific heat C_e was measured at a low temperature of 90 mK as a function of the field angle, where the direction of the magnetic field is rotated from [001] to [101] in the (010) plane. The specific heat under 5 kOe has a minimum for $H \| [101]$. This minimum is found to decrease with decreasing field and to be diminished under 2 kOe below 0.3 K. The minimum does not correspond to a node of the superconducting gap, but reveals its anisotropy; the gap is not uniform in magnitude over the Fermi surface possessing a minimum of $\Delta_{\rm min}$ for $< 110 >$ compared with $\Delta_{\rm max}$ for $< 100 >$, with $\Delta_{\rm min}/\Delta_{\rm max} = 0.2 - 0.3$.

An ultrahigh-resolution (360 μeV) photoemission spectroscopy (PES) study, using a laser as a photon source, also indicated that the superconducting gap of $\rm CeRu_2$ is highly anisotropic [183]. Figure 6.22 shows the PES data. The present superconducting spectrum at 3.8 K was analyzed using the Dynes function including a thermal broadening parameter. The superconducting gap was assumed to be anisotropic: $\Delta(\theta) = \Delta_{\rm min} + (\Delta_{\rm max} - \Delta_{\rm min}) \cos 2\theta$, where θ is the polar angle, and $\Delta_{\rm max}$ and $\Delta_{\rm min}$ are the maximum and minimum gap values, respectively, as shown in the inset of Fig. 6.22. The present

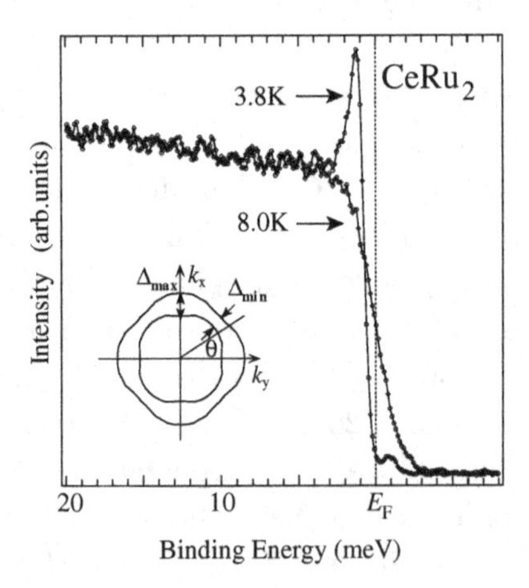

Figure 6.22: Photoemission spectra at 8.0 K (normal state) and 3.8 K (superconducting state) in $\rm CeRu_2$, cited from Ref. [183].

analyses indicate that the superconducting gap has an anisotropy with $\Delta_{min}/\Delta_{max} = 0.4$.

$CeRu_2$ is thus concluded to be an anisotropic s-wave superconductor with a three-dimensional electronic state, with an anisotropic gap of $\Delta_{min}/\Delta_{max} = 0.2 - 0.4$.

6.4.2 $CeCoIn_5$

We show the typical example used to determine the position of the node on the Fermi surface of $CeCoIn_5$ for which the electronic properties are described in Sec. 5.4.5. The Fermi surfaces in $CeCoIn_5$ are quasi-two dimensional, namely, nearly cylindrical Fermi surfaces. Here, we simply assume a cylindrical Fermi surface with four line nodes, as shown in Fig. 5.35(b) or Figs. 6.23(a) and (b), where four line nodes exist along the $\langle 110 \rangle$ direction, the renamed $(\pm\pi, \pm\pi)$ directions. Figures 6.23(a) and (b) show the (001) plane-cross-section of the Fermi surface. We discuss here the density of states in low magnetic fields ($H \ll H_{c2}$) and at low temperatures ($T \ll T_{sc}$). In the full-gap (s-wave type) superconductor, the density of states at the Fermi energy $D(H, \varepsilon = 0)$ is determined by the number of fluxoids, revealing $D(H, \varepsilon = 0) \sim H$. On the other hand, $D(H, \varepsilon = 0)$ is dominated by the extended quasiparticle states around the nodes, based on the Doppler shift or the Volvik effect [210]. In the presence of a shielding current $\boldsymbol{v}_s$, the energy ε of a quasiparticle with a momentum $\hbar\boldsymbol{k}$ is shifted by the Doppler effect. Specifically, $\varepsilon(\boldsymbol{k})$ is changed to $\varepsilon(\boldsymbol{k}) - \hbar\boldsymbol{k} \cdot \boldsymbol{v}_s$ or $\varepsilon(\boldsymbol{k}) - m^* \boldsymbol{v}_F \cdot \boldsymbol{v}_s$. In the s-wave superconductor, as shown in Fig. 6.23(c), the density of states $D(H, \varepsilon = 0)$ is unchanged, revealing no apparent effect due to the Doppler shift. On the other hand, $D(H, \varepsilon = 0)$ becomes dominant due to the effect of the Doppler shift because $D(H, \varepsilon = 0) \sim \sqrt{H}$, which is based on $D(\varepsilon) \sim |\varepsilon|$ for the line-node superconductor, as shown in Fig. 6.23(d). When the magnetic field is directed along the antinodal [100] direction, the Doppler-shifted quasiparticles around the four line nodes contribute to the density of the states $D(H, \varepsilon = 0)$ because $\boldsymbol{v}_s$ is perpendicular to $\boldsymbol{H}$ and then $\boldsymbol{v}_F \cdot \boldsymbol{v}_s \neq 0$. On the other hand, the quasiparticles around two line nodes do not contribute to the density of the states for the magnetic field along the nodal [110] direction

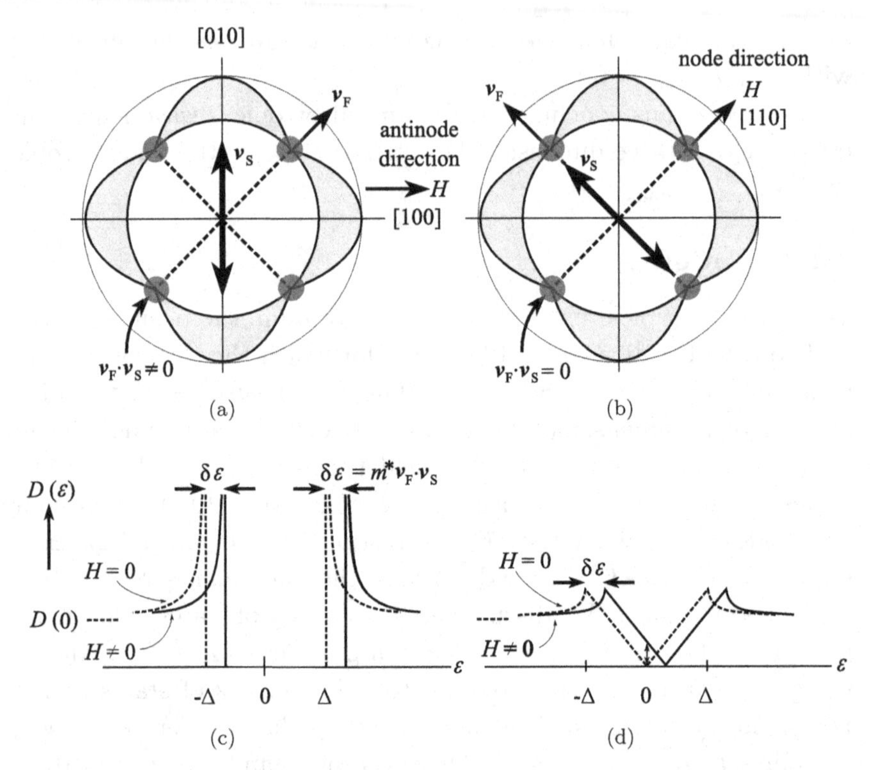

Figure 6.23: Tetragonal (001) plane-cross-section of a cylindrical Fermi surface and the superconducting gap with four line nodes along the [110] direction, where the four nodes exist along the $\langle 110 \rangle$ direction, so named $(\pm\pi, \pm\pi)$ directions. In (a), the four nodes contribute to the density of states $D(H, \varepsilon = 0)$ when the magnetic field is directed along the antinodal [100] direction, while in (b), namely for $H \|$ nodal [110] two nodes do not contribute to the density of states because $\boldsymbol{v}_{\mathrm{s}} \perp \boldsymbol{v}_{\mathrm{F}}$, showing a minimum of the density of states. The Volvik effect is not realized in the full-gap superconductor, as shown in (c), while it is effective in the line-node superconductor in (d).

because $\boldsymbol{v}_{\mathrm{F}} \cdot \boldsymbol{v}_{\mathrm{s}} = 0$. The corresponding density of states $D(H, \varepsilon = 0)$ for $H \| [110]$ is reduced compared to $D(H, \varepsilon = 0)$ for $H \| [100]$. When the magnetic field is rotated in the (001) plane, the density of states is expected to become maximum for $H \| [100]$ and minimum for $H \| [110]$, revealing the four-fold oscillatory symmetry [211]. This is seen in the $(C - C_{\mathrm{n}})/T$ data in Fig. 6.24.

Figure 6.24 shows the angular dependence of the specific heat in the form of $(C - C_{\mathrm{n}})/T$. Here, $C_{\mathrm{n}} = A(H)/T^2$ is due to the specific

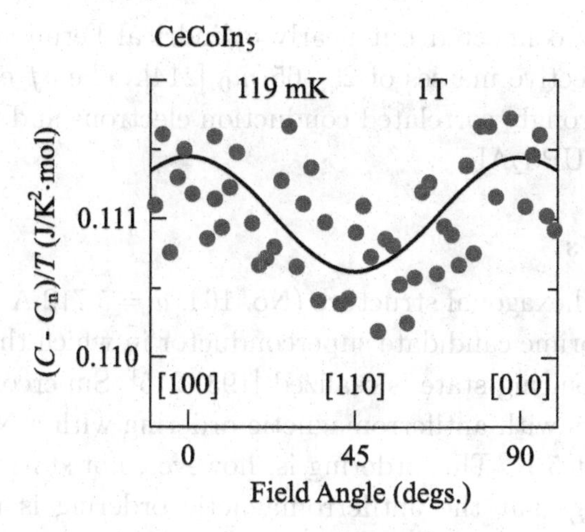

Figure 6.24: Angular dependence of the specific heat in the form of $(C - C_n)/T$ in $CeCoIn_5$, courtesy of T. Sakakibara, cited from Ref. [212].

heat nuclear Schottky contribution for ^{115}In ($I = 9/2$) and ^{59}Co ($I = 7/2$). At zero field, C/T above 300 mK is linear to temperature, as expected for a line-node superconductor of $CeCoIn_5$. Based on the Doppler shift effect, the nodes are determined for the [110] direction. $CeCoIn_5$ is a $d_{x^2-y^2}$-type superconductor [212].

The vertical line nodes in $CeCoIn_5$ are illustrated simply in Fig. 5.35(b). A typical example of the horizontal line nodes shown in Fig. 5.35(c) is of superconductivity in UPd_2Al_3 with the hexagonal structure. UPd_2Al_3 orders antiferromagnetically below $T_N = 14.5$ K with a relatively large ordered moment of $0.85\ \mu_B/U$. In UPd_2Al_3, the magnetic moments are ferromagnetically oriented along the $[11\bar{2}0]$ direction and are coupled antiferromagnetically along the [0001] direction with the propagation vector $\boldsymbol{q} = (0, 0, 1/2)$. Superconductivity in UPd_2Al_3 is observed below $T_{sc} = 1.8$ K in this stable antiferromagnetic state. The experimental results of thermal conductivity measurement with rotating magnetic field in various directions suggest the superconducting gap function has d-wave symmetry of the form $\Delta(k) = \Delta_0 \cos k_z c$, favoring the spin-singlet pairing state and a horizontal line node in the superconducting gap [213]. The Fermi surfaces in UPd_2Al_3 consist of a band 41-hole Fermi surface and a band 42-electron Fermi surface, which are corrugated and partially connected but nearly cylindrical Fermi surfaces, with cyclotron effective masses of 20–65 m_0 [214]. The $5f$ electrons contribute to strongly correlated conduction electrons and the magnetic moments in UPd_2Al_3.

6.4.3 UPt_3

UPt_3 with a hexagonal structure (No. 164, $a = 5.712$ Å, $c = 4.864$ Å, $Z = 2$) is a prime candidate superconductor in which the anisotropic spin-triplet pairing state is realized [193, 215]. Superconductivity in UPt_3 coexists with antiferromagnetic ordering with a Néel temperature of about 5 K. This ordering is, however, not static but dynamical, meaning that the antiferromagnetic ordering is not observed in the specific heat and magnetic susceptibility measurements. To understand the $5f$-electron nature, it is essential to clarify the Fermi surface property using high-quality single crystal samples grown by the Czochralski pulling method in a tetra-arc furnace [216, 217].

Figure 6.25(a) shows the angular dependence of the dHvA frequencies. The origin of the detected dHvA branches is identified on the basis of the $5f$-itinerant band model. Figure 6.25(b) shows the theoretical Fermi surfaces. The dHvA branches are identified as follows: (1) branch ω: band 37-electron, (2) branches τ, σ, ρ, λ, ε, and α: band 36-hole, and (3) branch δ: band 35-hole.

The dHvA frequencies, shown by circles in Fig 6.25(a), are in good agreement with theoretical results (solid lines) based on the $5f$-itinerant band model. The dHvA branch ω possesses the largest cross-sectional area and cyclotron mass. The cyclotron mass is determined as 80 m_0 in the field range from 150 to 175 kOe along the hexagonal [0001] direction (c-axis), 105 m_0 in the field range from 182 to 196 kOe for [$10\bar{1}0$], and 90 m_0 in the field range of 170 to 196 kOe for [$11\bar{2}0$]. The corresponding band masses are 5.09, 6.84 and 5.79 m_0, respectively. The cyclotron masses are about 15 times larger than the corresponding band masses. Note that all the bands contain an f-electron component of about 70 % and are flat in the dispersion.

UPt_3 exhibits the metamagnetic transition at about 200 kOe [218]. Figure 6.26(a) shows the high-field magnetization. This

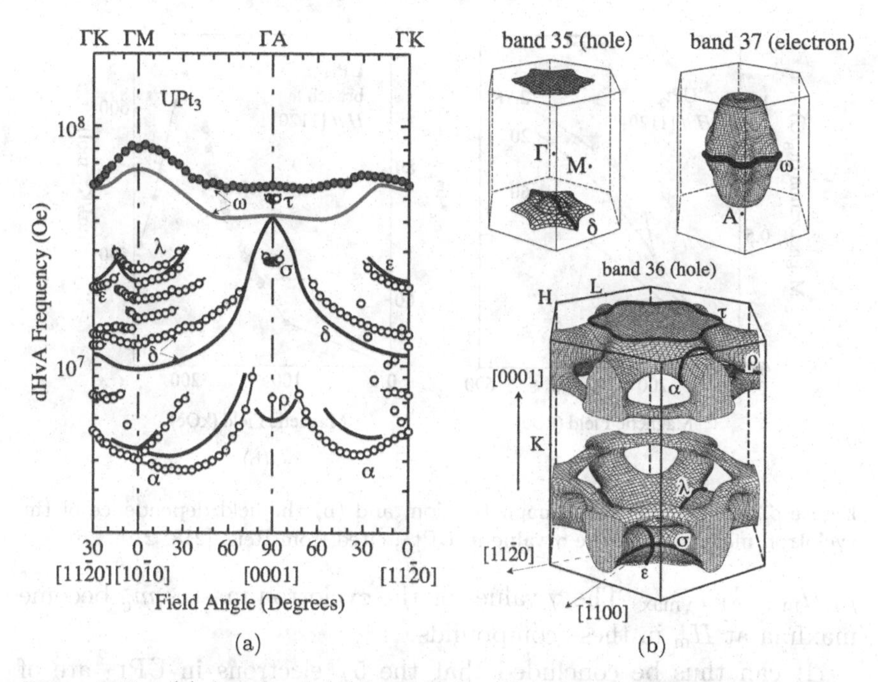

Figure 6.25: (a) Angular dependence of the dHvA frequencies in UPt_3. Theoretical results are shown in solid lines, and (b) the corresponding theoretical Fermi surfaces, cited from Refs. [216, 217].

metamagnetic transition persists up to about 30 K, which is close to the characteristic temperature $T_{\chi\max}$ ($\simeq$ 20 K) where the magnetic susceptibility has a maximum in the temperature dependence, as shown in Fig. 5.18(a). We note the growth of the cyclotron effective mass m_c^* at this metamagnetic transition, as shown in Fig. 6.26(b). The cyclotron mass in the main band 37-electron Fermi surface, branch ω, is strongly enhanced with an increase of the field from 63 m_0 at 125 kOe to 86 m_0 at 180 kOe [218]. This is consistent with the field dependence of γ shown in Fig. 6.26(b) [219].

It is noted that one of the characteristic properties of the heavy fermion compounds as in UPt_3 is metamagnetic behavior or an abrupt nonlinear increase of magnetization at the magnetic field H_m below $T_{\chi\max}$. Figure 5.18(d) shows the relation between $T_{\chi\max}$ and H_m in several Ce, Yb, and U compounds including UPt_3 [131]. The solid line represents the relation H_m (kOe) = 15 $T_{\chi\max}$ (K), namely

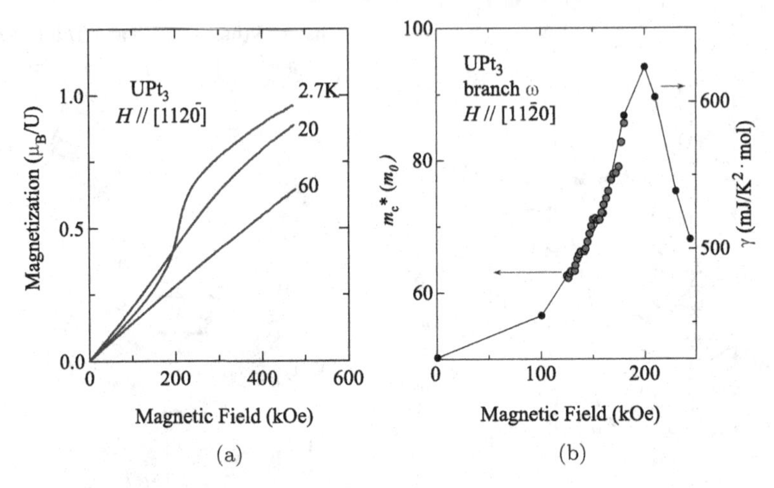

Figure 6.26: (a) High-field magnetization, and (b) the field dependence of the cyclotron mass m_c^* and the γ value in UPt_3, cited from Refs. [218, 219].

$\mu_B H_m = k_B T_{\chi max}$. The γ values or the cyclotron masses m_c^* become maxima at H_m in these compounds.

It can thus be concluded that the $5f$ electrons in UPt_3 are of itinerant nature — the dHvA results are well explained by the $5f$-itinerant band model. All the dHvA branches are heavy, with large cyclotron masses of 15–105 m_0, i.e. ten to twenty times larger than the corresponding band masses. The mass enhancement is caused by magnetic fluctuations, where the freedom of charge transfer of the $5f$ electrons appears in the form of the $5f$-itinerant band, namely the Fermi surface, but the freedom of spin fluctuations of the same $5f$ electrons reveals an unusual magnetic ordering and enhances the effective mass as in the many-body Kondo effect. These heavy conduction electrons or quasiparticles condense into Cooper pairs. To avoid a large overlap of the wave functions of the paired particles, the heavy fermion state chooses an anisotropic channel, like a spin triplet to form Cooper pairs in UPt_3 [193, 215, 220].

As shown in Fig. 6.27(a), three superconducting phases, A, B, and C exist in the temperature-field phase diagram [152, 221]. The power law dependence of the thermodynamic and transport quantities suggest two kinds of nodes in the superconducting gap: line and point nodes [215]. Moreover, a spin-triplet (odd-parity) pairing is suggested

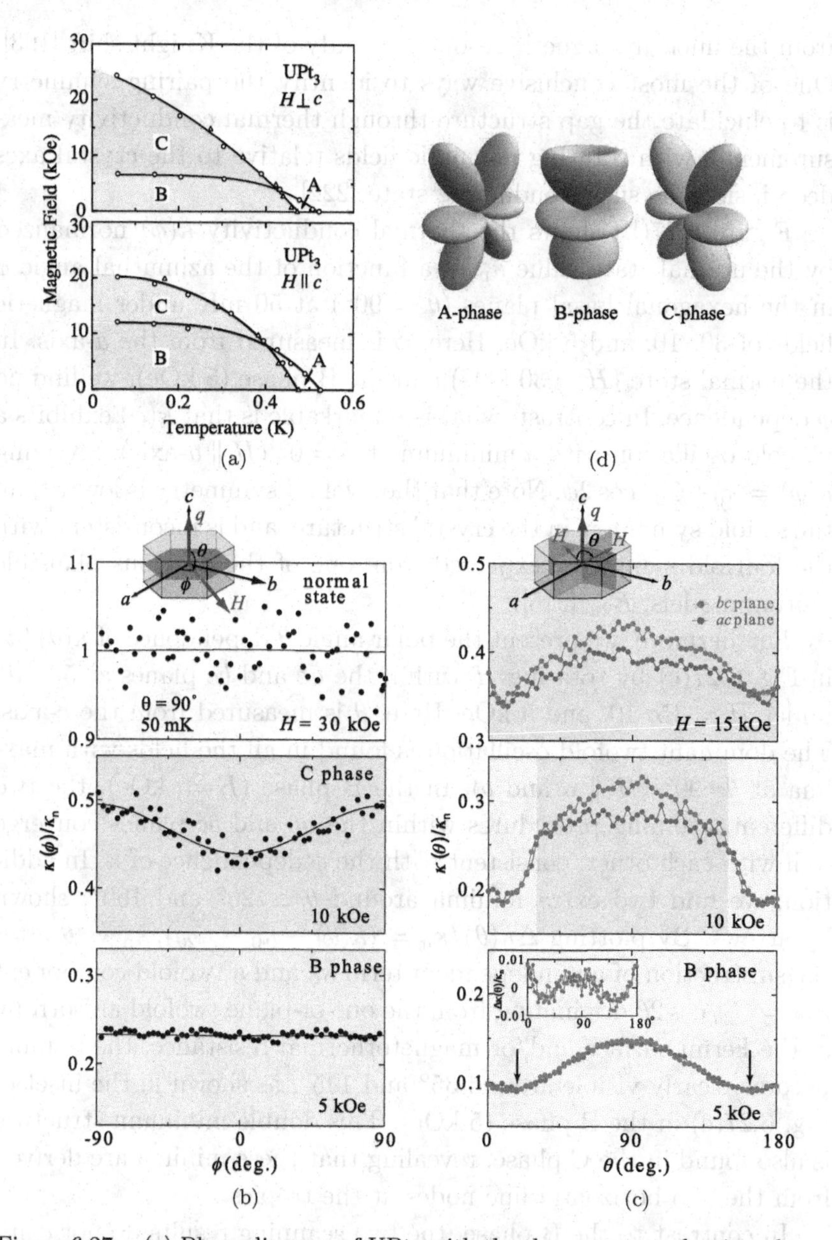

Figure 6.27: (a) Phase diagram of UPt_3 with the three superconducting phases, A, B, and C, (b) and (c) the angular dependences of the thermal conductivities, and (d) schematic shapes of the gap symmetries in the A, B, and C phases, cited from Refs. [152, 221, 222], courtesy of K. Izawa. The inset in (c) B phase ($H =$ 5 kOe) indicates $\Delta\kappa(\theta)/\kappa_{\rm n} = (\kappa(\theta) - \kappa_0 - \kappa_{2\theta})/\kappa_{\rm n}$ as a function of θ at 50 mK (see text).

from the nuclear magnetic resonance study of the Knight shift [193]. One of the most conclusive ways to identify the pairing symmetry is to elucidate the gap structure through thermal conductivity measurements, with rotating magnetic fields relative to the crystal axes deep inside the superconducting state [222].

Figure 6.27(b) shows the thermal conductivity $\kappa(\phi)$ normalized by the normal state value κ_n as a function of the azimuthal angle ϕ in the hexagonal basal planes ($\theta = 90°$) at 50 mK under magnetic fields of 30, 10, and 5 kOe. Here, ϕ is measured from the a-axis. In the normal state ($H = 30$ kOe) and the B phase (5 kOe), we find no ϕ dependence. In contrast, what is remarkable is that $\kappa(\phi)$ exhibits a twofold oscillation with a minimum at $\phi = 0°$ ($H \parallel a$-axis), revealing $\kappa(\phi) = \kappa_0 + C_{2\phi} \cos 2\phi$. Note that the twofold symmetry is lower than the sixfold symmetry in the crystal structure, and is inconsistent with the fourfold symmetry expected from one of the previous plausible pairing models, E_{2u} [215].

Furthermore, we present the polar angle θ dependence of $\kappa(\theta)/\kappa_n$ in Fig. 6.27(c) by rotating H within the ac and bc planes at 50 mK under $H = 15$, 10, and 5 kOe. Here, θ is measured from the c-axis. The dominant twofold oscillation is found in all the fields with maxima at θ=90° ($H \parallel a$ and b). In the B phase (H=5 kOe), the two different scanning procedures within the ac and bc planes converge well with each other, consistent with the ϕ dependence of κ. In addition, we find two extra minima around $\theta \simeq 20°$ and 160°, shown by arrows. By plotting $\Delta\kappa(\theta)/\kappa_n = (\kappa(\theta) - \kappa_0 - \kappa_{2\theta})/\kappa_n$ vs θ after the subtraction of a θ-independent term κ_0 and a twofold component $\kappa_{2\theta} = C_{2\theta} \cos 2\theta$ originating from the out-of-plane twofold anisotropy of the Fermi surface and/or magnetothermal resistance, the minima become clearly visible around 35° and 145°, as shown in the inset of Fig. 6.27(c) in the B phase (5 kOe). This double minimum structure is also found in the C phase, revealing that these minima are derived from the two horizontal line nodes at the tropics.

In contrast to the B phase, the two scanning results do not coincide in the C phase. The difference is diminished at $\theta = 0°$ ($H \parallel c$-axis) and maximized at $\theta = 90°$ ($H \parallel a, b$), being consistent with the in-plane two fold symmetry. Moreover, a significant appearance of the twofold symmetry across the BC transition can be seen at 10 kOe, as shown in Fig. 6.27(c) at 10 kOe.

These experimental results are summarized as follows: (1) the line node along the a-axis in the C phase, (2) the absence of in-phase gap anisotropy in the B phase, and (3) the two horizontal line nodes at the tropics in both B and C phases. Taking into account all these results and the $\boldsymbol{d}$ vector configurations assigned by the Knight shift, the pairing symmetry is concluded to be $\boldsymbol{b}k_a(5k_c^2-1)$ in the A phase, $(\boldsymbol{b}k_a+\boldsymbol{c}k_b)(5k_c^2-1)$ in the B phase, and $\boldsymbol{c}k_b(5k_c^2-1)$ $(H \parallel ab)$, and $\boldsymbol{a}k_b(5k_c^2-1)$ $(H \parallel c)$ in the C phase, where $\boldsymbol{a}$, $\boldsymbol{b}$, and $\boldsymbol{c}$ are unit vectors of the hexagonal axes representing the directions of $\boldsymbol{d}$ vectors. The pairing symmetry of UPt_3 is determined to be an E_{1u} representation in the f-wave category, and the spin part of the pair function is most likely of weak spin-orbit coupling, where the corresponding $\boldsymbol{d}$ vector can change its orientation against the magnetic field.

The present experimental result is, however, not consistent with the result of the angle-resolved specific heat measurement for UPt_3 [223]. The discrepancies between the two experiments for UPt_3 are not clear but the specific heat measurement mainly detects the conduction electrons (Fermi surfaces) with heavy masses, while the thermal conductivity measurement is sensitive to the conduction electrons (Fermi surfaces) with lighter masses. We stress here that the same results are obtained between two measurements for $CeCoIn_5$ and UPd_2Al_3 mentioned above.

6.4.4 UGe_2

UGe_2 crystallizes in the orthorhombic crystal structure (No. 65, $a = 4.0089$ Å, $b = 15.0889$ Å, $c = 4.0950$ Å, $Z = 4$) with a large lattice constant along the b-axis, as shown in Fig. 6.28(a), which corresponds to a short distance along the b-axis in the Brillouin zone [224]. The U zig-zag chain with the distance of the next nearest neighbor $d_{\text{U-U}}$=3.85 Å is formed along the a-axis. The ordered moment of $\mu_s \simeq 1.4\ \mu_B$/U orients along the a-axis, revealing a magnetic easy-axis, while b- and c-axes correspond to hard axes in the magnetization curves. UGe_2 is thus an Ising-type ferromagnet.

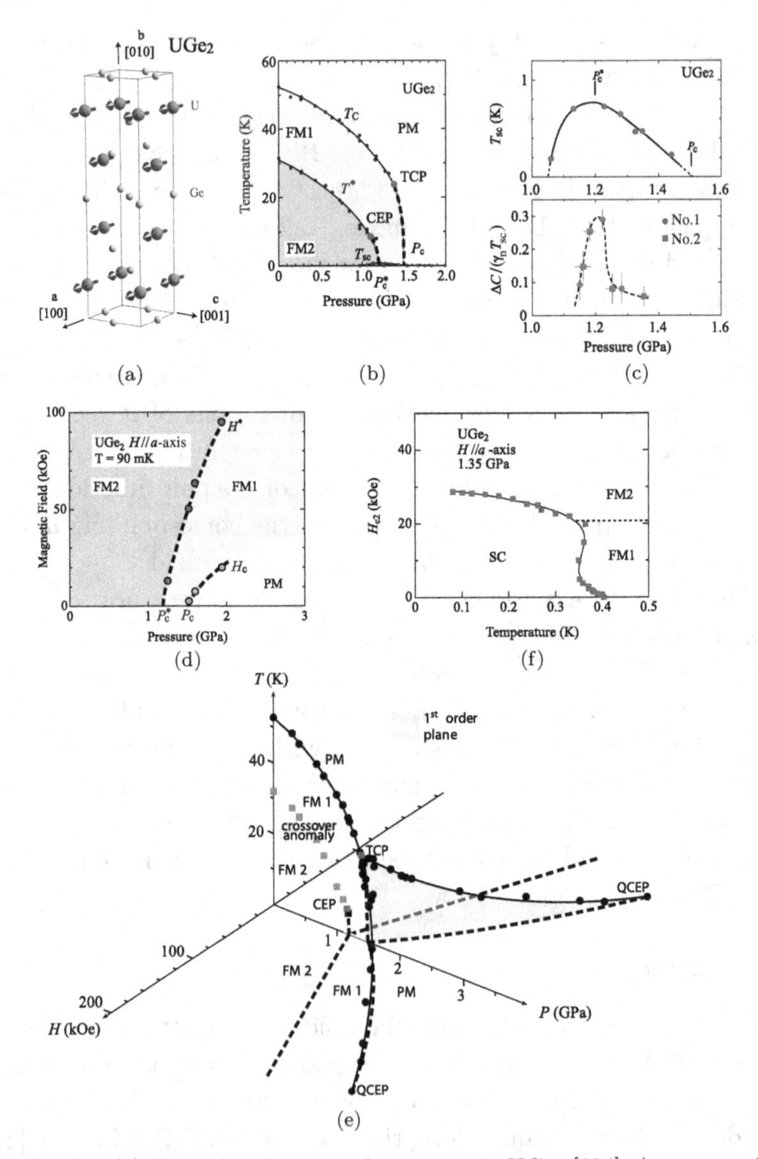

Figure 6.28: (a) Orthorhombic crystal structure in UGe_2 [224]. Arrows on the U sites denote the direction of the magnetic moment. (b) Temperature vs pressure phase diagram [227]. (c) Pressure dependence of T_{sc} determined by the resistivity measurement (upper panel) and the specific heat jump of $\Delta C/\gamma_n T_{sc}$ (lower panel) [230]. (d) Pressure dependence of the magnetic phase transition fields [231]. (e) Temperature–pressure–field phase diagram for $H \parallel a$-axis, and (f) temperature dependence of H_{c2} for $H \parallel a$-axis at 1.35 GPa, cited from Refs. [226, 232]. TCP, CEP, and QCEP are the tricritical point, critical end point, and quantum critical end point, respectively.

UGe_2 is a characteristic pressure-induced superconductor in the ferromagnetic state, suggesting the so-named parallel spin pairs of ↑↑ or ↓↓, as schematically shown in Fig. 6.8(c) [225, 226]. The coexistence of ferromagnetism and superconductivity is realized in this compound, together with the similar compounds URhGe, UCoGe, and most likely UIr. Up to date, this is limited only to $5f$-itinerant ferromagnets. In the pressure experiment of UGe_2 in Fig. 6.28(b) [227], it was found that the Curie temperature $T_C = 52$ K becomes zero at $P_c \simeq 1.5$ GPa. A second phase transition was found at $T^* \simeq 30$ K at ambient pressure, and becomes zero at $P_c^* \simeq 1.2$ GPa. The temperature region from T_C to T^* and/or the pressure region from P_c^* to P_c, named the weakly polarized phase, (FM1), has ordered moments of about 0.9 μ_B/U, as shown later in Fig. 6.29(a), while the lower temperature region ($T < T^*$) and/or the pressure region below P_c^*, named the strongly polarized phase (FM2), has ordered moments of 1.41–1.26 μ_B/U [228, 229]. Note that the phase transition at T^* is understood as a cross-over anomaly at ambient pressure because the magnetic moment increases smoothly between FM1 and FM2 and there exist no changes of the crystal structure and magnetic structure. On the other hand, a step-like change of the magnetic moment is observed between FM1 and FM2 with increasing pressure. In Fig. 6.28(b), there are two solid lines showing T_C and T^*. On these lines, there are also thick broken lines sharing a T_C line in the pressure region from a pressure corresponding to a tricritical point (TCP) to P_c, and a T^* line in the pressure region from a pressure corresponding to a critical end point (CEP) to P_c^*. These are of the first order [226].

When a magnetic field is applied along the a-axis, the FM1 phase changes into the FM2 phase at H^* in the pressure region from P_c^* to P_c, as shown in Fig. 6.28(d), and even in the pressure region where $P > P_c$, revealing a phase change from the paramagnetic (PM) phase to the FM1 phase at H_c. This corresponds to the metamagnetic transition, indicating step-like increases of the magnetization. The P–H phase diagram at 90 mK is shown in Fig. 6.28(d) [231]. Note that the two thick broken lines showing H^* and H_c are

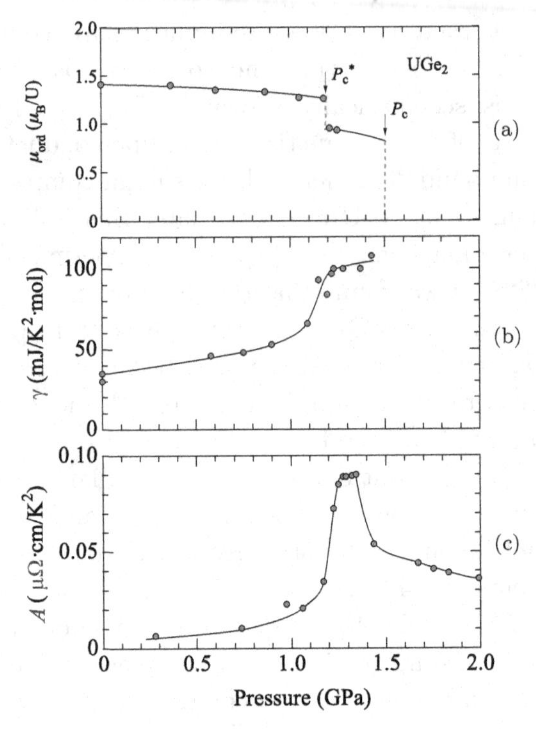

Figure 6.29: Pressure dependence of (a) ordered moment $\mu_{\rm ord}$ (b) electronic specific heat coefficient γ, and (c) A-value in UGe$_2$, cited from Refs. [227] and [228].

of the first order. The P–H–T phase diagram is also shown in Fig. 6.28(e) [226, 232].

Superconductivity appears in the ferromagnetic state. Figure 6.28(c) shows the pressure dependence of $T_{\rm sc}$ obtained through the electrical resistivity measurement and the pressure dependence of $\Delta C/\gamma_{\rm n}T_{\rm sc}$ (ΔC: jump of the specific heat at $T_{\rm sc}$, $\gamma_{\rm n}$: electronic specific heat coefficient at $T_{\rm sc}$) obtained through the specific heat measurement [230]. The superconducting region is wide in the resistivity measurement (upper figure) compared with that in the specific heat measurement (lower figure), but $T_{\rm sc}$ indicates a maximum at $P_{\rm c}^* \simeq 1.2$ GPa. The specific heat jump $\Delta C/\gamma_{\rm n}T_{\rm sc}$ is much smaller than 1.43 in the weak coupling BCS scheme. Note that the residual specific heat coefficient estimated at 0 K is quite large, at 70% of the γ value in the normal state, in spite of the use of a very high-quality single crystal sample with the residual resistivity ratio (ρ_{RT}/ρ_0, ρ_{RT}: resistivity at room temperature, ρ_0: residual resistivity) RRR $= 600$ and $\rho_0 = 0.2$ $\mu\Omega$·cm, as obtained by the Czochralski method. This might be related to the large magnetic moment of about 0.9 μ_B/U, shown in Fig. 6.29(a), and the corresponding self-induced vortex state. Note that the residual specific heat coefficient and the ordered moment are 50% and 0.4 μ_B/U in URhGe, and 15% and 0.05 μ_B/U in UCoGe [226]. The relation between the residual specific heat coefficient and the magnetic moment has been discussed theoretically [233].

A striking point on the superconducting properties is that the temperature dependence of H_{c2} for $H \parallel a$-axis indicates the field-enhanced superconducting phase when the pressure is tuned just above P_c^*, as shown in Fig. 6.28(f) [226, 232]. This peculiar "S"-shape of H_{c2} curve under $P = 1.35$ GPa is associated with the crossing of the metamagnetic transition at H^*, namely a change of the magnetic phase from FM1 to FM2.

As shown in Fig. 6.29(c), the A value of the electrical resistivity ρ $(= \rho_0 + AT^2)$ obtained from the resistivity experiment has a peak at about 1.3 GPa. The corresponding electronic specific heat coefficient γ reaches 100 mJ/(K^2·mol), as shown in Fig. 6.29(b). The main Fermi surfaces at ambient pressure, which were determined by the dHvA experiment and the energy band calculation, are nearly cylindrical along the b-axis [234–236]. The corresponding cyclotron masses are relatively large at 15–20 m_0. Thus, the $5f$ electrons in UGe_2 are itinerant, indicating band magnetism similar to $3d$-electron systems.

To further clarify the Fermi surface property, the dHvA experiment was carried out under pressure. A drastic change of the Fermi surfaces is found when pressure crosses P_c [227]. Figure 6.30 indicates the FFT spectra at several pressures. The FFT spectra are greatly different below and above about 1.5 GPa $(= P_c)$. In the present dHvA experiment, no dHvA signal is obtained in the pressure region from 1.25 to 1.50 GPa, namely, in the FM1 phase, but a clear signal at 1.54 GPa is obtained.

Figures 6.30(b) and (c) show the pressure dependence of the dHvA frequencies and the cyclotron effective masses, respectively.

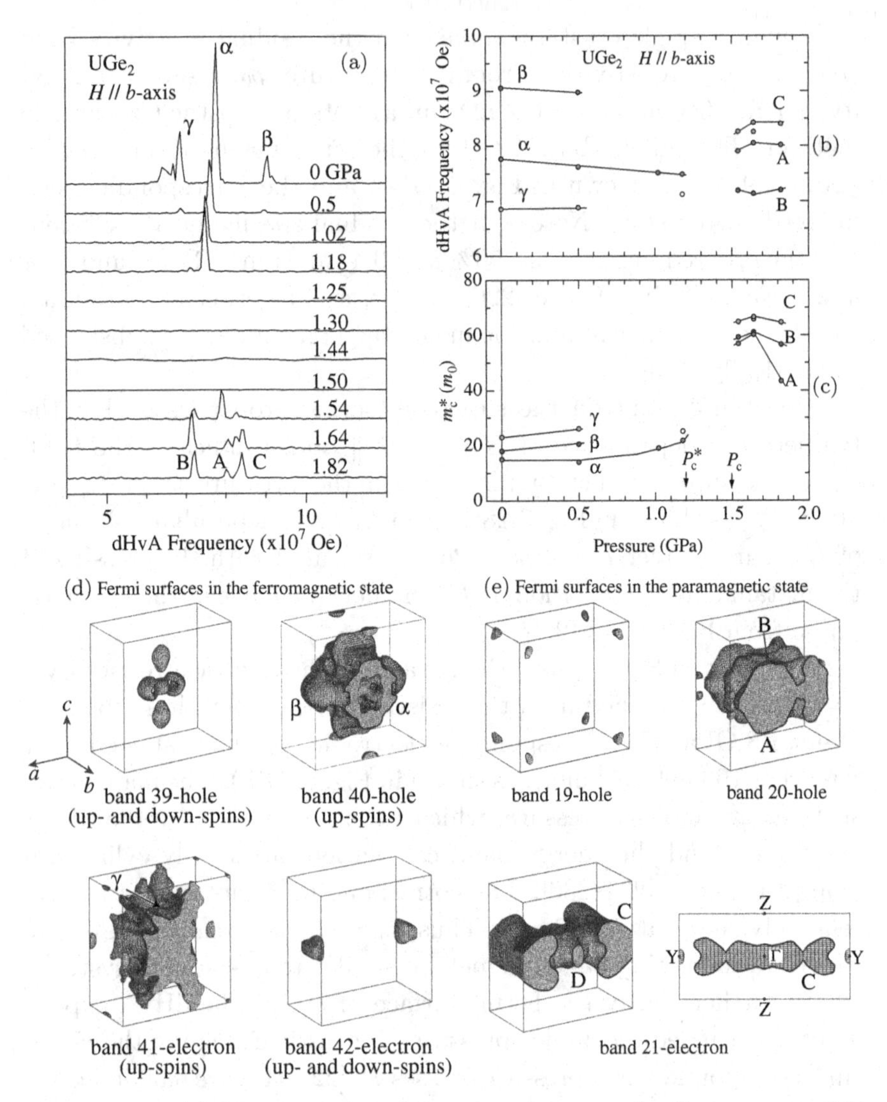

Figure 6.30: (a) FFT spectra under various pressures for the magnetic field along the b-axis (hard-axis), pressure dependences of (b) the dHvA frequencies and (c) the cyclotron masses, and theoretical Fermi surfaces in (d) the ferromagnetic (FM2 phase) and (e) paramagnetic states in UGe_2, cited from Ref. [227].

The dHvA frequency of branch α, together with that of branch β, decreases almost linearly with increasing pressure, as shown in Fig. 6.30(b). The cyclotron mass of 15 m_0 for branch α at ambient pressure increases up to 22 m_0 at 1.18 GPa with a tendency for steep increase above $P_c^* \simeq$1.2 GPa, as shown in Fig. 6.30(c). The cyclotron masses in the paramagnetic state are surprisingly large: 43, 57, 64 and 19 m_0 at 1.82 GPa for branches A, B, C and D, respectively. Note that branch D is not shown in Figs. 6.30(b) and (c) because of a small dHvA frequency $F = 1.59 \times 10^7$ Oe. The cyclotron masses in the paramagnetic state decrease with increasing pressure, which is approximately consistent with the A value in Fig. 6.29(c). It is noted that similar dHvA results are obtained [237, 238].

We will discuss a change of the Fermi surface under pressure with consideration to the theoretical results of energy band calculations in the ferromagnetic (FM2 phase) and paramagnetic states shown in Figs. 6.30(d) and (e), respectively. Fermi surfaces were calculated in the scheme of a fully-relativistic spin-polarized version of the linearized augmented-plane-wave (LAPW) method within a local spin-density approximation under the assumption that $5f$ electrons are itinerant and the magnetic moment is directed along the a-axis in the ferromagnetic state. Band 40-hole and 41-electron Fermi surfaces in Fig. 6.30(d) correspond to majority up-spin bands, while minority down-spin states are included partially in band 39-hole and 42-electron Fermi surfaces. In the present band calculations, a magnetic moment of 1.2 μ_B/U was obtained, in which spin and orbital moments are -1.3 μ_B/U and 2.5 μ_B/U, respectively.

The characteristic pressure region is classified into three regions (1) $P < P_c^*$, (2) $P_c^* < P < P_c$ and (3) $P > P_c$. In the first pressure region where $P < P_c^*$, the dHvA frequency of branch α, which most likely corresponds to the majority up-spin band 40-hole Fermi surface, decreases almost linearly with increasing pressure. This means that the volume of the majority up-spin band 40-hole Fermi surface decreases with increasing pressure. Correspondingly, the Curie temperature T_C decreases from 52 K at ambient pressure to about 30 K at P_c^*, and the ordered moment decreases from 1.41 to about 1.26 μ_B/U. Intuitively it is expected that with increasing pressure, volumes of both the band 40-hole and band 41-electron Fermi surfaces with majority up-spin states decrease, while volumes of both the band 39-hole and the band 42-electron Fermi surfaces with minority down-spin states increase correspondingly. Note that both the up- and down-spin states are involved in the band 39-hole and band 42-electron Fermi surfaces at ambient pressure. The ordered moment, which is proportional to the volume difference between Fermi surfaces with up- and down-spin states, thus decreases with increasing pressure. However, experimentally we have found no direct evidence for the Fermi surface with the down-spin state. A dHvA frequency of 7.12×10^7 Oe was observed at 1.18 GPa. It is not clear whether this branch corresponds to the Fermi surface with minority down-spin states or not.

Next, we will discuss the electronic state in the second pressure region of $P_c^* < P < P_c$. In this pressure region, the dHvA signal disappears completely. There are a few reasons for this. One is a large γ value in this region. The γ value reaches 100 mJ/(K^2·mol), as shown in Fig. 6.29(b), which is much larger than 35 mJ/(K^2·mol) at ambient pressure. A large effective mass of the conduction electron intensively reduces the dHvA amplitude.

In the paramagnetic state, clear dHvA oscillations are observed. Branches A, B, C and D are approximately identified in theoretical Fermi surfaces in Fig. 6.30(e), which were calculated by using the lattice parameter at ambient pressure in the paramagnetic state. As shown in Fig. 6.30(e), UGe_2 is a compensated metal because it possesses two molecules in the primitive cell. Band 19- and 20-hole Fermi surfaces are compensated by a band 21-electron Fermi surface, and up- and down-spin states are degenerate in the paramagnetic state. Four kinds of theoretical dHvA frequencies corresponding to branches A, B, C and D are: 9.21×10^7 Oe (8.74 m_0), 5.00×10^7 Oe (12.2 m_0), 11.7×10^7 Oe (14.2 m_0) and 0.38×10^7 Oe (6.01 m_0). These dHvA frequencies are quantitatively not in good agreement with the experimental values, although a rough correspondence between the theory and the experiment is obtained. This discrepancy is mainly due to low symmetry of the orthorhombic crystal structure.

The electronic states at low temperatures are thus changed as a function of pressure, from the strongly polarized ferromagnetic state (FM2) to the weakly polarized ferromagnetic state (FM1), and finally to the paramagnetic state (PM). Further, we will introduce one of the theoretical discussions on these phase changes as well as the ordered moments and the effective masses [239].

The theory is based on the periodic Anderson model. The theoretical results are shown simply in Fig. 6.31. In the paramagnetic phase shown in Fig. 6.31(a), the numbers of up- and down-spin electrons are the same. A large Fermi surface is obtained with Fermi momentum $k_{\mathrm{F}}^{(\mathrm{L})}$ because the f-electron state contributes to the band. With weak polarization, the band will be changed into a FM0 phase in Fig. 6.31(b). When the polarization increases, the up-sign electrons will fill up the lower hybridized band as shown in Fig. 6.31(c). This state might correspond to FM1. In this state, the Fermi surface for the up-spin states disappears, revealing a half-metallic state. When the polarization increases further, the up-spin electrons start to fill the upper hybridized band, as shown in Fig. 6.31(d), to form a state which might correspond to FM2. In the FM2 state, the f electrons will be polarized almost completely. The corresponding electronic state is the small Fermi surface state with momentum $k_{\mathrm{F}}^{(\mathrm{S})}$. When the f-electron level ϵ_f is shifted upward from the FM2 state, the electronic states are changed into FM2 → FM1 → FM0 → PM, as shown in Fig. 6.31(e). The corresponding magnetization or the ordered moment $\mu_{\mathrm{ord}} = n_{f\uparrow} - n_{f\downarrow}$ is changed as shown in Fig. 6.31(f). Here, V is the hybridization matrix element, and W is the band width. The effective mass was also calculated (not shown here), revealing a large mass in FM1.

6.4.5 $PrTi_2Al_{20}$

Superconductivity is achieved on the basis of the quadrupole moment in $PrOs_4Sb_{12}$, $PrTi_2Al_{20}$, and PrV_2Al_{20} with the cubic caged structure (No. 227, a = 14.725 Å, Z = 8 in $PrTi_2Al_{20}$) [240, 241]. $PrOs_4Sb_{12}$ is one of the so-called skutterudite compounds. $PrTi_2Al_{20}$ and PrV_2Al_{20} crystallize in the same crystal structure as the heavy

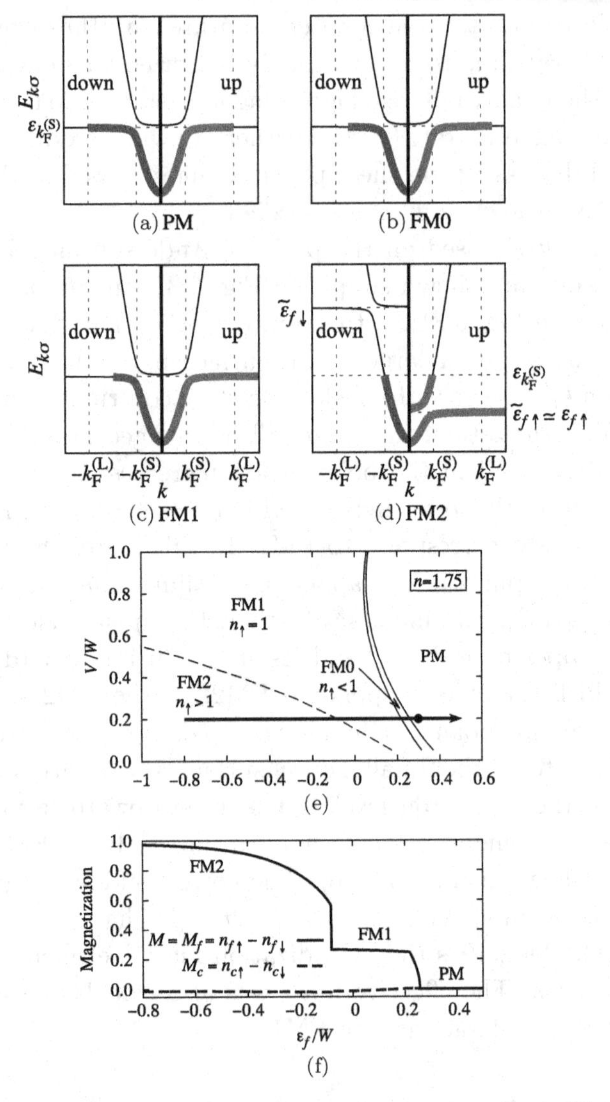

Figure 6.31: Schematic band structures of paramagnetic and ferromagnetic phases in the periodic Anderson model. (a) Paramagnetic phase (PM), (b) weakly polarized ferromagnetic phase (FM0), (c) half-metallic phase (FM1), and (d) ferromagnetic phase with an almost completely polarized f-electron state (FM2), cited from Ref. [239]. $E_{\kappa\sigma}$ denotes the quasiparticle energy. Right (left) part of each figure shows the up- (down-) spin band. The occupied states are represented by the bold lines. (e) Theoretical phase diagrams for the total number $n = n_\uparrow + n_\downarrow$ of electrons per site, $n = 1.75$. Solid lines indicate the second-order phase transitions, and a dashed line indicates the first-order phase transition. A bold arrow indicates the parameter set for which the magnetization was calculated in (f).

fermion compounds $YbIr_2Zn_{20}$ and $YbCo_2Zn_{20}$, shown in Fig. 5.25, where the unit cell contains 8 formula units and each Pr atom is encapsulated in the cage formed by 16 Al atoms. This brings about a small CEF splitting energy of 100–150 K. In the case of $PrTi_2Al_{20}$, the CEF ground state is a Γ_3 non-Kramers doublet, and the excited states (and the splitting energy from the ground state) are Γ_4 (triplet, 65 K), Γ_5 (triplet, 107 K), and Γ_1 (singlet, 156 K). Note that the CEF ground state of $PrOs_4Sb_{12}$ is the Γ_1 singlet but the quadrupole ordering is induced due to the level crossing with an excited state under a magnetic field [242]. The superconducting properties of $PrOs_4Sb_{12}$ are $T_{sc} = 1.85$ K, $-dH_{c2}/dT(\text{at } T_{sc}) =$ 18 kOe/K, and $H_{c2}(0) = 22$ kOe, indicating heavy fermion superconductivity [243].

On the other hand, $PrTi_2Al_{20}$ and PrV_2Al_{20} display quadrupole ordering with a ferroquadrupole ordering temperature $T_{FQ} = 2.0$ K and an antiferroquadrupole one $T_{AFQ} = 0.6$ K, respectively [244, 245]. Moreover, superconductivity is observed in the quadrupole ordered phase, with $T_{sc} = 0.2$ K and 0.05 K, respectively. It is interesting that T_{FQ} in $PrTi_2Al_{20}$ increases slightly with increasing pressure, has a maximum at around 5–6 GPa, and decreases rather steeply with further increasing pressure, as shown in Fig. 6.32. The superconducting transition temperature starts to increase from 5–6 GPa and most likely has a maximum around 10 GPa: $T_{sc} =$ 0.97 K, $-dH_{c2}/dT = 62$ kOe/K, and $H_{c2}(0) = 45$ kOe at $P =$ 9.1 GPa [244]. Heavy fermion superconductivity is thus realized. The

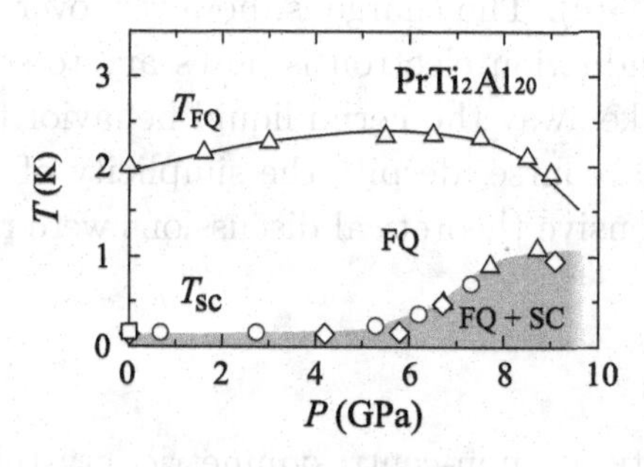

Figure 6.32: Pressure dependence of the ferroquadrupole ordering temperature T_{FQ} and the superconducting transition temperature T_{sc} in $PrTi_2Al_{20}$, cited from Ref. [244].

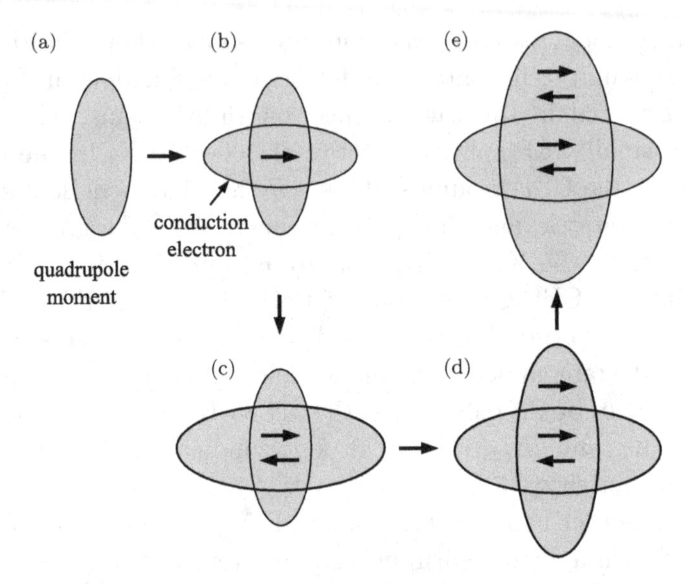

Figure 6.33: Interaction between the quadrupole moment and the conduction electron.

present heavy fermion is most likely due to the quadrupole Kondo effect, revealing non-Fermi liquid properties [246].

Hence, the non-Fermi liquid behavior in this system can be simply explained as follows. Figure 6.33(a) shows the charge distribution of the quadrupole moment which an electron screens so as to be spherical in charge distribution, as shown in Fig. 6.33(b). The conduction electron, however, possesses a spin. Therefore, another electron whose spin direction is opposite is necessary for the spin compensation, as shown in Fig. 6.33(c). The charge is, however, over-screened. Therefore, another conduction electron is necessary to screen the charge. These repeats take away the Fermi liquid behavior from the conduction electrons. Of course, despite the simplicity of this quantitative explanation, extensive theoretical discussions were required to arrive at it [247].

6.4.6 $CeIrSi_3$

Superconductivity in non-centrosymmetric crystal structures has been reported in $CePt_3Si$ [248, 249] with the tetragonal structure ($P4mm$), UIr [250, 251] with the monoclinic structure ($P2_1$),

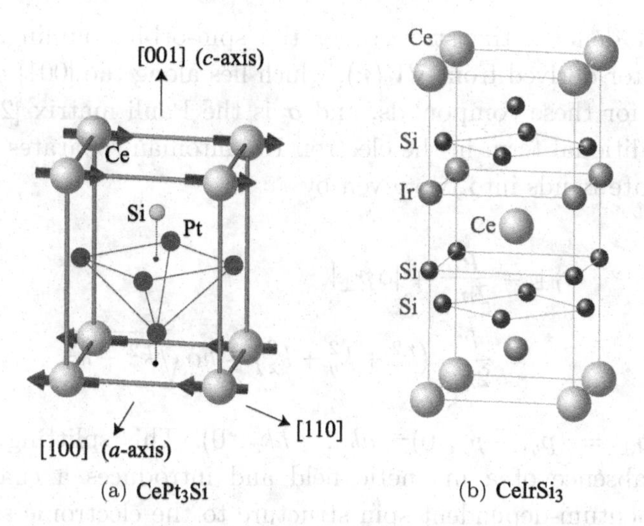

Figure 6.34: (a) Crystal and magnetic structure of $CePt_3Si$, and the crystal structure of $CeIrSi_3$, which also lacks inversion symmetry along the [001] direction.

$CeRhSi_3$ [252–254], $CeIrSi_3$ [168, 255, 256], $CeCoGe_3$ [257, 258] and $CeIrGe_3$ [259] with the tetragonal $BaNiSn_3$-type structure ($I4mm$). We show in Fig. 6.34 the crystal structures of $CePt_3Si$ and $CeIrSi_3$, which lack inversion symmetry along the tetragonal [001] direction (c-axis). In $CeIrSi_3$ (No. 107, a = 4.238 Å, c = 9.784 Å, Z = 2), the Ce atoms occupy the four corners and the body center of the tetragonal structure, similar to the well-known tetragonal $CeCu_2Si_2$ ($ThCr_2Si_2$-type), but the Ir and Si atoms lack inversion symmetry along the [001] direction.

Inversion is an essential symmetry for the formation of Cooper pairs. In non-centrosymmetric metals as in $CePt_3Si$ and $CeIrSi_3$, a splitting of Fermi surfaces with different spins occurs due to the presence of a Rashba-type antisymmetric spin-orbit coupling

$$\begin{aligned} \mathcal{H}_{\text{so}} &= -\frac{\hbar}{4m^{*2}c^2}(\nabla V(\boldsymbol{r}) \times \boldsymbol{p}) \cdot \boldsymbol{\sigma} \qquad (6.69) \\ &= \alpha(\boldsymbol{n} \times \boldsymbol{p}) \cdot \boldsymbol{\sigma} \\ &= \alpha \boldsymbol{p}_\perp \cdot \boldsymbol{\sigma} \end{aligned}$$

where α denotes the strength of the spin-orbit coupling, $\boldsymbol{n}$ is a unit vector derived from $\nabla V(\boldsymbol{r})$, which lies along the [001] direction (c-axis) for these compounds, and $\boldsymbol{\sigma}$ is the Pauli matrix [260–263]. This additional term in the electron Hamiltonian separates the spin degenerate bands into two given by

$$\varepsilon_{p\pm} = \frac{\boldsymbol{p}^2}{2m^*} \mp |\alpha \boldsymbol{p}_\perp| \tag{6.70}$$

$$= \frac{\hbar^2}{2m^*}\left(k_x^2 + k_y^2 + k_z^2\right) \mp \hbar\alpha\sqrt{k_x^2 + k_y^2} \tag{6.71}$$

where $\boldsymbol{p}_\perp = (p_y,\ -p_x,\ 0)=(\hbar k_y,\ -\hbar k_x,\ 0)$. This splitting appears in the absence of a magnetic field and introduces a characteristic momentum-dependent spin structure to the electronic states, as shown in Fig. 6.35(c), which is compared with a spherical Fermi surface with degenerate up- and down-spin states in Fig. 6.35(a). Note that the spins of the conduction electrons are rotated in the direction of the effective magnetic field, $\boldsymbol{n}\times\boldsymbol{p}$, clockwise or counterclockwise, depending on the up- and down-spin states. For comparison, in Fig. 6.35(b), we show the well-known Zeeman splitting, where the degenerate Fermi surface is split into two Fermi surfaces corresponding to a majority and minority spin, respectively, for a given quantization axis parallel to an applied magnetic field.

Due to the Fermi surface splitting in Fig. 6.35(c), the dHvA frequency F is split into two dHvA frequencies, F_+ and F_-. Using $m_{\mathrm{c}}^* = (\hbar^2/2\pi)\partial S/\partial\varepsilon$ and $S_{\mathrm{F}} = (2\pi e/c\hbar)F$, the following relations are obtained:

$$\begin{aligned}\Delta\varepsilon &= 2\,|\alpha \boldsymbol{p}_\perp| \\ &= \frac{\hbar e}{m_{\mathrm{c}}^* c}\Delta F \\ &= \frac{\hbar e}{m_{\mathrm{c}}^* c}\,|F_+ - F_-|\,.\end{aligned} \tag{6.72}$$

We can thus determine the magnitude of the antisymmetric spin-orbit interaction $2|\alpha\boldsymbol{p}_\perp|$ via the dHvA experiment where we use the average m_{c}^* of two orbits for m_{c}^* in Eq. (6.72).

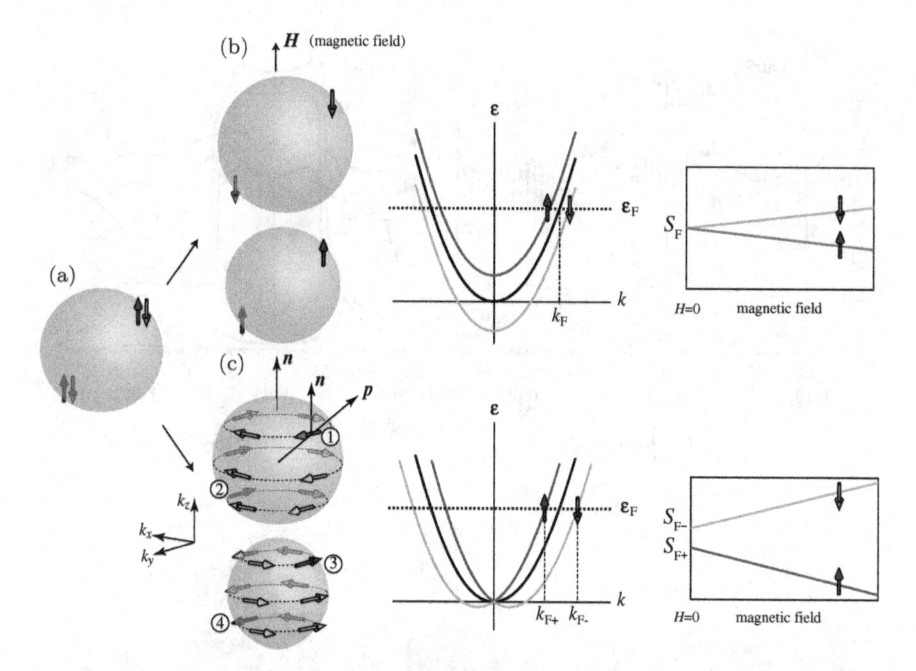

Figure 6.35: (a) Spherical Fermi surface with degenerate up- (↑) and down- (↓) spin states. Arrows do not correspond to the directions of the magnetic moment, but to the directions of the spins. (b) The Fermi surface and the corresponding energy bands are split into two components depending on the up- and down-spin states when the magnetic field H is applied to the material. The maximum cross-sectional area S_F is also split into two components as a function of the magnetic field, well known as Zeeman splitting. (c) The Fermi surface and the corresponding energy band are split into two components depending on the up and down spin states due to the antisymmetric spin-orbit coupling even when $H = 0$. The field dependence of the maximum cross-sectional areas S_{F-} and S_{F+} are also shown in the non-centrosymmetric structure.

Figure 6.36 shows the experimental and theoretical angular dependence of dHvA frequency in $LaIrSi_3$, together with the theoretical Fermi surfaces [255]. The detected dHvA branches are well explained by the result of the FLAPW energy band calculation. The split Fermi surfaces are very similar to each other in topology but different in volume. The magnitude of the antisymmetric spin-orbit coupling $2|\alpha \boldsymbol{p}_\perp|$ is 1100 K in branch α and 1250 K in branch β. Interestingly, these values are 460(α) and 420 K(β) in $LaCoGe_3$, 510(α) and 510 K(β) in $LaRhGe_3$, and 1100(α) and 1100 K(β) in $LaIrGe_3$. The antisymmetric spin-orbit coupling in $LaIrSi_3$ and $LaIrGe_3$ is

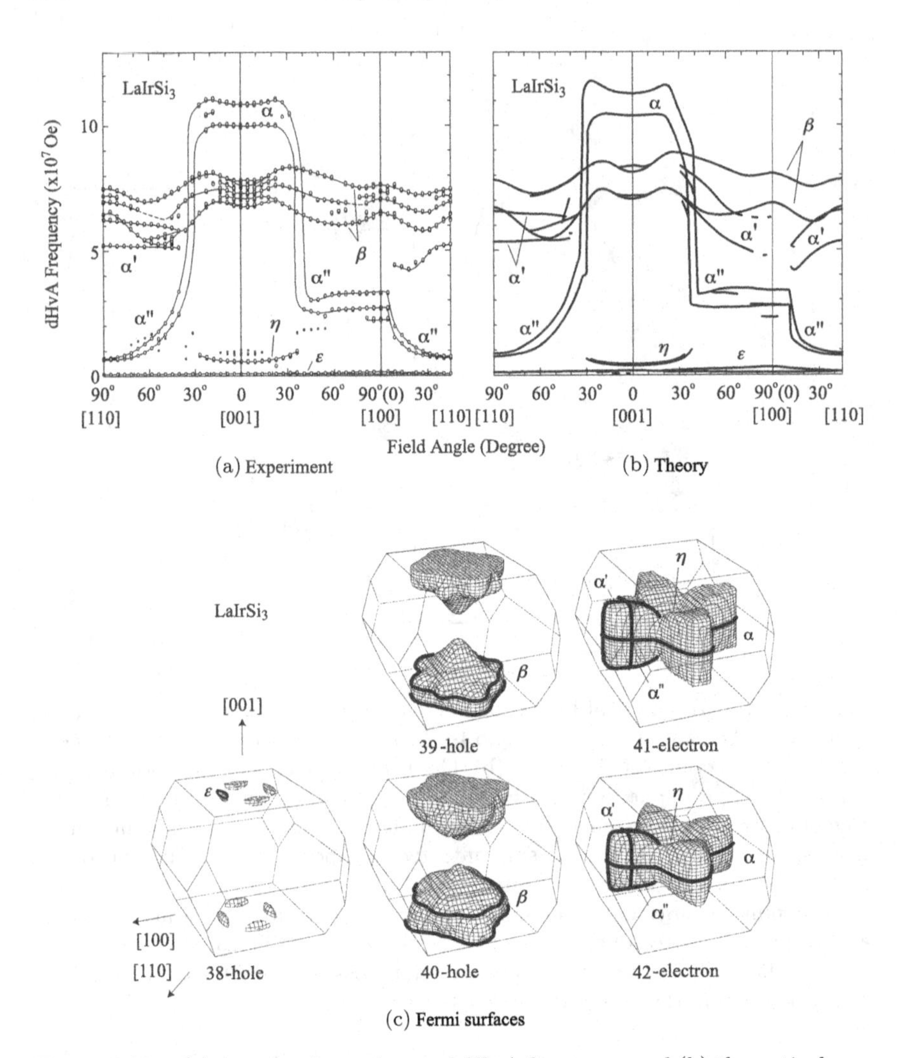

Figure 6.36: (a) Angular dependence of dHvA frequency and (b) theoretical one, together with (c) theoretical Fermi surfaces in $LaIrSi_3$, cited from Ref. [255].

relatively large compared with those in $LaCoGe_3$ and $LaRhGe_3$. The difference is related to the characteristic radial wave function $\phi(r)$ of Ir-5d electrons and the relatively large effective atomic number Z_{eff} in Ir close to the nuclear center [264].

As noted above, the existence of inversion symmetry in the crystal structure is believed to be an important factor for the formation of Cooper pairs, particularly for the spin-triplet configuration.

For example, parallel spin pairing is prohibited in the non-centrosymmetric structure because one conduction electron with a momentum $\boldsymbol{p}$ and an up-spin state, named ①, and the other conduction electron with a momentum $-\boldsymbol{p}$ and an up-spin state, named ④, belong to two different Fermi surfaces, as shown in Fig. 6.35(c). However, the possible existence of a spin-triplet pairing state in the non-centrosymmetric tetragonal crystal structure was proposed theoretically [265, 266].

We explain the pressure-induced superconducting state of $CeIrSi_3$ [255, 256]. The effect of pressure on the electronic state was studied through resistivity measurements. Figure 6.37 shows the low-temperature resistivity at $P = 0$, 1.95, and 2.65 GPa. With increasing pressure, the Néel temperature, shown with arrows, decreases monotonically, although it is not clearly defined at pressures higher than 2 GPa, where pressure-induced superconductivity appears, as shown in Fig. 6.37 for 1.95 GPa. The antiferromagnetic ordering disappears completely at $P = 2.65$ GPa. The superconducting transition temperature T_{sc}, also shown with arrows, increases as a function of pressure and finally attains a value of $T_{sc} = 1.6$ K at 2.65 GPa. Note that the resistivity at 2.65 GPa does not show a T^2-dependence, but indicates a linear T-dependence, which persists up to 18 K.

Figure 6.38 shows the temperature dependence of the ac-specific heat at pressures 1.31, 2.19 and 2.58 GPa. At 1.31 GPa, the antiferromagnetic ordering is clearly observed at $T_N = 4.5$ K, but at 2.19 GPa

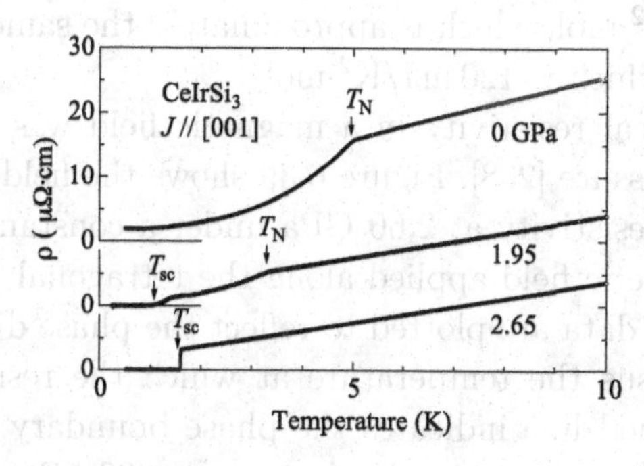

Figure 6.37: Temperature dependence of the electrical resistivity in $CeIrSi_3$ at 0, 1.95 and 2.65 GPa, cited from Refs. [255, 256].

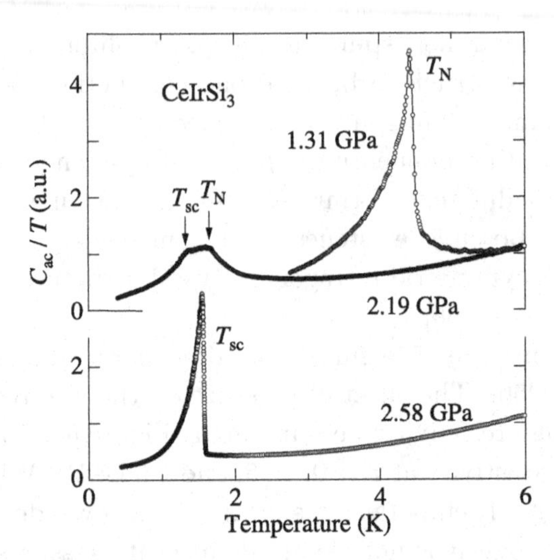

Figure 6.38: Temperature dependence of the ac-specific heat in the form of C_{ac}/T at 1.31, 2.19 and 2.58 GPa in $CeIrSi_3$, Ref. [267].

antiferromagnetism with $T_N = 1.7\,K$ coexists with superconductivity, with $T_{sc} = 1.4\,K$. An exclusively superconducting phase is observed only above the critical pressure $P_c = 2.25\,GPa$ [267]. The specific heat has a huge jump at the superconducting transition above P_c, and $\Delta C_{ac}/C_{ac}(T_{sc})$ at 2.58 GPa is 5.75 at $T_{sc} = 1.6\,K$, which is extremely large when compared with the BCS value of $\Delta C_{ac}/\gamma T_{sc} = 1.43$. This value is the largest in all the superconductors discussed. The antiferromagnet $CeIrSi_3$ is thus changed by pressure into a strong-coupling superconductor. The γ value at 2.58 GPa is roughly estimated to be $100 \pm 20\,mJ/K^2{\cdot}mol$, which is approximately the same as γ at ambient pressure, which is $120\,mJ/K^2{\cdot}mol$.

The electrical resistivity in a magnetic field was measured as a function of pressure [268]. Figure 6.39 shows the field dependence of the electrical resistivity at 2.60 GPa under a constant temperature, with the magnetic field applied along the tetragonal [001] direction. The resistivity data are plotted to reflect the phase diagram of magnetic field versus the temperature at which the resistivity reaches zero. The dashed line indicates the phase boundary which, extrapolated to zero temperature, clearly exceeds 300 kOe, and is roughly estimated to be about 450 kOe ($= 450 \pm 100$ kOe).

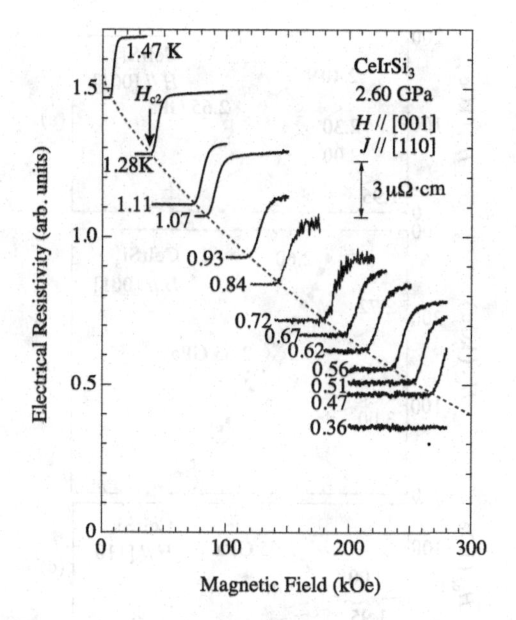

Figure 6.39: Field dependence of the electrical resistivity at a constant temperature for $H \parallel [0\,0\,1]$ in $CeIrSi_3$, Ref. [268]. The zero-resistivity data at each constant temperature are shifted to a vertical scale corresponding to the temperature so that the dashed line indicates the temperature vs upper critical field relation.

The upper critical field H_{c2} is determined for a wide pressure range from 1.95 to 3.00 GPa, as shown in Fig. 6.40. $H_{c2}(0) \simeq 50$ kOe at 1.95 GPa is almost the same for both $H \parallel [0\,0\,1]$ and $[1\,1\,0]$. The temperature dependence of H_{c2}, however, differs strongly between $H \parallel [0\,0\,1]$ and $[1\,1\,0]$. The upper critical field for $H \parallel [0\,0\,1]$ displays an increasing feature with decreasing temperature, while for $H \parallel [1\,1\,0]$, saturation is observed at low temperatures.

With further increasing pressure, the upper critical field between $H \parallel [0\,0\,1]$ and $[1\,1\,0]$ deviates substantially. The superconducting properties become highly anisotropic: $-dH_{c2}/dT = 170$ kOe/K at $T_{sc} = 1.56$ K, and $H_{c2}(0) \simeq 450$ kOe for $H \parallel [0\,0\,1]$, and $-dH_{c2}/dT = 145$ kOe/K at $T_{sc} = 1.59$ K, and $H_{c2}(0) = 95$ kOe for $H \parallel [1\,1\,0]$ at 2.65 GPa, as shown in Fig. 6.41. The upper critical field H_{c2} for $H \parallel [1\,1\,0]$ shows strong signs of Pauli paramagnetic suppression with decreasing temperature because the orbital limiting field $H_{orb}(= -0.7(dH_{c2}/dT)T_{sc})$ in Eq. (6.61) is estimated to

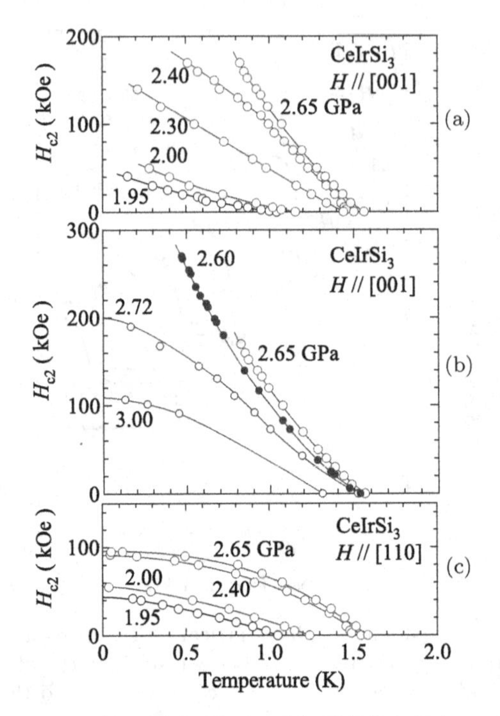

Figure 6.40: Upper critical field H_{c2} for (a) $H \parallel [0\,0\,1]$ in pressure range from 1.95 GPa to 2.65 GPa, (b) 2.65 GPa to 3.0 GPa, and (c) $H \parallel [1\,1\,0]$ in pressure range from 1.95 GPa to 2.65 GPa in CeIrSi$_3$, Ref. [268].

be 170 kOe [195, 269], which is larger than $H_{c2}(0) = 95$ kOe for $H \parallel [1\,1\,0]$. On the other hand, the upper critical field $H \parallel [0\,0\,1]$ is not destroyed by spin polarization based on Zeeman coupling but possesses an upward curvature below 1 K. A similar result is also obtained for CeRhSi$_3$ [252, 253].

It is interesting to compare the results of H_{c2} with the upper critical field of the non-heavy-fermion reference compound LaIrSi$_3$. This is a conventional superconductor with $T_{sc} \simeq 0.9$ K, and has an exponential temperature dependence of specific heat [255]. Figure 6.42 shows the temperature dependence of the upper critical field H_{c2} in LaIrSi$_3$, which was obtained by resistivity measurements in a magnetic field [268]. The anisotropy of H_{c2} is small between $H \parallel [0\,0\,1]$ and $[1\,1\,0]$. Note that the upper critical field for $H \parallel [1\,1\,0]$ is slightly larger than that for $H \parallel [0\,0\,1]$: $-dH_{c2}/dT = 2.6$ kOe/K and $H_{c2}(0) \simeq 1.7$ kOe

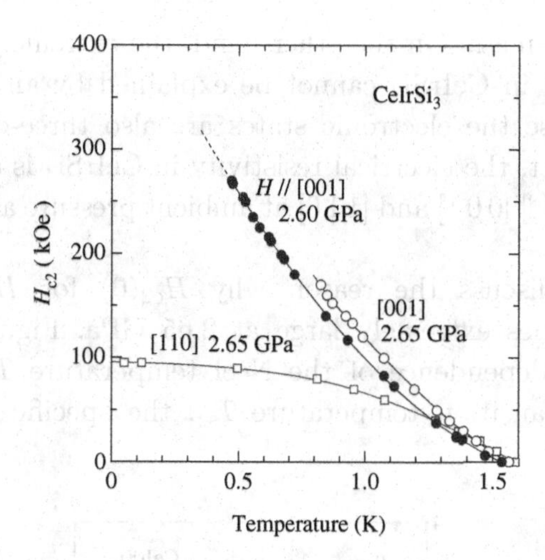

Figure 6.41: Temperature dependence of upper critical field H_{c2} for the magnetic field along [0 0 1] at 2.60 GPa, together with those at 2.65 GPa in $CeIrSi_3$, Ref. [268].

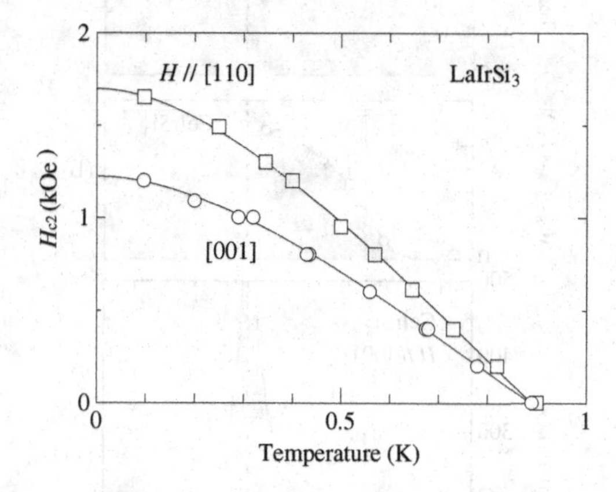

Figure 6.42: Temperature dependence of upper critical field H_{c2} for $H \parallel [0\,0\,1]$ and [1 1 0] in $LaIrSi_3$, Ref. [268].

for $H \parallel [1\,1\,0]$, and $-dH_{c2}/dT = 1.9$ kOe/K and $H_{c2}(0) \simeq 1.25$ kOe for $H \parallel [0\,0\,1]$. The solid lines connecting the data indicate guidelines based on the WHH theory mentioned in Sec. 6.3 [269].

As shown above, the electronic structure inferred from the Fermi surface of $LaIrSi_3$ is three-dimensional [255]. The small anisotropy of H_{c2} in $LaIrSi_3$ is most likely due to the corresponding anisotropy of its effective mass. On the other hand, the extremely large $H_{c2}(0)$ for $H \parallel [0\,0\,1]$ in $CeIrSi_3$ cannot be explained by an effective-mass model, because the electronic states are also three-dimensional in $CeIrSi_3$. In fact, the electrical resistivity in $CeIrSi_3$ is approximately the same for $J \parallel [0\,0\,1]$ and $[1\,1\,0]$ at ambient pressure as well as under pressure.

We will discuss the reason why $H_{c2}(0)$ for $H \parallel [0\,0\,1]$ in $CeIrSi_3$ becomes extremely large at 2.65 GPa. Figure 6.43 shows the pressure dependence of the Néel temperature T_N, the superconducting transition temperature T_{sc}, the specific heat jump at

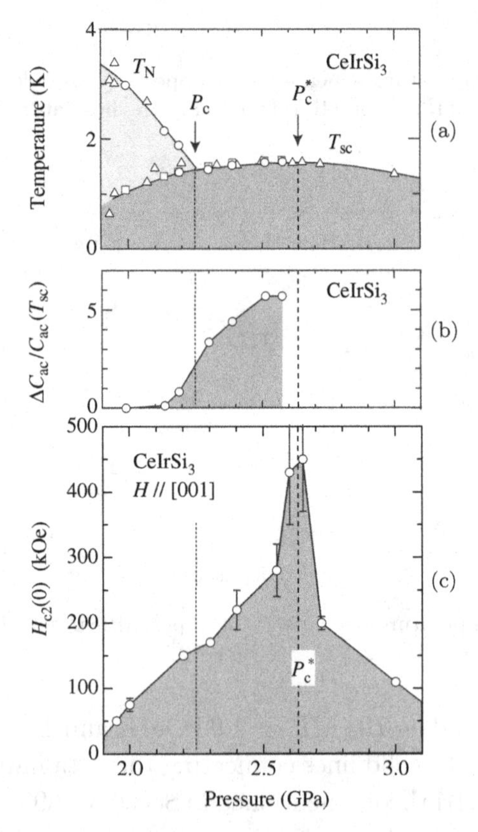

Figure 6.43: Pressure dependence of (a) the Néel temperature T_N and superconducting transition temperature T_{sc}, (b) specific heat jump $\Delta C_{ac}/C_{ac}(T_{sc})$, and (c) the upper critical field $H_{c2}(0)$ for $H \parallel [0\,0\,1]$ in $CeIrSi_3$, Refs. [267] and [268].

T_{sc}, $\Delta C_{ac}/C_{ac}(T_{sc})$, and the upper critical field at 0 K $H_{c2}(0)$ for $H \parallel [0\,0\,1]$. The ac-specific heat measurements indicate both the antiferromagnetic ordering at $T_N = 1.88\,K$ and the superconducting transition at $T_{sc} = 1.40\,K$ at 2.19 GPa [267]. The critical pressure, where the Néel temperature becomes zero, is estimated to be $P_c = 2.25\,GPa$. Above $P_c = 2.25\,GPa$, the antiferromagnetic ordering is absent, and only the superconducting transition is observed in the ac-specific heat measurement.

The superconducting transition temperature becomes maximum at about 2.6 GPa, as shown in Fig. 6.43(a). Simultaneously, the specific heat at T_{sc} 'jumps' and becomes extremely large. This might be reflected in the upward curvature of H_{c2}, as noted above. As shown in Fig. 6.43(c), the upper critical field $H_{c2}(0)$ at 0 K for $H \parallel [0\,0\,1]$ becomes maximum at $P_c^* \simeq 2.63$ GPa.

Pressure-induced superconductivity in f-electron systems is not simple because magnetism does not disappear completely, especially in external magnetic fields. The electrical resistivity measurement is clear for superconductivity because of $\rho = 0$ but is not definite for the magnetic ordering, whereas the specific heat measurement is sensitive to the magnetic ordering even in magnetic fields. From the resistivity and ac-specific heat experiments, the antiferromagnetic (AF) and superconducting (SC) phase diagram was constructed, as shown in Fig. 6.44(a) [270]. The present phase diagram for $CeIrSi_3$ is slightly different from that for $CeRhIn_5$ [166, 271], as shown in Fig. 6.44(b). In $CeRhIn_5$, the field-induced AF phase line crosses the H_{c2} line and the AF phase exists in the SC phase. On the other hand, the AF phase line touches the H_{c2} line and disappears at lower magnetic fields in $CeIrSi_3$.

The antiferromagnetic phase was thus investigated under various magnetic fields and pressures in $CeIrSi_3$. We show in Fig. 6.45 the AF and SC phase diagram under various pressures. The antiferromagnetic phase is robust in magnetic fields. In other words, superconductivity is realized in the antiferromagnetic state. It is concluded that the H_{c2} value for $H \parallel [0\,0\,1]$ is maximum when the antiferromagnetic phase disappears completely in magnetic fields. This pressure corresponds to $P_c^* \simeq 2.63$ GPa.

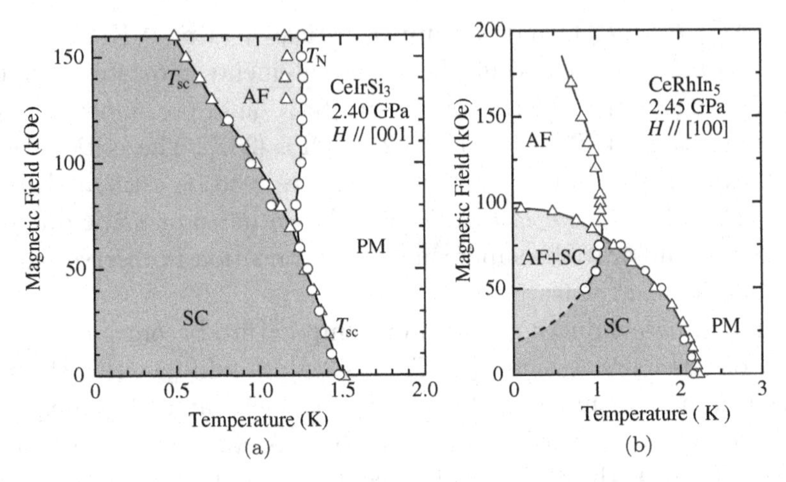

Figure 6.44: Antiferromagnetic and superconducting phase diagram in (a) $CeIrSi_3$, Ref. [270] and (b) $CeRhIn_5$, Refs. [271] and [166].

Also note that the result of the ^{29}Si-NMR experiment at 2.7 GPa for $CeIrSi_3$ supports the present result [272]. In the normal state, the nuclear spin-lattice relaxation rate $1/T_1$ shows a $\sqrt{T}$ dependence, as shown in Fig. 6.46. When the system is close to an antiferromagnetic quantum critical point, the isotropic antiferromagnetic spin-fluctuation model predicts the relation of $1/T_1 \propto T/\sqrt{T+\theta_p}$ [273], where θ_p is a measure of the proximity of the system to the quantum critical point. If $\theta_p = 0$, $1/T_1$ shows a $\sqrt{T}$ dependence. In this context, the NMR experiment shows that the electronic state at 2.7 GPa is very close to the quantum critical point. Moreover, the temperature dependence of $1/T_1$ below T_{sc} is T^3-dependent without a coherence peak just below T_{sc}, revealing the presence of line nodes in the superconducting energy gap. Here, we note that the isotropic antiferromagnetic spin-fluctuation model is based on SCR theory, shown in Table 4.2. In Table 4.2, it is described that $\chi(Q)^{-1}$ follows $T^{3/2}$ dependence at low temperatures. In the present temperature region, namely at relatively higher temperatures, $\chi(Q)$ follows the Curie–Weiss law, namely $\chi(Q) \sim 1/(T+\theta_p)$. This means $1/T_1 \sim T\chi(Q)^{1/2} \sim T/\sqrt{T+\theta_p}$.

Moreover, the ^{29}Si Knight shift experiment was carried out under pressure of 2.8 GPa and magnetic field of 13.26 kOe, revealing the

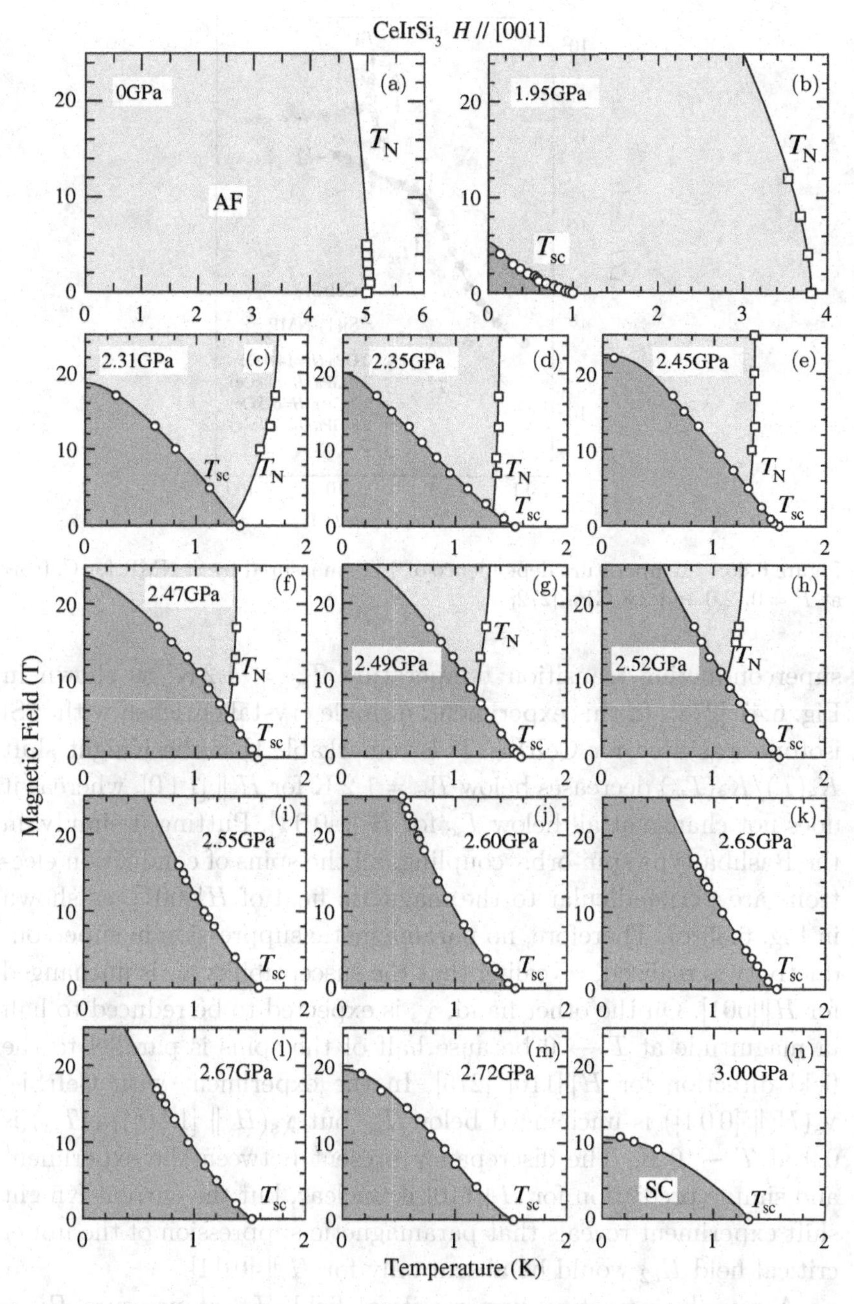

Figure 6.45: Antiferromagnetic and superconducting phase diagrams in magnetic fields under various pressures in CeIrSi$_3$, Ref. [270].

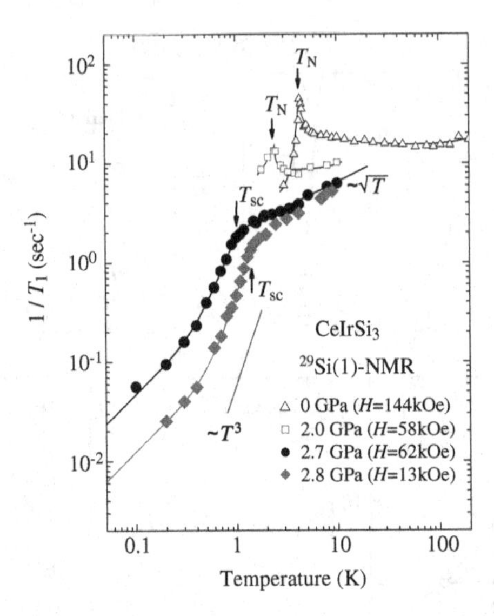

Figure 6.46: Temperature dependence of $1/T_1$ measured by Si NMR for $CeIrSi_3$ at $P = 0$, 2.0 and 2.8 GPa [272].

superconducting transition temperature $T_{sc} = 1.2\,K$, as shown in Fig. 6.47 [274]. In this experiment, a single crystal enriched with ^{29}Si isotope was used for $CeIrSi_3$. It is remarkable that the Knight shift $K_s(T)/K_s(T_{sc})$ decreases below $T_{sc} = 1.2$ K for $H \parallel [1\,1\,0]$, whereas it does not change at all below T_{sc} for $H \parallel [0\,0\,1]$. Putting it simply, in the Rashba-type spin-orbit coupling, all the spins of conduction electrons are perpendicular to the magnetic field of $H\|[001]$, as shown in Fig. 6.35(c). Therefore, no paramagnetic suppression in superconductivity is realized, revealing that the susceptibility χ_s is unchanged for $H\|[001]$. On the other hand, χ_s is expected to be reduced to half in magnitude at $T \to 0$ because half of the spins is parallel to the field direction for $H\|[110]$ [275]. In the experiment with $CeIrSi_3$, $\chi_s(H \parallel [0\,0\,1])$ is unchanged below T_{sc} but $\chi_s(H \parallel [1\,1\,0])/\chi(T_{sc})$ is 0.9 at $T \to 0$ K. The discrepancy present between the experiment and simple prediction for $H\|[110]$ is unclear, but the current Knight shift experiment reveals that paramagnetic suppression of the upper critical field H_{c2} would be absent only for $H \parallel [0\,0\,1]$.

A rapidly increasing upper critical field H_{c2} at pressure $P_c^* \simeq$ 2.63 GPa is observed for $H \parallel [0\,0\,1]$, indicating a huge $H_{c2}(0) \simeq$

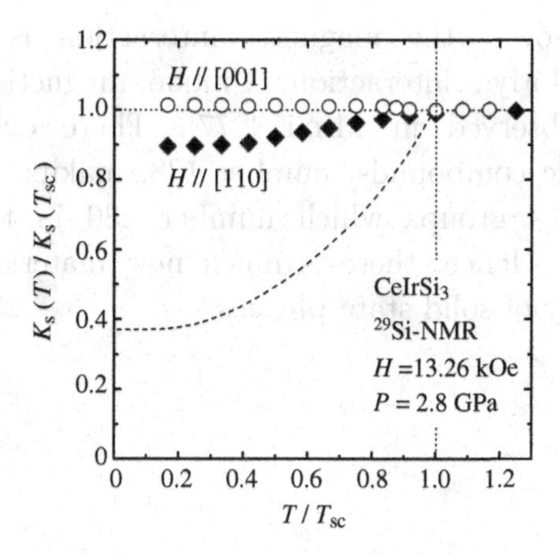

Figure 6.47: Temperature dependence of $K_s(T)/K_s(T_{sc})$ versus T/T_{sc} for $H \parallel [1\,1\,0]$ and $[0\,0\,1]$ in the superconducting state with $T_{sc} = 1.2$ K in CeIrSi$_3$. A broken curve indicates the calculation for a case of $\chi \to 0$ at low-T limit, assuming the residual density of states to be 37% at the Fermi energy, Ref. [274].

450 kOe, while the upper critical field H_{c2} for $H \parallel [1\,1\,0]$ indicates Pauli paramagnetic suppression. The electronic instability, non-centrosymmetry and strong-coupling superconductivity are combined into the huge $H_{c2}(0)$ value for $H \parallel [0\,0\,1]$ in CeIrSi$_3$. The electronic instability produces a large slope for the upper critical field, $-dH_{c2}/dT$ at T_{sc}, where $P_c^* = 2.63$ GPa is the quantum critical point. It has been predicted on theoretical grounds, together with the results of the ^{29}Si Knight shift experiment, that paramagnetic suppression of H_{c2} would be absent only for $H \parallel [0\,0\,1]$, based on the temperature dependence of the spin susceptibility below T_{sc} for this type of non-centrosymmetric superconductor. It turns out that CeIrSi$_3$ is a strong-coupling superconductor, concluded by looking at the results of specific heat measurements. These characteristics result in the huge H_{c2} values for $H \parallel [0\,0\,1]$ and a large anisotropy of $H_{c2}(0)$ between $H \parallel [1\,1\,0]$ and $[0\,0\,1]$. The present experimental results are discussed in the theoretical work [276].

Finally we note that superconductivity and magnetism in non-centrosymmetric compounds have been developed into a new field in condensed matter physics. The corresponding Fermi surface splits into two Fermi surfaces due to antisymmetric spin-orbit coupling. Moreover, the magnetic interaction is based on the Dzyloshinskii-Moriya interaction. Unique magnetism of so-called skyrmion is observed in MnSi [277]. There exist many non-centrosymmetric compounds, number 138, making up a large proportion of space groups which number 230 in total (shown in Appendix A.1). Hence, there is much new material to explore in our future study of solid state physics.

References

[1] W. A. Harrison: Phys. Rev. **118** (1960) 1190.
[2] P. Hohenberg and W. Kohn: Phys. Rev. **136** (1964) B864.
[3] W. Kohn and L. J. Sham: Phys. Rev. **140** (1965) A1133.
[4] O. Gunnarsson and B. I. Lundgvist: Phys. Rev. B **13** (1976) 4274.
[5] Y. Ōnuki and A. Hasegawa: in *Handbook on the Physics and Chemistry of Rare Earths*, ed. J. K. A. Gschneidner and L. Eyring (North-Holland, Amsterdam, 1995), Vol. 20, p. 1.
[6] Y. Ōnuki and R. Settai: Low. Temp. Phys. **38** (2012) 89.
[7] J. Wilson, F. D. Salvo, and S. Mahajan: Adv. Phys. **24** (1975) 117.
[8] L. F. Mattheiss: Phys. Rev. B **8** (1973) 3719.
[9] R. Inada, Y. Ōnuki, and S. Tanuma: Phys. Lett. A **69** (1979) 453.
[10] A. Teruya, M. Takeda, A. Nakamura, H. Harima, Y. Haga, K. Uchima, M. Hedo, T. Nakama, and Y. Ōnuki: J. Phys. Soc. Jpn. **84** (2015) 054703.
[11] P. Schroeder, B. Blumenstock, V. Heinen, W. Pratt, and S. Steenwyk: Physica B+ C **107** (1981) 137.
[12] L. D. Landau: Sov. Phys. JETP **3** (1957) 920; **5** (1957) 101; **8** (1959) 70.
[13] Y. Ōnuki, T. Goto, and T. Kasuya: in *Materials Science and Technology*, ed. K. H. J. Buschow (Wiley-VCH, Weinheim, 1991), Vol. 3A.
[14] P. A. Heisenberg: Z. Phys. **49** (1928) 619.
[15] M. A. Ruderman and C. Kittel: Phys. Rev. **96** (1954) 99.
[16] T. Kasuya: Prog. Theor. Phys. **16** (1956) 45.
[17] K. Yosida: Phys. Rev. **106** (1957) 893.
[18] L. Néel: Ann. Phys. **5** (1936) 232.
[19] D. Aoki, Y. Katayama, R. Settai, N. Suzuki, K. Sugiyama, K. Kindo, H. Harima, and Y. Ōnuki: J. Phys. Soc. Jpn. **67** (1998) 4251.
[20] M. Kakihana, H. Akamine, K. Tomori, K. Nishimura, A. Teruya, A. Nakamura, F. Honda, D. Aoki, M. Nakashima, Y. Amako, *et al.*: J. Alloys Comp. **694** (2017) 439.
[21] M. Nakashima, Y. Amako, K. Matsubayashi, Y. Uwatoko, M. Nada, K. Sugiyama, M. Hagiwara, Y. Haga, T. Takeuchi, A. Nakamura, H. Akamine, K. Tomori, T. Yara, Y. Ashitomi, M. Hedo, T. Nakama, and Y. Ōnuki: J. Phys. Soc. Jpn. **86** (2017) 034708.
[22] Y. Takikawa, S. Ebisu, and S. Nagata: J. Phys. Chem. Solids **71** (2010) 1592.

[23] J. H. Van Vleck: *The Theory of Electric and Magnetic Susceptibilities* (Oxford University Press, 1932).
[24] W. E. Gardner, J. Penfold, T. F. Smith, and I. R. Harris: J. Phys. F: Met. Phys. **2** (1972) 133.
[25] T. Takeuchi, A. Nakamura, M. Hedo, T. Nakama, and Y. Ōnuki: J. Phys. Soc. Jpn. **83** (2014) 114001.
[26] Y. Hiranaka, A. Nakamura, M. Hedo, T. Takeuchi, A. Mori, Y. Hirose, K. Mitamura, K. Sugiyama, M. Hagiwara, T. Nakama, and Y. Ōnuki: J. Phys. Soc. Jpn. **82** (2013) 083708.
[27] K. Ueda and Y. Ōnuki: *Physics of Heavy Electron Systems* (Shokabo, Tokyo, 1998) [in Japanese].
[28] K. W. H. Stevens: Proc. Phys. Soc., Sect. A **65** (1952) 209.
[29] M. T. Hutchings: in *Solid State Physics, Advances in Research and Applications*, ed. F. Seitz and B. Turnbull (Academic, New York, 1965), Vol. 16, p. 227.
[30] P. Morin: J. Magn. Magn. Mater. **71** (1988) 151.
[31] R. Settai, S. Araki, P. Ahmet, M. Abliz, K. Sugiyama, Y. Ōnuki, T. Goto, H. Mitamura, T. Goto, and S. Takayanagi: J. Phys. Soc. Jpn. **71** (1998) 151.
[32] M. Abliz, R. Settai, P. Ahmet, D. Aoki, K. Sugiyama, and Y. Ōnuki: Phil. Mag. B **75** (1998) 443.
[33] P. Ahmet, M. Abliz, R. Settai, K. Sugiyama, Y. Ōnuki, T. Takeuchi, K. Kindo, and S. Takayanagi: J. Phys. Soc. Jpn. **65** (1996) 1077.
[34] K. Sugiyama, M. Nakashima, Y. Yoshida, R. Settai, T. Takeuchi, K. Kindo, and Y. Ōnuki: Physica B **256–261** (1999) 896.
[35] K. Sugiyama, M. Nakashima, Y. Yoshida, Y. Kimura, K. Kindo, T. Takeuchi, R. Settai, and Y. Ōnuki: J. Phys. Soc. Jpn. **67** (1998) 3244.
[36] Y. Yoshida, K. Sugiyama, T. Takeuchi, Y. Kimura, D. Aoki, M. Kozaki, R. Settai, K. Kindo, and Y. Ōnuki: J. Phys. Soc. Jpn. **67** (1998) 1421.
[37] R. Shiina, O. Sakai, H. Shiba, and P. Thalmeier: J. Phys. Soc. Jpn. **67** (1998) 941.
[38] H. Kusunose and Y. Kuramoto: J. Phys. Soc. Jpn. **74** (2005) 3139.
[39] Y. Haga, T. Honma, E. Yamamoto, H. Ohkuni, Y. Ōnuki, M. Ito, and N. Kimura: Jpn. J. Appl. Phys. **37** (1998) 3604.
[40] Y. Haga, T. D. Matsuda, N. Tateiwa, E. Yamamoto, Y. Ōnuki, and Z. Fisk: Philoso. Mag. **94** (2014) 3672.
[41] N. D. Dung, Y. Ota, K. Sugiyama, T. D. Matsuda, Y. Haga, K. Kindo, M. Hagiwara, T. Takeuchi, R. Settai, and Y. Ōnuki: J. Phys. Soc. Jpn. **78** (2009) 024712.
[42] K. Kadowaki and S. Woods: Solid State Commun. **58** (1986) 507.
[43] K. Miyake, T. Matsuura, and C. Varma: Solid State Commun. **71** (1989) 1149.
[44] N. Tsujii, H. Kontani, and K. Yoshimura: Phys. Rev. Lett. **94** (2005) 057201.
[45] N. V. Hieu, T. Takeuchi, H. Shishido, C. Tonohiro, T. Yamada, H. Nakashima, K. Sugiyama, R. Settai, T. D. Matsuda, Y. Haga,

M. Hagiwara, K. Kindo, S. Araki, Y. Nozue, and Y. Ōnuki: J. Phys. Soc. Jpn. **76** (2007) 064702.
[46] N. V. Hieu, H. Shishido, T. Takeuchi, A. Thamizhavel, H. Nakashima, K. Sugiyama, R. Settai, T. D. Matsuda, Y. Haga, M. Hagiwara, K. Kindo, and Y. Ōnuki: J. Phys. Soc. Jpn. **75** (2006) 074708.
[47] S. Chang, P. Pagliuso, W. Bao, J. Gardner, I. Swainson, J. Sarrao, and H. Nakotte: Phys. Rev. B **66** (2002) 132417.
[48] Y. Takeda, N. Dung, Y. Nakano, T. Ishikura, S. Ikeda, T. D. Matsuda, E. Yamamoto, Y. Haga, T. Takeuchi, R. Settai, and Y. Ōnuki: J. Phys. Soc. Jpn. **77** (2008) 104710.
[49] J. Kondo: J. Phys. Soc. Jpn. **16** (1961) 1690.
[50] E. Goremychkin, A. Y. Muzychka, and R. Osborn: J. Exp. Theor. Phys. **83** (1996) 738.
[51] H. Pinto, M. Melamud, M. Kuznietz, and H. Shaked: Phys. Rev. B **31** (1985) 508.
[52] H. Sato, Y. Hoshina, K. Yonemitsu, Y. Onuki, and T. Komatsubara: J. Phys. Soc. Jpn. **55** (1986) 2471.
[53] J. Benz, C. Pfleiderer, O. Stockert, and H. Löhneysen: Physica B **259–261** (1999) 380.
[54] K. Behnia, D. Jaccard, and J. Flouquet: J. Phys.: Condens. Matter **16** (2004) 5187.
[55] A. Lacerda, A. de Visser, P. Haen, P. Lejay, and J. Flouquet: Phys. Rev. B **40** (1989) 8759.
[56] Y. Kitaoka, K. Ueda, T. Kohara, Y. Kohori, and K. Asayama: in *Theoretical and Experimental Aspects of Valence Fluctuations and Heavy Fermions*, ed. L. C. Gupta and S. K. Malik (Plenum Press, New York, 2012).
[57] T. Moriya: J. Phys. Soc. Jpn. **18** (1963) 516.
[58] J. Korringa: Physica **16** (1950) 601.
[59] H. Tou, Y. Kitaoka, K. Asayama, N. Kimura, Y. Ōnuki, E. Yamamoto, and K. Maezawa: Phys. Rev. Lett. **77** (1996) 1374.
[60] J. Rossat-Mignod, P. Burlet, and G. Lander: Neutron Elastic Scattering of the Actinides. in *Handbook on the Physics and Chemistry of the Actinides*, A. Freeman and G. Lander (eds), Vol. 1 (Elsevier Science Publishers B. V, 1984).
[61] T. Honma, H. Amitsuka, S. Yasunami, K. Tenya, T. Sakakibara, H. Mitamura, T. Goto, G. Kido, S. Kawarazaki, Y. Miyako, K. Sugiyama, and M. Date: J. Phys. Soc. Jpn. **67** (1998) 1017.
[62] K. Iwasa, H. Kobayashi, T. Onimaru, K. T. Matsumoto, N. Nagasawa, T. Takabatake, S. Ohira-Kawamura, T. Kikuchi, Y. Inamura, and K. Nakajima: J. Phys. Soc. Jpn. **82** (2013) 043707.
[63] K. Lea, M. Leask, and W. Wolf: J. Phys. Chem. Solids **23** (1962) 1381.
[64] A. Schenck: in *Selected Topics in Magnetism*, ed. L. Gupta and M. Multani (World Scientific, 1993), Vol. 2 of *Frontiers in Solid State Sciences.*
[65] S. Barth, H. R. Ott, F. N. Gygax, B. Hitti, E. Lippelt, A. Schenck, C. Baines, B. van den Brandt, T. Konter, and S. Mango: Phys. Rev. Lett. **59** (1987) 2991.

[66] H. Nakamura, Y. Kitaoka, K. Asayama, and J. Flouquet: J. Magn. Magn. Mater. **76–77** (1988) 465.
[67] K. Yamaji: J. Phys. Soc. Jpn. **58** (1989) 1520.
[68] Y. Yoshida, A. Mukai, R. Settai, Y. Ōnuki, and H. Takei: J. Phys. Soc. Jpn. **67** (1998) 2551.
[69] Y. Yoshida, A. Mukai, R. Settai, K. Miyake, Y. Inada, Y. Ōnuki, Y. Aoki, and H. Sato: J. Phys. Soc. Jpn. **68** (1999) 3041.
[70] P. Coleman, P. W. Anderson, and T. V. Ramakrishnan: Phys. Rev. Lett. **55** (1985) 414.
[71] A. Fert, A. Hamzić, and P. Levy: J. Magn. Magn. Mater. **63** (1987) 535.
[72] Y. Ōnuki, T. Yamazaki, T. Omi, I. Ukon, A. Kobori, and T. Komatsubara: J. Phys. Soc. Jpn. **58** (1989) 2126.
[73] T. Namiki, H. Sato, H. Sugawara, Y. Aoki, R. Settai, and Y. Onuki: J. Phys. Soc. Jpn. **76** (2007) 054708.
[74] D. Shoenberg: *Magnetic Oscillations in Metals* (Cambridge Monographs on Physics. Cambridge University Press, Cambridge, 1984), Cambridge Monographs on Physics.
[75] H. Yamagami and A. Hasegawa: J. Phys. Soc. Jpn. **62** (1993) 592.
[76] M. Takashita, H. Aoki, T. Terashima, S. Uji, K. Maezawa, R. Settai, and Y. Ōnuki: J. Phys. Soc. Jpn. **65** (1996) 515.
[77] S. Fujimori: J. Phys.: Condens. Matter **28** (2016) 153002.
[78] S. Fujimori, Y. Takeda, T. Okane, Y. Saitoh, A. Fujimori, H. Yamagami, Y. Haga, E. Yamamoto, and Y. Ōnuki: J. Phys. Soc. Jpn. **85** (2016) 062001.
[79] M. Seah and W. Dench: Surf. Interface Anal. **1** (1979) 2.
[80] T. Ishii, K. Soda, K. Naito, T. Miyahara, H. Kato, S. Sato, T. Mori, M. Taniguchi, A. Kakizaki, Y. Onuki, and T. Komatsubara: Physica Scripta **35** (1987) 603.
[81] J.-G. Cheng, K. Matsubayashi, S. Nagasaki, A. Hisada, T. Hirayama, M. Hedo, H. Kagi, and Y. Uwatoko: Rev. Sci. Instrum. **85** (2014) 093907.
[82] T. Kobayashi, H. Hidaka, H. Kotegawa, K. Fujiwara, and M. Eremets: Rev. Sci. Instrum. **78** (2007) 023909.
[83] G. Giriat, Z. Ren, P. Pedrazzini, and D. Jaccard: Solid State Commun. **209** (2015) 55.
[84] B. Salce, J. Thomasson, A. Demuer, J. Blanchard, J. Martinod, L. Devoille, and A. Guillaume: Rev. Sci. Instrum. **71** (2000) 2461.
[85] I. R. Walker: Rev. Sci. Instrum. **70** (1999) 3402.
[86] T. F. Smith, C. W. Chu, and M. B. Maple: Cryogenics **9** (1969) 53.
[87] T. Osakabe and K. Kakurai: Jpn. J. Appl. Phys. **47** (2008) 6544.
[88] G. J. Piermarini and S. Block: Rev. Sci. Instrum. **46** (1975) 973.
[89] H. K. Mao and P. M. Bell: Science **200** (1978) 1145.
[90] A. Drozdov, M. Eremets, I. Troyan, V. Ksenofontov, and S. Shylin: Nature **525** (2015) 73.
[91] E. C. Stoner: Proc. Phys. Soc. A **165** (1938) 372.
[92] J. W. D. Connolly: Phys. Rev. **159** (1967) 415.
[93] H. Yamada: Physica B+C **149** (1988) 390.
[94] E. Burzo, E. Gratz, and V. Pop: J. Magn. Magn. Mater. **123** (1993) 159.

[95] T. Goto, K. Fukamichi, T. Sakakibara, and H. Komatsu: Solid State Commun. **72** (1989) 945.
[96] E. Wohlfarth and P. Rhodes: Philos. Mag. **7** (1962) 1817.
[97] M. Shimizu: J. Phys. (Paris) **43** (1982) 155.
[98] H. Yamada: Phys. Rev. B **47** (1993) 11211.
[99] T. Sakakibara, H. Mitamura, and T. Goto: Physica B: Cond. Matt. **201** (1994) 127.
[100] M. Imai, C. Michioka, H. Ohta, A. Matsuo, K. Kindo, H. Ueda, and K. Yoshimura: Phys. Rev. B **90** (2014) 014407.
[101] T. Moriya: J. Magn. Magn. Mater. **14** (1979) 1.
[102] T. Moriya: *Spin Fluctuations in Itinerant Electron Systems* (Springer Berlin/Heidelberg, 1985).
[103] M. Shiga: *Introduction to Magnetism* (Uchida Rokakuho, 2007) [in Japanese].
[104] B. T. Matthias and R. M. Bozorth: Phys. Rev. **109** (1958) 604.
[105] B. T. Matthias, A. M. Clogston, H. J. Williams, E. Corenzwit, and R. C. Sherwood: Phys. Rev. Lett. **7** (1961) 7.
[106] A. Teruya, F. Suzuki, D. Aoki, F. Honda, A. Nakamura, M. Nakashima, Y. Amako, H. Harima, M. Hedo, T. Nakama, and Y. Ōnuki: J. Phys. Soc. Jpn. **85** (2016) 064716.
[107] A. Teruya, A. Nakamura, T. Takeuchi, H. Harima, K. Uchima, M. Hedo, T. Nakama, and Y. Ōnuki: J. Phys. Soc. Jpn. **83** (2014) 113702.
[108] A. Teruya, A. Nakamura, T. Takeuchi, F. Honda, D. Aoki, H. Harima, K. Uchima, M. Hedo, T. Nakama, and Y. Ōnuki: Physics Procedia **75** (2015) 876.
[109] A. Mewis: Z. Naturforsch. **35b** (1980) 141.
[110] T. Moriya and K. Ueda: The Phys. Soc. Japan **52** (1997) 422 [in Japanese].
[111] J. Kondo: Prog. Theor. Phys. **32** (1964) 37.
[112] K. G. Wilson: Rev. Mod. Phys. **47** (1975) 773.
[113] A. A. Abrikosov: Physics **1** (1965) 5.
[114] K. Yosida: Phys. Rev. **147** (1966) 223.
[115] K. Yosida and A. Yoshimori: in *Magnetism: Magnetic Properties of Metallic Alloys*, ed. G. T. Rado and H. Suhl (Academic Press, New York & London, 1973), Vol. 5, p. 253.
[116] M. D. Daybell: in *Magnetism: Magnetic Properties of Metallic Alloys*, ed. G. T. Rado and H. Suhl (Academic Press, New York & London, 1973), Vol. 5, p. 121.
[117] K. Yamada, K. Yosida, and K. Hanzawa: Prog. Theor. Phys. **71** (1984) 450.
[118] Y. Ōnuki, Y. Furukawa, and T. Komatsubara: J. Phys. Soc. Jpn. **53** (1984) 2734.
[119] A. Sumiyama, Y. Oda, H. Nagano, Y. Ōnuki, K. Shibutani, and T. Komatsubara: J. Phys. Soc. Jpn. **55** (1986) 1294.
[120] Y. Ōnuki and T. Komatsubara: J. Magn. Magn. Mater. **63 & 64** (1987) 281.
[121] K. Satoh, T. Fujita, Y. Maeno, Y. Ōnuki, and T. Komatsubara: J. Phys. Soc. Jpn. **58** (1989) 1012.

[122] H. Asano, M. Umino, Y. Ōnuki, T. Komatsubara, F. Izumi, and N. Watanabe: J. Phys. Soc. Jpn. **55** (1986) 454.
[123] M. D. Daybell: in *Magnetism: Magnetic Properties of Metallic Alloys*, ed. G. T. Rado and H. Suhl (Academic Press, New York & London, 1973), Vol. 5, p. 133.
[124] V. T. Rajan, J. H. Lowenstein, and N. Andrei: Phys. Rev. Lett. **49** (1982) 497.
[125] V. T. Rajan: Phys. Rev. Lett. **51** (1983) 308.
[126] S. Chapman, M. Hunt, P. Meeson, P. H. P. Reinders, M. Springford, and M. Norman: J. Phys.: Condens. Matter **2** (1990) 8123.
[127] Y. Hirose, J. Sakaguchi, M. Ohya, M. Matsushita, F. Honda, R. Settai, and Y. Ōnuki: J. Phys. Soc. Jpn. **81** (2012) SB009.
[128] A. Amato, D. Jaccard, J. Flouquet, F. Lapierre, J. L. Tholence, R. A. Fisher, S. E. Lacy, J. A. Olsen, and N. E. Phillips: J. Low Temp. Phys. **68** (1987) 371.
[129] L. Regnault, W. Erkelens, J. Rossat-Mignod, J. Flouquet, E. Walker, D. Jaccard, A. Amato, and B. Hennion: J. Magn. Magn. Mater. **63** (1987) 289.
[130] J. Rossat-Mignod, L. Regnault, J. Jacoud, C. Vettier, P. Lejay, J. Flouquet, E. Walker, D. Jaccard, and A. Amato: J. Magn. Magn. Mater. **76–77** (1988) 376.
[131] Y. Ōnuki, R. Settai, T. Takeuchi, K. Sugiyama, F. Honda, Y. Haga, E. Yamamoto, T. D. Matsuda, N. Tateiwa, D. Aoki, I. Sheikin, H. Harima, and H. Yamagami: J. Phys. Soc. Jpn. **81** (2012) SB001.
[132] S. Doniach: in *Valence Instabilities and Related Narrow Band Phenomena*, ed. R. D. Parks (Plenum, New York, 1977), p. 169.
[133] H. H. Hill: in *Plutonium and Other Actinides*, ed. W. N. Miner (AIME, New York, 1970), p. 2.
[134] A. Galatanu, Y. Haga, T. D. Matsuda, S. Ikeda, E. Yamamoto, D. Aoki, T. Takeuchi, and Y. Ōnuki: J. Phys. Soc. Jpn. **74** (2005) 1582.
[135] H. Winkelmann, M. M. Abd-Elmeguid, H. Micklitz, J. P. Sanchez, P. Vulliet, K. Alami-Yadri, and D. Jaccard: Phys. Rev. B **60** (1999) 3324.
[136] H. v. Löhneysen, A. Rosch, M. Vojta, and P. Wölfle: Rev. Mod. Phys. **79** (2007) 1015.
[137] Q. Si and F. Steglich: Science **329** (2010) 1161.
[138] Y. Ōnuki, S. Yasui, M. Matsushita, S. Yoshiuchi, M. Ohya, Y. Hirose, N. D. Dung, F. Honda, T. Takeuchi, R. Settai, K. Sugiyama, E. Yamamoto, T. D. Matsuda, Y. Haga, T. Tanaka, Y. Kubo, and H. Harima: J. Phys. Soc. Jpn. **80** (2011) SA003.
[139] T. Takeuchi, S. Yasui, M. Toda, M. Matsushita, S. Yoshiuchi, M. Ohya, K. Katayama, Y. Hirose, N. Yoshitani, F. Honda, K. Sugiyama, M. Hagiwara, K. Kindo, E. Yamamoto, Y. Haga, T. Tanaka, Y. Kubo, R. Settai, and Y. Ōnuki: J. Phys. Soc. Jpn. **79** (2010) 064609.
[140] M. S. Torikachvili, S. Jia, E. D. Mun, S. T. Hannahs, R. C. Black, W. K. Neils, D. Martien, S. L. Bud'ko, and P. C. Canfield: Proc. Natl. Acad. Sci. USA **104** (2007) 9960.

[141] S. Yoshiuchi, M. Toda, M. Matsushita, S. Yasui, Y. Hirose, M. Ohya, K. Katayama, F. Honda, K. Sugiyama, M. Hagiwara, K. Kindo, T. Takeuchi, E. Yamamoto, Y. Haga, R. Settai, T. Tanaka, Y. Kubo, and Y. Ōnuki: J. Phys. Soc. Jpn. **78** (2009) 123711.
[142] M. Ohya, M. Matsushita, S. Yoshiuchi, T. Takeuchi, F. Honda, R. Settai, T. Tanaka, Y. Kubo, and Y. Ōnuki: J. Phys. Soc. Jpn. **79** (2010) 083601.
[143] Y. Hirose, M. Toda, S. Yoshiuchi, S. Yasui, K. Sugiyama, F. Honda, M. Hagiwara, K. Kindo, R. Settai, and Y. Ōnuki: J. Phys.: Conf. Ser. **273** (2011) 012003.
[144] F. Honda, S. Yasui, S. Yoshiuchi, T. Takeuchi, R. Settai, and Y. Ōnuki: J. Phys. Soc. Jpn. **79** (2010) 083709.
[145] K. Sugiyama and Y. Ōnuki: in *High Magnetic Fields*, ed. F. Herlach and N. Miura (World Scientific, 2003), Vol. 2, p. 139.
[146] T. Ebihara, I. Umehara, A. K. Albessard, K. Satoh, and Y. Ōnuki: J. Phys. Soc. Jpn. **61** (1992) 1473.
[147] I. Umehara, N. Nagai, and Y. Ōnuki: J. Phys. Soc. Jpn. **60** (1991) 1294.
[148] I. Umehara, Y. Kurosawa, N. Nagai, M. Kikuchi, K. Satoh, and Y. Ōnuki: J. Phys. Soc. Jpn. **59** (1990) 2848.
[149] A. Hasegawa and H. Yamagami: J. Phys. Soc. Jpn. **60** (1991) 1654.
[150] A. Hasegawa, H. Yamagami, and H. Johbettoh: J. Phys. Soc. Jpn. **59** (1990) 2457.
[151] K. Miyake and H. Maebashi: J. Phys. Soc. Jpn. **71** (2002) 1007.
[152] Y. Haga, H. Sakai, and S. Kambe: J. Phys. Soc. Jpn. **76** (2007) 051012.
[153] J. L. Sarrao, L. A. Morales, J. D. Thompson, B. L. Scott, G. R. Stewart, F. Wastin, J. Rebizant, P. Boulet, E. Colineau, and G. H. Lander: Nature (London) **420** (2002) 297 .
[154] H. Shishido, R. Settai, D. Aoki, S. Ikeda, H. Nakawaki, N. Nakamura, T. Iizuka, Y. Inada, K. Sugiyama, T. Takeuchi, K. Kindo, T. C. Kobayashi, Y. Haga, H. Harima, Y. Aoki, T. Namiki, H. Sato, and Y. Ōnuki: J. Phys. Soc. Jpn. **71** (2002) 162.
[155] Y. Haga, D. Aoki, T. D. Matsuda, K. Nakajima, Y. Arai, E. Yamamoto, A. Nakamura, Y. Homma, Y. Shiokawa, and Y. Ōnuki: J. Phys. Soc. Jpn. **74** (2005) 1698.
[156] D. Aoki, Y. Haga, T. D. Matsuda, N. Tateiwa, S. Ikeda, Y. Homma, H. Sakai, Y. Shiokawa, E. Yamamoto, A. Nakamura, R. Settai, and Y. Ōnuki: J. Phys. Soc. Jpn. **76** (2007) 063701.
[157] S. Ikeda, H. Shishido, M. Nakashima, R. Settai, D. Aoki, Y. Haga, H. Harima, Y. Aoki, T. Namiki, H. Sato, and Y. Ōnuki: J. Phys. Soc. Jpn. **70** (2001) 2248.
[158] C. Petrovic, P. G. Pagliuso, M. F. Hundley, R. Movshovich, J. L. Sarrao, J. D. Thompson, Z. Fisk, and P. Monthoux: J. Phys.: Condens. Matter **13** (2001) L337.
[159] S. Ikeda, Y. Tokiwa, T. Okubo, Y. Haga, E. Yamamoto, Y. Inada, R. Settai, and Y. Ōnuki: J. Nucl. Sci. Technol. Suppl. **3** (2002) 206.

[160] K. Izawa, H. Yamaguchi, Y. Matsuda, H. Shishido, R. Settai, and Y. Ōnuki: Phys. Rev. Lett. **87** (2001) 057002.
[161] T. Tayama, A. Harita, T. Sakakibara, Y. Haga, H. Shishido, R. Settai, and Y. Ōnuki: Phys. Rev. B **65** (2002) 180504.
[162] A. Bianchi, R. Movshovich, C. Capan, P. G. Pagliuso, and J. L. Sarrao: Phys. Rev. Lett. **91** (2003) 187004.
[163] K. Kakuyanagi, M. Saitoh, K. Kumagai, S. Takashima, M. Nohara, H. Takagi, and Y. Matsuda: Phys. Rev. Lett. **94** (2005) 047602.
[164] M. Kenzelmann, T. Strässle, C. Niedermayer, M. Sigrist, B. Padmanabhan, M. Zolliker, A. Bianchi, R. Movshovich, E. Bauer, J. Sarrao, and J. Thompson: Science **321** (2008) 1652.
[165] Y. Matsuda and H. Shimahara: J. Phys. Soc. Jpn. **76** (2007) 051005.
[166] Y. Ida, R. Settai, Y. Ota, F. Honda, and Y. Ōnuki: J. Phys. Soc. Jpn. **77** (2008) 084708.
[167] R. Settai, H. Shishido, S. Ikeda, Y. Murakawa, M. Nakashima, D. Aoki, Y. Haga, H. Harima, and Y. Ōnuki: J. Phys.: Condens. Matter **13** (2001) L627.
[168] R. Settai, T. Takeuchi, and Y. Ōnuki: J. Phys. Soc. Jpn. **76** (2007) 051003.
[169] H. Shishido, R. Settai, H. Harima, and Y. Ōnuki: J. Phys. Soc. Jpn. **74** (2005) 1103.
[170] M. Yashima, S. Kawasaki, Y. Kitaoka, H. Shishido, R. Settai, and Y. Ōuki: Physica B: Cond. Matt. **378–380** (2006) 94.
[171] G. Knebel, D. Aoki, D. Braithwaite, B. Salce, and J. Flouquet: Phys. Rev. B **74** (2006) 020501.
[172] T. Park, F. Ronning, H. Q. Yuan, M. B. Salamon, R. Movshovich, J. L. Sarrao, and J. D. Thompson: Nature (London) **440** (2006) 65.
[173] H. Shishido, R. Settai, H. Harima, and Y. Ōnuki: Physica B **378–380** (2006) 92.
[174] S. Araki, R. Settai, T. C. Kobayashi, H. Harima, and Y. Ōnuki: Phys. Rev. B **64** (2001) 224417.
[175] S. Araki, R. Settai, M. Nakashima, H. Shishido, S. Ikeda, H. Nakawaki, Y. Haga, N. Tateiwa, T. C. Kobayashi, H. Harima, H. Yamagami, Y. Aoki, T. Namiki, H. Sato, and Y. Ōnuki: J. Phys. Chem. Solids **63** (2002) 1133.
[176] R. Settai, T. Kubo, T. Shiromoto, D. Honda, H. Shishido, K. Sugiyama, Y. Haga, T. D. Matsuda, K. Betsuyaku, H. Harima, T. C. Kobayashi, and Y. Ōnuki: J. Phys. Soc. Jpn. **74** (2005) 3016.
[177] J. Bardeen, L. N. Cooper, and J. R. Schrieffer: Phys. Rev. **108** (1957) 1175.
[178] L. Cooper: Phys. Rev. **104** (1956) 1189.
[179] C. A. Reynolds, B. Serin, and L. B. Nesbitt: Phys. Rev. **84** (1951) 691.
[180] L. C. Hebel and C. P. Slichter: Phys. Rev. **113** (1959) 1504.
[181] K. Yosida: Phys. Rev. **110** (1958) 769.
[182] H. Mukuda, K. Ishida, Y. Kitaoka, and K. Asayama: J. Phys. Soc. Jpn. **67** (1998) 2101.
[183] T. Kiss, F. Kanetaka, T. Yokoya, T. Shimojima, K. Kanai, S. Shin, Y. Onuki, T. Togashi, C. Zhang, C. T. Chen, and S. Watanabe: Phys. Rev. Lett. **94** (2005) 057001.

[184] M. Hedo, Y. Inada, E. Yamamoto, Y. Haga, Y. Ōnuki, Y. Aoki, T. D. Matsuda, H. Sato, and S. Takahashi: J. Phys. Soc. Jpn. **67** (1998) 272.

[185] S. Kittaka, T. Sakakibara, M. Hedo, Y. Ōnuki, and K. Machida: J. Phys. Soc. Jpn. **82** (2013) 123706.

[186] Z. Pribulova, T. Klein, J. Marcus, C. Marcenat, F. Levy, M. S. Park, H. G. Lee, B. W. Kang, S. I. Lee, S. Tajima, and S. Lee: Phys. Rev. Lett. **98** (2007) 137001.

[187] X. X. Xi: Reports on Progress in Physics **71** (2008) 116501.

[188] R. Liu, B. W. Veal, A. P. Paulikas, J. W. Downey, P. J. Kostić, S. Fleshler, U. Welp, C. G. Olson, X. Wu, A. J. Arko, and J. J. Joyce: Phys. Rev. B **46** (1992) 11056.

[189] Y. Ōnuki, R. Settai, Y. Haga, Y. Machida, K. Izawa, F. Honda, and D. Aoki: Comptes Rendus Physique **15** (2014) 616. Emergent Phenomena in Actinides.

[190] M. Sigrist and K. Ueda: Rev. Mod. Phys. **63** (1991) 239.

[191] Y. Kuramoto and Y. Kitaoka: *Dynamics of Heavy Electrons* (Oxford Science Publication, Oxford, 2000).

[192] M. Kyogaku, Y. Kitaoka, K. Asayama, C. Geibel, C. Schank, and F. Steglich: J. Phys. Soc. Jpn. **62** (1993) 4016.

[193] H. Tou, Y. Kitaoka, K. Ishida, K. Asayama, N. Kimura, Y. Ōnuki, E. Yamamoto, Y. Haga, and K. Maezawa: Phys. Rev. Lett. **80** (1998) 3129.

[194] V. L. Ginzburg: Zh. Eksp. Teor. Fiz **20** (1950) 1064.

[195] N. R. Werthamer, E. Helfand, and P. C. Hohenberg: Phys. Rev. **147** (1966) 295.

[196] A. M. Clogston: Phys. Rev. Lett. **9** (1962) 266.

[197] B. S. Chandrasekhar: Appl. Phys. Lett. **1** (1962) 7.

[198] K. Yagasaki, M. Hedo, and T. Nakama: J. Phys. Soc. Jpn. **62** (1993) 3825.

[199] A. D. Huxley, C. Paulson, O. Laborde, J. L. Tholence, D. Sanchez, A. Junod, and R. Calemczuk: J. Phys.: Condens. Matter **5** (1993) 7709.

[200] M. Hedo, Y. Inada, K. Sakurai, E. Yamamoto, Y. Haga, Y. Ōnuki, S. Takahashi, M. Higuchi, T. Maehira, and A. Hasegawa: Philoso. Mag. Part B **77** (1998) 975.

[201] K. Matsuda, Y. Kohori, and T. Kohara: J. Phys. Soc. Jpn. **64** (1995) 2750.

[202] K. Ishida, H. Mukuda, Y. Kitaoka, K. Asayama, and Y. Ōnuk: Z. Naturforsch. **51a** (1996) 793.

[203] K. Maki: Physics **1** (1964) 21–30: K. Maki and T. Tsuzuki: Phys. Rev. **139** (1965) A868.

[204] G. Eilenberger: Phys. Rev. **153** (1967) 584.

[205] K. Tenya, S. Yasunami, T. Tayama, H. Amitsuka, T. Sakakibara, M. Hedo, Y. Inada, E. Yamamoto, Y. Haga, and Y. Ōnuki: J. Phys. Soc. Jpn. **68** (1999) 224.

[206] M. Naito, A. Matsuda, K. Kitazawa, S. Kambe, I. Tanaka, and H. Kojima: Phys. Rev. B **41** (1990) 4823.

[207] A. L. Fetter and P. C. Hohenberg: in *Superconductivity*, ed. R. D. Parks (Marcel Dekker, New York, 1995), p. 852.

[208] K. A. Moler, D. J. Baar, J. S. Urbach, R. Liang, W. N. Hardy, and A. Kapitulnik: Phys. Rev. Lett. **73** (1994) 2744.
[209] Y. Nakazawa and K. Kanoda: Phys. Rev. B **55** (1997) R8670.
[210] G. E. Volovik: JETP Lett. **58** (1993) 469.
[211] I. Vekhter, P. Hirschfeld, J. Carbotte, and E. Nicol: Phys. Rev. B **59** (1999) R9023.
[212] K. An, T. Sakakibara, R. Settai, Y. Ōnuki, M. Hiragi, M. Ichioka, and K. Machida: Phys. Rev. Lett. **104** (2010) 037002.
[213] T. Watanabe, K. Izawa, Y. Kasahara, Y. Haga, Y. Ōnuki, P. Thalmeier, K. Maki, and Y. Matsuda: Physica C **426** (2005) 234.
[214] Y. Inada, H. Yamagami, Y. Haga, K. Sakurai, Y. Tokiwa, T. Honma, E. Yamamoto, Y. Ōnuki, and T. Yanagisawa: J. Phys. Soc. Jpn. **68** (1999) 3643.
[215] R. Joynt and L. Taillefer: Rev. Mod. Phys. **74** (2002) 235.
[216] N. Kimura, T. Komatsubara, D. Aoki, Y. Ōnuki, Y. Haga, E. Yamamoto, H. Aoki, and H. Harima: J. Phys. Soc. Jpn. **67** (1998) 2185.
[217] N. Kimura, T. Tani, H. Aoki, T. Komatsubara, S. Uji, D. Aoki, Y. Inada, Y. Ōnuki, Y. Haga, E. Yamamoto, and H. Harima: Physica B: Cond. Matt. **281** (2000) 710.
[218] K. Sugiyama, M. Nakashima, D. Aoki, K. Kindo, N. Kimura, H. Aoki, T. Komatsubara, S. Uji, Y. Haga, E. Yamamoto, H. Harima, and Y. Ōnuki: Phys. Rev. B **60** (1999) 9248.
[219] H. P. van der Meulen, Z. Tarnawski, A. de Visser, J. J. M. Franse, J. A. A. J. Perenboom, D. Althof, and H. van Kempen: Phys. Rev. B **41** (1990) 9352.
[220] G. R. Stewart and B. L. Brandt: Phys. Rev. B **29** (1984) 3908.
[221] T. Sakakibara, K. Tenya, M. Ikeda, T. Tayama, H. Amitsuka, E. Yamamoto, K. Maezawa, N. Kimura, R. Settai, and Y. Ōnuki: J. Phys. Soc. Jpn. Suppl. B **65** (1996) 202.
[222] Y. Machida, A. Itoh, Y. So, K. Izawa, Y. Haga, E. Yamamoto, N. Kimura, Y. Onuki, Y. Tsutsumi, and K. Machida: Phys. Rev. Lett. **108** (2012) 157002.
[223] S. Kittaka, K. An, T. Sakakibara, Y. Haga, E. Yamamoto, N. Kimura, Y. Ōnuki, and K. Machida: J. Phys. Soc. Jpn. **82** (2013) 024707.
[224] K. Oikawa, T. Kamiyama, H. Asano, Y. Ōnuki, and M. Kohgi: J. Phys. Soc. Jpn. **65** (1996) 3229.
[225] S. S. Saxena, P. Agarwal, K. Ahilan, F. M. Grosche, R. K. W. Haselwimmer, M. J. Steiner, E. Pugh, I. R. Walker, S. R. Julian, P. Monthoux, G. G. Lonzarich, A. Huxley, I. Sheikin, D. Braithwaite, and J. Flouquet: Nature **406** (2000) 587.
[226] D. Aoki and J. Flouquet: J. Phys. Soc. Jpn. **81** (2012) 011003.
[227] R. Settai, M. Nakashima, S. Araki, Y. Haga, T. C. Kobayashi, N. Tateiwa, H. Yamagami, and Y. Onuki: J. Phys.: Condens. Matter **14** (2002) L29.
[228] N. Tateiwa, K. Hanazono, T. C. Kobayashi, K. Amaya, T. Inoue, K. Kindo, Y. Koike, N. Metoki, Y. Haga, R. Settai, and Y. Ōnuki: J. Phys. Soc. Jpn. **70** (2001) 2876.

[229] C. Pfleiderer and A. D. Huxley: Phys. Rev. Lett. **89** (2002) 147005.
[230] N. Tateiwa, T. C. Kobayashi, K. Amaya, Y. Haga, R. Settai, and Y. Ōnuki: Phys. Rev. B **69** (2004) 180513.
[231] Y. Haga, M. Nakashima, R. Settai, S. Ikeda, T. Okubo, S. Araki, T. C. Kobayashi, N. Tateiwa, and Y. Onuki: J. Phys.: Condens. Matter **14** (2002) L125.
[232] I. Sheikin, A. Huxley, D. Braithwaite, J. P. Brison, S. Watanabe, K. Miyake, and J. Flouquet: Phys. Rev. B **64** (2001) 220503.
[233] H. Kusunose and Y. Kimoto: J. Phys. Soc. Jpn. **82** (2013) 094711.
[234] K. Satoh, S. W. Yun, I. Umehara, Y. Ōnuki, S. Uji, T. Shimizu, and H. Aoki: J. Phys. Soc. Jpn. **61** (1992) 1827.
[235] A. B. Shick and W. E. Pickett: Phys. Rev. Lett. **86** (2001) 300.
[236] H. Yamagami: J. Phys.: Condens. Matter **15** (2003) S2271.
[237] T. Terashima, T. Matsumoto, C. Terakura, S. Uji, N. Kimura, M. Endo, T. Komatsubara, and H. Aoki: Phys. Rev. Lett. **87** (2001) 166401.
[238] T. Terashima, T. Matsumoto, C. Terakura, S. Uji, N. Kimura, M. Endo, T. Komatsubara, H. Aoki, and K. Maezawa: Phys. Rev. B **65** (2002) 174501.
[239] K. Kubo: Phys. Rev. B **87** (2013) 195127.
[240] M. B. Maple: J. Phys. Soc. Jpn. **74** (2005) 222.
[241] T. Onimaru and H. Kusunose: J. Phys. Soc. Jpn. **85** (2016) 082002.
[242] M. Kohgi, K. Iwasa, M. Nakajima, N. Metoki, S. Araki, N. Bernhoeft, J.-M. Mignot, A. Gukasov, H. Sato, Y. Aoki, et al.: J. Phys. Soc. Jpn. **72** (2003) 1002.
[243] Y. Aoki, T. Tayama, T. Sakakibara, K. Kuwahara, K. Iwasa, M. Kohgi, W. Higemoto, D. E. MacLaughlin, H. Sugawara, and H. Sato: J. Phys. Soc. Jpn. **76** (2007) 051006.
[244] K. Matsubayashi, T. Tanaka, J. Suzuki, A. Sakai, S. Nakatsuji, K. Kitagawa, Y. Kubo, and Y. Uwatoko: JPS Conf. Proc. **3** (2014) 011077.
[245] M. Tsujimoto, Y. Matsumoto, T. Tomita, A. Sakai, and S. Nakatsuji: Phys. Rev. Lett. **113** (2014) 267001.
[246] A. Tsuruta and K. Miyake: J. Phys. Soc. Jpn. **84** (2015) 114714.
[247] D. Cox: Physica B: Cond. Matt. **186** (1993) 312.
[248] E. Bauer, G. Hilscher, H. Michor, C. Paul, E. W. Scheidt, A. Gribanov, Y. Seropegin, H. Noël, M. Sigrist, and P. Rogl: Phys. Rev. Lett. **92** (2004) 027003.
[249] E. Bauer, H. Kaldarar, A. Prokofiev, E. Royanian, A. Amato, J. Sereni, W. Brämer-Escamilla, and I. Bonalde: J. Phys. Soc. Jpn. **76** (2007) 051009.
[250] T. Akazawa, H. Hidaka, T. Fujiwara, T. C. Kobayashi, E. Yamamoto, Y. Haga, R. Settai, and Y. Ōnukinuki: J. Phys.: Condens. Matter **16** (2004) L29.
[251] T. C. Kobayashi, A. Hori, S. Fukushima, H. Hidaka, H. Kotegawa, T. Akazawa, K. Takeda, Y. Ohishi, and E. Yamamoto: J. Phys. Soc. Jpn. **76** (2007) 051007.
[252] N. Kimura, K. Ito, K. Saitoh, Y. Umeda, H. Aoki, and T. Terashima: Phys. Rev. Lett. **95** (2005) 247004.

[253] N. Kimura, K. Ito, H. Aoki, S. Uji, and T. Terashima: Phys. Rev. Lett. **98** (2007) 197001.
[254] N. Kimura, Y. Muro, and H. Aoki: J. Phys. Soc. Jpn. **76** (2007) 051010.
[255] Y. Okuda, Y. Miyauchi, Y. Ida, Y. Takeda, C. T., Y. Oduchi, T. Yamada, N. D. Dung, T. D. Matsuda, Y. Haga, T. Takeuchi, M. Hagiwara, K. Kindo, H. Harima, K. Sugiyama, R. Settai, and Y. Ōnuki: J. Phys. Soc. Jpn. **76** (2007) 044708.
[256] I. Sugitani, Y. Okuda, H. Shishido, T. Yamada, A. Thamizhavel, E. Yamamoto, T. D. Matsuda, Y. Haga, T. Takeuchi, R. Settai, and Y. Ōnuki: J. Phys. Soc. Jpn. **75** (2006) 043703.
[257] R. Settai, Y. Okuda, I. Sugitani, Y. Ōnuki, T. D. Matsuda, Y. Haga, and H. Harima: Int. J. Mod. Phys. B **21** (2007) 3238.
[258] T. Kawai, H. Muranaka, T. Endo, N. D. Dung, Y. Doi, S. Ikeda, T. D. Matsuda, Y. Haga, H. Harima, R. Settai, and Y. Ōnuki: J. Phys. Soc. Jpn. **77** (2008) 064717.
[259] F. Honda, I. Bonalde, K. Shimizu, S. Yoshiuchi, Y. Hirose, T. Nakamura, R. Settai, and Y. Ōnuki: Phys. Rev. B **81** (2010) 140507.
[260] E. I. Rashba: Sov. Phys. Solid State. **2** (1960) 1109.
[261] P. A. Frigeri, D. F. Agterberg, A. Koga, and M. Sigrist: Phys. Rev. Lett. **92** (2004) 097001.
[262] K. V. Samokhin: Phys. Rev. B **70** (2004) 104521.
[263] V. Mineev and K. Samokhin: Phys. Rev. B **72** (2005) 212504.
[264] T. Kawai, H. Muranaka, T. Endo, N. D. Dung, Y. Doi, S. Ikeda, T. D. Matsuda, Y. Haga, H. Harima, R. Settai, and Y. Ōnuki: J. Phys. Soc. Jpn. **77** (2008) 064717.
[265] P. Frigeri, D. Agterberg, and M. Sigrist: New J. Phys. **6** (2004) 115.
[266] S. Fujimoto: J. Phys. Soc. Jpn. **76** (2007) 051008.
[267] N. Tateiwa, Y. Haga, T. D. Matsuda, S. Ikeda, E. Yamamoto, Y. Okuda, Y. Miyauchi, R. Settai, and Y. Ōnuki: J. Phys. Soc. Jpn. **76** (2007) 083706.
[268] R. Settai, Y. Miyauchi, T. Takeuchi, F. Lévy, I. Sheikin, and Y. Ōnuki: J. Phys. Soc. Jpn. **77** (2008) 073705.
[269] E. Helfand and N. R. Werthamer: Phys. Rev. **147** (1966) 288.
[270] R. Settai, K. Katayama, D. Aoki, I. Sheikin, G. Knebel, J. Flouquet, and Y. Ōnuki: J. Phys. Soc. Jpn. **80** (2011) 094703.
[271] G. Knebel, D. Aoki, D. Braithwaite, N. Cherroret, B. Salce, and J. Flouquet: J. Phys. Soc. Jpn. **76** (2007) 124.
[272] H. Mukuda, T. Fujii, T. Ohara, A. Harada, M. Yashima, Y. Kitaoka, Y. Okuda, R. Settai, and Y. Ōnuki: Phys. Rev. Lett. **100** (2008) 107003.
[273] T. Moriya and K. Ueda: Adv. Phys. **49** (2000) 555.
[274] H. Mukuda, T. Ohara, M. Yashima, Y. Kitaoka, R. Settai, Y. Ōnuki, K. M. Itoh, and E. E. Haller: Phys. Rev. Lett. **104** (2010) 017002.
[275] P. A. Frigeri, D. F. Agterberg, and M. Sigrist: New J. Phys. **6** (2004) 115.
[276] Y. Tada, N. Kawakami, and S. Fujimoto: Phys. Rev. Lett. **101** (2008) 267006.
[277] S. Mühlbauer, B. Binz, F. Jonietz, C. Pfleiderer, A. Rosch, A. Neubauer, R. Georgii, and P. Böni: Science **323** (2009) 915.

Appendix

A.1 230 space groups

Crystal structures of compounds are classified on the basis of 230 space groups. They are divided into 92 centrosymmetric ones (their space group numbers are shown in solid type) and 138 non-centrosymmetric ones (in bold solid type), together with short and full international symbols and also with the Schenflies symbol. The non-centrosymmetric crystal structure does not possess a center of inversion. Suffix c of the space group number, for example, **198**^{c} refers to the chiral structure, where the 138 non-centrosymmetric structures contain the chiral structures based on 65 space groups. The following sets of chiral space groups (total 22 in number) are mirrorsymmetric: No.76 vs No.78, 91 vs 95, 92 vs 96, 144 vs 145, 151 vs 153, 152 vs 154, 169 vs 170, 171 vs 172, 178 vs 179, 180 vs 181, and 212 vs 213, courtesy of A. Teruya.

Triclinic, Monoclinic							
1^{c}	$P1$	$(P1)$	C_1^1	**9**	Cc	$(C1c1)$	C_s^4
2	$P\bar{1}$	$(P\bar{1})$	C_i^1	10	$P2/m$	$(P12/m1)$	C_{2h}^1
3^{c}	$P2$	$(P121)$	C_2^1	11	$P2_1/m$	$(P12_1/m1)$	C_{2h}^2
4^{c}	$P2_1$	$(P12_11)$	C_2^2	12	$C2/m$	$(C12/m1)$	C_{2h}^3
5^{c}	$C2$	$(C121)$	C_2^3	13	$P2/c$	$(P12/c1)$	C_{2h}^4
6	Pm	$(P1m1)$	C_s^1	14	$P2_1/c$	$(P12_1/c1)$	C_{2h}^5
7	Pc	$(P1c1)$	C_s^2	15	$C2/c$	$(C12/c1)$	C_{2h}^6
8	Cm	$(C1m1)$	C_s^3				

Orthorhombic							
16[c]	$P222$	$(P222)$	D_2^1	**46**	$Ima2$	$(Ima2)$	C_{2v}^{22}
17[c]	$P222_1$	$(P222_1)$	D_2^2	47	$Pmmm$	$(P2/m2/m2/m)$	D_{2h}^1
18[c]	$P2_12_12$	$(P2_12_12)$	D_2^3	48	$Pnnn$	$(P2/n2/n2/n)$	D_{2h}^2
19[c]	$P2_12_12_1$	$(P2_12_12_1)$	D_2^4	49	$Pccm$	$(P2/c2/c2/m)$	D_{2h}^3
20[c]	$C222_1$	$(C222_1)$	D_2^5	50	$Pban$	$(P2/b2/a2/n)$	D_{2h}^4
21[c]	$C222$	$(C222)$	D_2^6	51	$Pmma$	$(P2_1/m2/m2/a)$	D_{2h}^5
22[c]	$F222$	$(F222)$	D_2^7	52	$Pnna$	$(P2/n2_1/n2/a)$	D_{2h}^6
23[c]	$I222$	$(I222)$	D_2^8	53	$Pmna$	$(P2/m2/n2_1/a)$	D_{2h}^7
24[c]	$I2_12_12_1$	$(I2_12_12_1)$	D_2^9	54	$Pcca$	$(P2_1/c2/c2/a)$	D_{2h}^8
25	$Pmm2$	$(Pmm2)$	C_{2v}^1	55	$Pbam$	$(P2_1/b2_1/a2/m)$	D_{2h}^9
26	$Pmc2_1$	$(Pmc2_1)$	C_{2v}^2	56	$Pccn$	$(P2_1/c2_1/c2/n)$	D_{2h}^{10}
27	$Pcc2$	$(Pcc2)$	C_{2v}^3	57	$Pbcm$	$(P2/b2_1/c2_1/m)$	D_{2h}^{11}
28	$Pma2$	$(Pma2)$	C_{2v}^4	58	$Pnnm$	$(P2_1/n2_1/n2/m)$	D_{2h}^{12}
29	$Pca2_1$	$(Pca2_1)$	C_{2v}^5	59	$Pmmn$	$(P2_1/m2_1/m2/n)$	D_{2h}^{13}
30	$Pnc2$	$(Pnc2)$	C_{2v}^6	60	$Pbcn$	$(P2_1/b2/c2_1/n)$	D_{2h}^{14}
31	$Pmn2_1$	$(Pmn2_1)$	C_{2v}^7	61	$Pbca$	$(P2_1/b2_1/c2_1/a)$	D_{2h}^{15}
32	$Pba2$	$(Pba2)$	C_{2v}^8	62	$Pnma$	$(P2_1/n2_1/m2_1/a)$	D_{2h}^{16}
33	$Pna2_1$	$(Pna2_1)$	C_{2v}^9	63	$Cmcm$	$(C2/m2/c2_1/m)$	D_{2h}^{17}
34	$Pnn2$	$(Pnn2)$	C_{2v}^{10}	64	$Cmca$	$(C2/m2/c2_1/a)$	D_{2h}^{18}
35	$Cmm2$	$(Cmm2)$	C_{2v}^{11}	65	$Cmmm$	$(C2/m2/m2/m)$	D_{2h}^{19}
36	$Cmc2_1$	$(Cmc2_1)$	C_{2v}^{12}	66	$Cccm$	$(C2/c2/c2/m)$	D_{2h}^{20}
37	$Ccc2$	$(Ccc2)$	C_{2v}^{13}	67	$Cmma$	$(C2/m2/m2/a)$	D_{2h}^{21}
38	$Amm2$	$(Amm2)$	C_{2v}^{14}	68	$Ccca$	$(C2/c2/c2/a)$	D_{2h}^{22}
39	$Abm2$	$(Abm2)$	C_{2v}^{15}	69	$Fmmm$	$(F2/m2/m2/m)$	D_{2h}^{23}
40	$Ama2$	$(Ama2)$	C_{2v}^{16}	70	$Fddd$	$(F2/d2/d2/d)$	D_{2h}^{24}
41	$Aba2$	$(Aba2)$	C_{2v}^{17}	71	$Immm$	$(I2/m2/m2/m)$	D_{2h}^{25}
42	$Fmm2$	$(Fmm2)$	C_{2v}^{18}	72	$Ibam$	$(I2/b2/a2/m)$	D_{2h}^{26}
43	$Fdd2$	$(Fdd2)$	C_{2v}^{19}	73	$Ibca$	$(I2/b2/c2/a)$	D_{2h}^{27}
44	$Imm2$	$(Imm2)$	C_{2v}^{20}	74	$Imma$	$(I2/m2/m2/a)$	D_{2h}^{28}
45	$Iba2$	$(Iba2)$	C_{2v}^{21}				

Tetragonal							
75[c]	$P4$	$(P4)$	C_4^1	**109**	$I4_1md$	$(I4_1md)$	C_{4v}^{11}
76[c]	$P4_1$	$(P4_1)$	C_4^2	**110**	$I4_1cd$	$(I4_1cd)$	C_{4v}^{12}
77[c]	$P4_2$	$(P4_2)$	C_4^3	**111**	$P\bar{4}2m$	$(P\bar{4}2m)$	D_{2d}^1
78[c]	$P4_3$	$(P4_3)$	C_4^4	**112**	$P\bar{4}2c$	$(P\bar{4}2c)$	D_{2d}^2
79[c]	$I4$	$(I4)$	C_4^5	**113**	$P\bar{4}2_1m$	$(P\bar{4}2_1m)$	D_{2d}^3
80[c]	$I4_1$	$(I4_1)$	C_4^6	**114**	$P\bar{4}2_1c$	$(P\bar{4}2_1c)$	D_{2d}^4
81	$P\bar{4}$	$(P\bar{4})$	S_4^1	**115**	$P\bar{4}m2$	$(P\bar{4}m2)$	D_{2d}^5
82	$I\bar{4}$	$(I\bar{4})$	S_4^2	**116**	$P\bar{4}c2$	$(P\bar{4}c2)$	D_{2d}^6

(*Continued*)

(*Continued*)

83	$P4/m$	$(P4/m)$	C_{4h}^1	**117**	$P\bar{4}b2$	$(P\bar{4}b2)$	D_{2d}^7
84	$P4_2/m$	$(P4_2/m)$	C_{4h}^2	**118**	$P\bar{4}n2$	$(P\bar{4}n2)$	D_{2d}^8
85	$P4/n$	$(P4/n)$	C_{4h}^3	**119**	$I\bar{4}m2$	$(I\bar{4}m2)$	D_{2d}^9
86	$P4_2/n$	$(P4_2/n)$	C_{4h}^4	**120**	$I\bar{4}c2$	$(I\bar{4}c2)$	D_{2d}^{10}
87	$I4/m$	$(I4/m)$	C_{4h}^5	**121**	$I\bar{4}2m$	$(I\bar{4}2m)$	D_{2d}^{11}
88	$I4_1/a$	$(I4_1/a)$	C_{4h}^6	**122**	$I\bar{4}2d$	$(I\bar{4}2d)$	D_{2d}^{12}
89[c]	$P422$	$(P422)$	D_4^1	123	$P4/mmm$	$(P4/m2/m2/m)$	D_{4h}^1
90[c]	$P42_12$	$(P42_12)$	D_4^2	124	$P4/mcc$	$(P4/m2/c2/c)$	D_{4h}^2
91[c]	$P4_122$	$(P4_122)$	D_4^3	125	$P4/nbm$	$(P4/n2/b2/m)$	D_{4h}^3
92[c]	$P4_12_12$	$(P4_12_12)$	D_4^4	126	$P4/nnc$	$(P4/n2/n2/c)$	D_{4h}^4
93[c]	$P4_222$	$(P4_222)$	D_4^5	127	$P4/mbm$	$(P4/m2_1/bm)$	D_{4h}^5
94[c]	$P4_22_12$	$(P4_22_12)$	D_4^6	128	$P4/mnc$	$(P4/m2_1/nc)$	D_{4h}^6
95[c]	$P4_322$	$(P4_322)$	D_4^7	129	$P4/nmm$	$(P4/n2_1/mm)$	D_{4h}^7
96[c]	$P4_32_12$	$(P4_32_12)$	D_4^8	130	$P4/ncc$	$(P4/n2_1/cc)$	D_{4h}^8
97[c]	$I422$	$(I422)$	D_4^9	131	$P4_2/mmc$	$(P4_2/m2/m2/c)$	D_{4h}^9
98[c]	$I4_122$	$(I4_122)$	D_4^{10}	132	$P4_2/mcm$	$(P4_2/m2/c2/m)$	D_{4h}^{10}
99	$P4mm$	$(P4mm)$	C_{4v}^1	133	$P4_2/nbc$	$(P4_2/n2/b2/c)$	D_{4h}^{11}
100	$P4bm$	$(P4bm)$	C_{4v}^2	134	$P4_2/nnm$	$(P4_2/n2/n2/m)$	D_{4h}^{12}
101	$P4_2cm$	$(P4_2cm)$	C_{4v}^3	135	$P4_2/mbc$	$(P4_2/m2_1/b2/c)$	D_{4h}^{13}
102	$P4_2nm$	$(P4_2nm)$	C_{4v}^4	136	$P4_2/mnm$	$(P4_2/m2_1/n2/m)$	D_{4h}^{14}
103	$P4cc$	$(P4cc)$	C_{4v}^5	137	$P4_2/nmc$	$(P4_2/n2_1/m2/c)$	D_{4h}^{15}
104	$P4nc$	$(P4nc)$	C_{4v}^6	138	$P4_2/ncm$	$(P4_2/n2_1/c2/m)$	D_{4h}^{16}
105	$P4_2mc$	$(P4_2mc)$	C_{4v}^7	139	$I4/mmm$	$(I4/m2/m2/m)$	D_{4h}^{17}
106	$P4_2bc$	$(P4_2bc)$	C_{4v}^8	140	$I4/mcm$	$(I4/m2/c2/m)$	D_{4h}^{18}
107	$I4mm$	$(I4mm)$	C_{4v}^9	141	$I4_1/amd$	$(I4_1/a2/m2/d)$	D_{4h}^{19}
108	$I4cm$	$(I4cm)$	C_{4v}^{10}	142	$I4_1/acd$	$(I4_1/a2/c2/d)$	D_{4h}^{20}

Trigonal (or Rhombohedral)

143[c]	$P3$	$(P3)$	C_3^1	**156**	$P3m1$	$(P3m1)$	C_{3v}^1
144[c]	$P3_1$	$(P3_1)$	C_3^2	**157**	$P31m$	$(P31m)$	C_{3v}^2
145[c]	$P3_2$	$(P3_2)$	C_3^3	**158**	$P3c1$	$(P3c1)$	C_{3v}^3
146[c]	$R3$	$(R3)$	C_3^4	**159**	$P31c$	$(P31c)$	C_{3v}^4
147	$P\bar{3}$	$(P\bar{3})$	C_{3i}^1	**160**	$R3m$	$(R3m)$	C_{3v}^5
148	$R\bar{3}$	$(R\bar{3})$	C_{3i}^2	**161**	$R3c$	$(R3c)$	C_{3v}^6
149[c]	$P312$	$(P312)$	D_3^1	162	$P\bar{3}1m$	$(P\bar{3}12/m)$	D_{3d}^1
150[c]	$P321$	$(P321)$	D_3^2	163	$P\bar{3}1c$	$(P\bar{3}12/c)$	D_{3d}^2
151[c]	$P3_112$	$(P3_112)$	D_3^3	164	$P\bar{3}m1$	$(P\bar{3}2/m1)$	D_{3d}^3
152[c]	$P3_121$	$(P3_121)$	D_3^4	165	$P\bar{3}c1$	$(P\bar{3}2/c1)$	D_{3d}^4
153[c]	$P3_212$	$(P3_212)$	D_3^5	166	$R\bar{3}m$	$(R\bar{3}2/m)$	D_{3d}^5
154[c]	$P3_221$	$(P3_221)$	D_3^6	167	$R\bar{3}c$	$(R\bar{3}2/c)$	D_{3d}^6
155[c]	$R32$	$(R32)$	D_3^7				

Hexagonal

168[c]	$P6$	$(P6)$	C_6^1	**182**[c]	$P6_322$	$(P6_322)$	D_6^6
169[c]	$P6_1$	$(P6_1)$	C_6^2	**183**	$P6mm$	$(P6mm)$	C_{6v}^1
170[c]	$P6_5$	$(P6_5)$	C_6^3	**184**	$P6cc$	$(P6cc)$	C_{6v}^2
171[c]	$P6_2$	$(P6_2)$	C_6^4	**185**	$P6_3cm$	$(P6_3cm)$	C_{6v}^3
172[c]	$P6_4$	$(P6_4)$	C_6^5	**186**	$P6_3mc$	$(P6_3mc)$	C_{6v}^4
173[c]	$P6_3$	$(P6_3)$	C_6^6	**187**	$P\bar{6}m2$	$(P\bar{6}m2)$	D_{3h}^1
174	$P\bar{6}$	$(P\bar{6})$	C_{3h}^1	**188**	$P\bar{6}c2$	$(P\bar{6}c2)$	D_{3h}^2
175	$P6/m$	$(P6/m)$	C_{6h}^1	**189**	$P\bar{6}2m$	$(P\bar{6}2m)$	D_{3h}^3
176	$P6_3/m$	$(P6_3/m)$	C_{6h}^2	**190**	$P\bar{6}2c$	$(P\bar{6}2c)$	D_{3h}^4
177[c]	$P622$	$(P622)$	D_6^1	191	$P6/mmm$	$(P6/m2/m2/m)$	D_{6h}^1
178[c]	$P6_122$	$(P6_122)$	D_6^2	192	$P6/mcc$	$(P6/m2/c2/c)$	D_{6h}^2
179[c]	$P6_522$	$(P6_522)$	D_6^3	193	$P6_3/mcm$	$(P6_3/m2/c2/m)$	D_{6h}^3
180[c]	$P6_222$	$(P6_222)$	D_6^4	194	$P6_3/mmc$	$(P6_3/m2/m2/c)$	D_{6h}^4
181[c]	$P6_422$	$(P6_422)$	D_6^5				

Cubic

195[c]	$P23$	$(P23)$	T^1	**213**[c]	$P4_132$	$(P4_132)$	O^7
196[c]	$F23$	$(F23)$	T^2	**214**[c]	$I4_132$	$(I4_132)$	O^8
197[c]	$I23$	$(I23)$	T^3	**215**	$P\bar{4}3m$	$(P\bar{4}3m)$	T_d^1
198[c]	$P2_13$	$(P2_13)$	T^4	**216**	$F\bar{4}3m$	$(F\bar{4}3m)$	T_d^2
199[c]	$I2_13$	$(I2_13)$	T^5	**217**	$I\bar{4}3m$	$(I\bar{4}3m)$	T_d^3
200	$Pm\bar{3}$	$(P2/m\bar{3})$	T_h^1	**218**	$P\bar{4}3n$	$(P\bar{4}3n)$	T_d^4
201	$Pn\bar{3}$	$(P2/n\bar{3})$	T_h^2	**219**	$F\bar{4}3c$	$(F\bar{4}3c)$	T_d^5
202	$Fm\bar{3}$	$(F2/m\bar{3})$	T_h^3	**220**	$I\bar{4}3d$	$(I\bar{4}3d)$	T_d^6
203	$Fd\bar{3}$	$(F2/d\bar{3})$	T_h^4	221	$Pm\bar{3}m$	$(P4/m\bar{3}2/m)$	O_h^1
204	$Im\bar{3}$	$(I2/m\bar{3})$	T_h^5	222	$Pn\bar{3}n$	$(P4/n\bar{3}2/n)$	O_h^2
205	$Pa\bar{3}$	$(P2_1/a\bar{3})$	T_h^6	223	$Pm\bar{3}n$	$(P4_2/m\bar{3}2/n)$	O_h^3
206	$Ia\bar{3}$	$(I2_1/a\bar{3})$	T_h^7	224	$Pn\bar{3}m$	$(P4_2/n\bar{3}2/m)$	O_h^4
207[c]	$P432$	$(P432)$	O^1	225	$Fm\bar{3}m$	$(F4/m\bar{3}2/m)$	O_h^5
208[c]	$P4_232$	(P_2432)	O^2	226	$Fm\bar{3}c$	$(F4/m\bar{3}2/c)$	O_h^6
209[c]	$F432$	$(F432)$	O^3	227	$Fd\bar{3}m$	$(F4_1/d\bar{3}2/m)$	O_h^7
210[c]	$F4_132$	$(F4_132)$	O^4	228	$Fd\bar{3}c$	$(F4_1/d\bar{3}2/c)$	O_h^8
211[c]	$I432$	$(I432)$	O^5	229	$Im\bar{3}m$	$(I4/m\bar{3}2/m)$	O_h^9
212[c]	$P4_332$	$(P4_332)$	O^6	230	$Ia\bar{3}d$	$(I4_1/a\bar{3}2/d)$	O_h^{10}

A.2 Brillouin zones

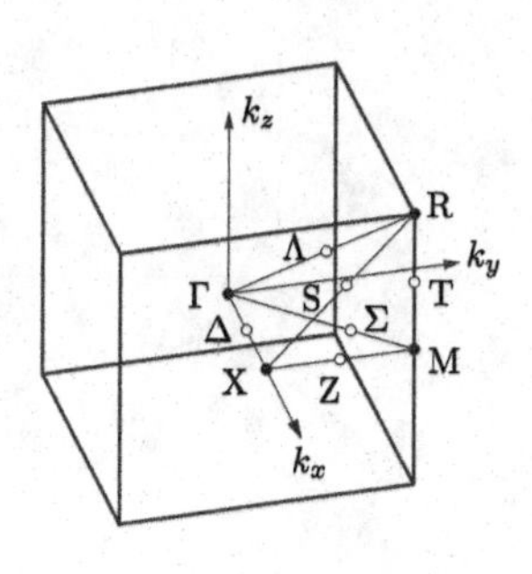

(a) simple cubic

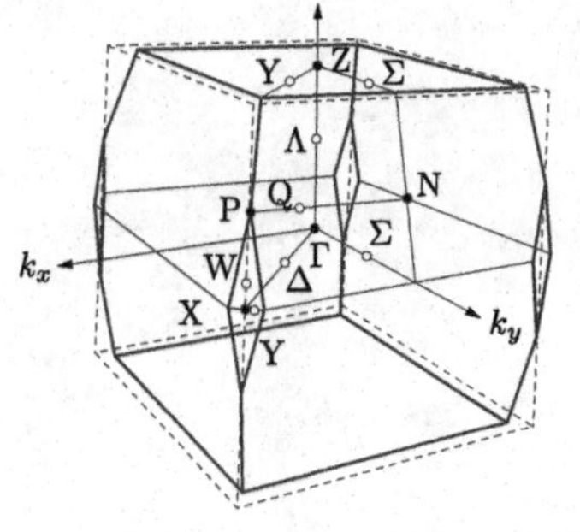

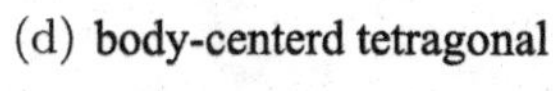

(d) body-centerd tetragonal

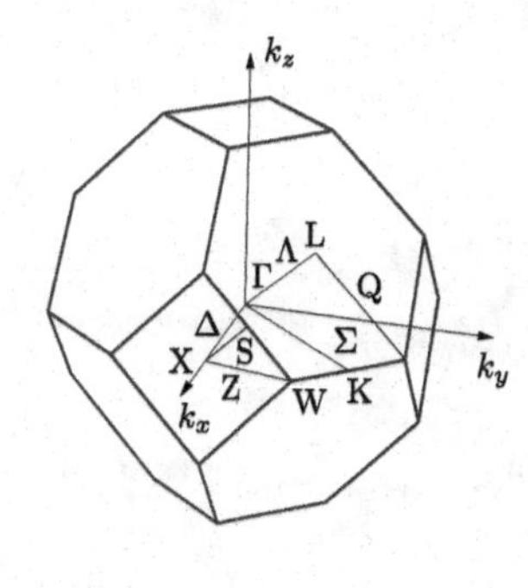

(b) fcc

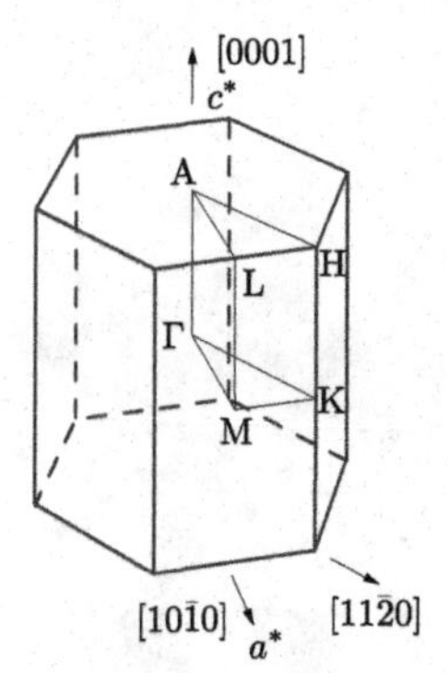

(e) hexagonal

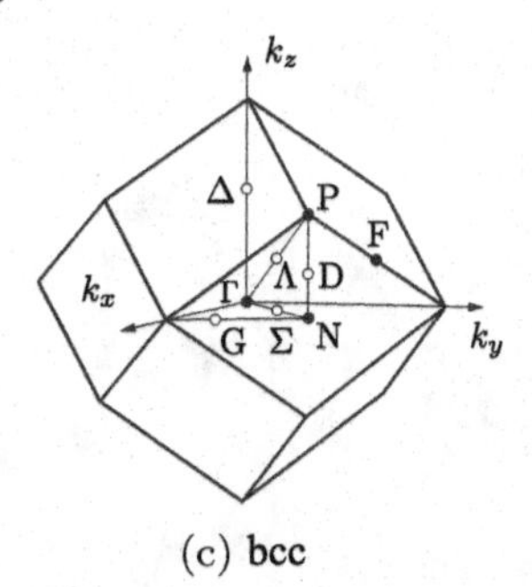

(c) bcc

Index

Printed in the United States
By Bookmasters